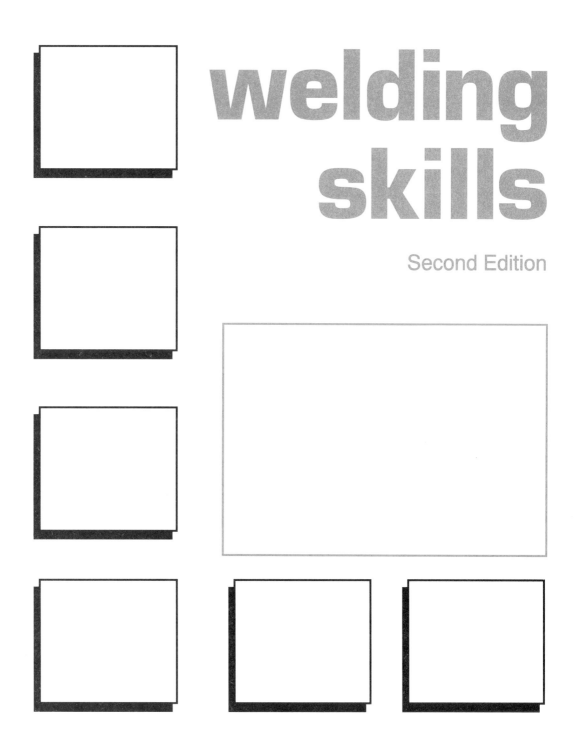

welding skills

Second Edition

 AMERICAN TECHNICAL PUBLISHERS, INC.
HOMEWOOD, ILLINOIS 60430

R. T. Miller

2 3 4 5 6 7 8 9 – 97 – 9 8 7 6 5 4 3 2 1

Printed in the United States of America

ISBN 0-8269-3007-7

INTRODUCTION

Welding Skills, 2nd Edition is a comprehensive text which addresses one of the most important manufacturing processes used in industry today. While comprehensive in scope, the book begins with a basic introductory approach, and builds on the knowledge and skills presented. While welding technology continues to evolve, basic principles that are fundamental to the joining of metals still remain.

Welding Skills, 2nd Edition blends new welding technology with fundamental principles in a direct, concise approach. An open, graphic format is used throughout the book. Numerous charts and tables supplement information in the text and serve as an invaluable reference for the learner. Photographs and large two-color drawings detail welding processes and techniques.

Welding Skills, 2nd Edition provides a basic background in welding processes. Current welding processes and American Welding Society (AWS) terminology are detailed throughout the text. Additionally, worker health and safety is emphasized as these areas have increasingly become an integral part of the manufacturing processes. Potential health and safety hazards are discussed with appropriate cautions and warnings throughout the text. A separate chapter also discusses welding safety.

Many welding processes require mastery of essential manipulative skills. Welding exercises are provided in selected sections of the text to aid the learner in acquiring and retaining these skills. Material requirements, fit-up, equipment set-up, and welding procedures are listed in concise steps within the exercises. Specific welding tasks to be practiced can be selected from the weld types, weld joints, and positions detailed in the respective exercises.

Welding Skills, 2nd Edition is divided into six sections:
- Introduction to Welding
- Oxyacetylene Welding — OAW
- Shielded Metal-Arc Welding — SMAW
- Gas Shielded-Arc Welding
- Special Welding Processes
- Supplementary Welding Data

Introduction to Welding provides a foundation of theory and knowledge relating to the welding process. Industrial applications and occupational opportunities are introduced. Facility, equipment, and worker safety aspects related to welding are included.

As heat is directed to joint members, changes in the internal structure of the metals begin. The behavior of metals in the welding process and the design and configurations of welded parts are detailed. Procedures which compensate for heat distortion are covered.

Oxyacetylene Welding covers all facets of oxyacetylene welding from equipment and set-up, to welding positions on various metals. Other fuels and required equipment, in addition to acetylene, are included.

Shielded Metal-Arc Welding details machines and accessories, electrodes, and methods used to achieve quality welds. Beginning welding procedures are introduced, with more advanced welding techniques covered in chapters on flat, horizontal, vertical, and overhead positions.

Gas Shielded-Arc Welding details welding processes which utilize shielding gases for production efficiency and quality. Special attention is given to the joining of aluminum and alloys.

Special Welding Processes covers related welding and material-joining processes. Adhesive bonding of materials is presented in the production welding chapter. The area of automated welding processes is addressed in the robotics and welding chapter.

Supplementary Welding Data includes information on weld testing, welding symbols, and welder certification. The joining of metals is performed after careful interpretation of welding symbols on weld prints. Weld quality is maintained by the ability of the welder to complete the weld according to specifications and through procedures for testing the weld following AWS testing procedures.

Welding Skills, 2nd Edition can be used as a stand-alone text or in conjunction with *Welding Skills Workbook,* 2nd Edition. *Printreading for Welders*, *Printreading for Welders Transparencies*, and *Printreading for Welders Tests* and their respective instructor's guides can be used with *Welding Skills,* 2nd Edition as part of a comprehensive instructional program containing maximum content exposure, skill development, and testing.

The Publisher

ACKNOWLEDGMENTS

The author and publisher are grateful to the following companies and organizations for providing technical information and assistance.

- Advanced Robotics Corporation
- Alcoa Aluminum Company
- American Optical Corporation
- American Welding Society
- Arcair Company
- Automation International
  (formerly Berkeley-Davis Corporation)
- Bernard Welding Company
- Boeing Commercial Airplane Group
- Caterpillar Industrial Inc.
  (formerly Caterpillar Tractor Company)
- Chemetron Corporation
- Cincinnati Milacron Company
- Clarage Fan Company
- CRC Crose International
- Crutcher Resources Corporation
- DND Corporation
- ESAB Welding and Cutting Products
  (formerly Airco Welding Products)
- Fel-Pro Chemical Products
- Fibre-Metal Products
- G.A.L. Gage Company
- GM — North American Operations
- Hobart Brothers Company
- Industrial Plastic Fabricators, Inc.
- Kamweld Products Company
- Kidde Automated Systems, Inc.
- Linde Division, Union Carbide Corporation
- LORS Machinery, Inc.
- LTV Steel Corporation
- Magnaflux Corporation
- Mapp Industrial Gas Company
- Mathey/Leland
- McDonnell Douglas Corporation
- Metallizing Company of America
- Miller Electric Manufacturing Company
- Motoman, Inc.
- National Electrical Manufacturers Association
- Nederman, Inc.
- Nelson Stud Welding Division, TRW Corporation
- Peer Division, Landis Machine Company
- Prairie State College
- Pratt & Whitney Corporation
- Robot Institute of America
- Rockwell International
- Sciaky Brothers, Inc.
- Seelye Plastics, Inc.
- Sellstrom Mfg. Co.
- Sonobond Ultrasonics, Inc.
- Standard United Aircraft Corporation
- Taylor-Winfield Corporation
- Tempil Division, Big Three Industries
- The Lincoln Electric Company
- The Stanley Works
- Thermal Dynamics Corporation
- Tinius Olsen Testing Machine Company, Inc.
- U. S. Department of Labor
- Wall Colmonoy Corporation
- Weston Instruments

CONTENTS

INTRODUCTION TO WELDING

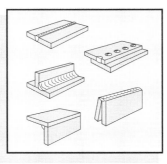

OXYACETYLENE WELDING — OAW

SHIELDED METAL-ARC WELDING — SMAW

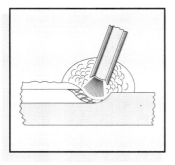

GAS SHIELDED-ARC WELDING

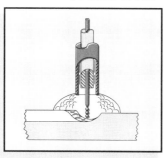

SPECIAL WELDING PROCESSES

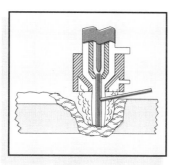

SUPPLEMENTARY WELDING DATA

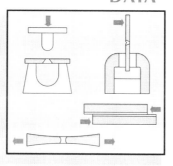

(The Lincoln Electric Company)

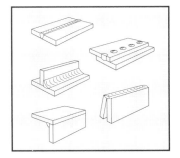

AN ESSENTIAL SKILL

Introduction to Welding

Welding is essential to the expansion and productivity of our industries. Welding has become one of the principal means of fabricating and repairing metal products. It is almost impossible to name an industry, large or small, that does not employ some type of welding. Industry has found that welding is an efficient, dependable, and economical means of joining metal in practically all metal fabricating operations and in most construction, Figure 1-1.

Figure 1-1. Welding is used in industry to join metals efficiently and economically. (*The Lincoln Electric Company*)

Figure 1-2. Many parts of airplanes are joined by various welding processes. *(Boeing Commercial Airplane Group)*

WHERE WELDING IS USED

In tooling-up for a new model automobile, a manufacturer may spend upward of a million dollars on welding equipment. Many buildings, bridges, and ships are fabricated by welding. Where construction noise must be kept at a minimum, such as in the building of hospital additions, the value of welding as the chief means of joining steel sections is particularly significant.

Without welding, the aircraft industries would never be able to meet the enormous demands for planes, rockets, and missiles, Figure 1-2. Rapid progress in the space program, has been made possible by new methods and knowledge of welding metallurgy.

Probably the most sizable contribution welding has made to society is the manufacture of special products for household use. Welding processes are employed in the construction of such items as television sets, refrigerators, kitchen cabinets, dishwashers, and other similar products.

As a means of fabrication, welding has proved fast, dependable, and flexible. It lowers production costs by simplifying design and eliminates costly patterns and machining operations.

Welding is used extensively for the manufacture and repair of farm equipment, mining and oil machinery, machine tools, jigs and fixtures, and in the construction of boilers, furnaces, and railway cars. With improved techniques for adding new metal to worn parts, welding has also resulted in economy for highly competitive industries, Figures 1-3 through 1-10.

Figure 1-3. The shielded metal-arc welding process is used in fabricating many industrial products. *(Hobart Brothers Company)*

Figure 1-4. Welding is often indispensable in assembling the steel structure of a building.

Figure 1-5. Robotic welding machines are programmed to perform repetitive welds on mass-produced products. (*The Lincoln Electric Company*)

Figure 1-6. A machine-operated gas metal-arc process is used to manufacture this water heater. (*Berkeley-Davis Inc.*)

Figure 1-7. Welding is commonly used in the maintenance and repair of farm equipment. (*The Lincoln Electric Company*)

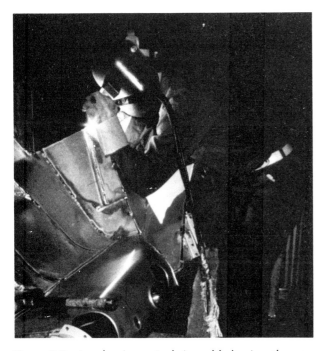

Figure 1-8. An aluminum tank is welded using the gas tungsten-arc welding process. (*Miller Electric Manufacturing Company*)

Figure 1-9. A stainless steel restaurant kettle is fabricated with a gas tungsten-arc spot welder.

Figure 1-10. Welding plays an important role in the construction and repair of heavy machinery and equipment. *(Miller Electric Manufacturing Company)*

DEVELOPMENT OF WELDING PROCESSES

Modern welding processes used today evolved from discoveries and inventions dating back to the year 2000 B.C. Forge welding was first used as a means of joining metals together. It was a crude process of joining metal by heating and hammering until the metal was fused together. Today, forge welding is still practiced but in limited applications.

The welding processes as we know them today were developing at the beginning of the twentieth century. Demands of a growing industrial economy spurred the development of three new efficient welding processes. The discovery of acetylene, later combined with oxygen, produced a flame that could be used for welding and cutting. The application of heat generated from an electric arc between carbon electrodes was the basis for the carbon arc welding

process. Resistance welding, which also uses electricity, was developed and first used in industry during this period.

Oxyacetylene welding, carbon arc welding, and resistance welding were further developed, resulting in several new and improved ways of welding metals together. One of the most significant developments was the discovery of an electrode that is consumed into the weld while providing heat from an arc. Modifications to the coating applied on consumable electrodes allowed greater applications of the shielded metal arc process.

Another improvement in the arc welding process was the addition of an inert shielding gas to protect the weld area from atmospheric contamination. This was an especially important process needed for welding magnesium and aluminum on World War II fighter planes. The electrode used was made out of tungsten and was not consumed in the weld. Originally, helium was used as a shielding gas, but helium was later replaced by less expensive argon.

Gas metal arc welding, a process also using shielding gas and an electric arc, was developed to increase the speed at which weld metal could be deposited. Different from gas tungsten arc welding, gas metal arc welding used a consumable wire electrode. The use of carbon dioxide as a shielding gas reduced the cost of this process, increasing its popularity in the manufacturing industry.

New developments in the field of welding continue to address new requirements and applications in today's industry. The welding processes discussed in this book are the product of continued refinements and variations of the welding processes discovered at the turn of the century.

Types of Welding Processes

Of the many processes of welding in use today, oxyfuel gas welding, arc, and resistance dominate the field. These processes are best explained from the standpoint of the operator's duties.

The principal duty of the operator employing oxyfuel gas welding equipment is to control and direct the heat on the edges of metal to be joined, while applying a suitable metal filler to the molten pool. The intense heat is obtained from the combustion of gas, usually acetylene and oxygen. For this reason, this process is called oxyacetylene welding. In some applications other gases—such as Mapp, propane, or natural gas—may be utilized to generate the intense heat necessary for welding.

The skills required for this job are adjustment of the regulators, selection of proper tips and filler rod, preparation of the metal edges to be joined, and the technique of flame and rod manipulation. The gas welder may also be called upon to do flame cutting with a cutting attachment and extra oxygen pressure. Flame or oxygen cutting is employed to cut various metals to a desired size or shape, or to remove excess metal from castings.

The three main types of arc welding processes used today are shielded metal-arc welding, gas tungsten-arc welding, and gas metal-arc welding. Shielded metal-arc welders perform their skill by first striking an arc at the starting point of a weld and maintaining this electric arc to fuse the metal joints. The molten metal from the tip of the electrode is then deposited in the joint, together with the molten metal of the edges, and solidifies to form a sound and uniform connection. The arc welding operator is expected to select the proper electrodes for the job or be able to follow instructions

as stated in the job specifications, to read welding symbols, and to weld any type of joint using the technique required.

Gas-shielded arc processes have gained recognition as being superior to the metal arc process. With gas-shielded arc both the arc and molten puddle are covered by a shield of inert gas. The shield of inert gas prevents atmospheric contamination, thereby producing a sounder weld. The processes known as gas tungsten arc welding and gas metal arc welding, sometimes called Tig and Mig, are either manually or automatically operated.

Resistance welding operators are responsible for the control of machines which fuse metals together by heat and pressure. If two pieces of metal are placed between electrodes which become conductors for a low voltage and high amperage current, the materials, because of their own resistance, will become heated to a plastic state. To complete the weld, the current is interrupted before pressure is released, thereby allowing the weld metal to cool for solid strength.

The operator's duty is to properly adjust the machine current, pressure, and feed settings suitable for the material to be welded. The welder usually will be responsible for the alignment of parts to be assembled and for controlling the passage of parts through the welding machine.

Selection of the Proper Welding Process

There are no hard and fast rules which govern the type of welding that is to be used for a particular job. In general, the controlling factors are kind of metals to be joined, costs involved, nature of products to be fabricated, and production techniques. Some welding jobs are best completed using the oxyfuel welding process as compared to shielded metal arc welding process.

Oxyacetylene welding, the most popular oxy-fuel gas welding process, is used in all metal working industries and in the field as well as for plant maintenance. Because of its flexibility and mobility, it is widely used in maintenance and repair work. The welding unit can be moved on a two-wheeled cart or transported by truck to any field job where breakdowns occur. Its adaptability makes the oxyacetylene process suitable for welding, brazing, cutting, and heat treating.

The chief advantages of shielded metal arc welding are the rapidity with which a high quality weld can be made at a relatively low cost and the variety of applications. Specific applications of this process are found in the manufacture of structural steel for buildings, bridges, and machinery. Shielded metal arc welding is considered ideal for making storage and pressure tanks as well as for production line products using standard commercial metals.

Since the development of gas-shielded arc processes, there are indications that they will be used extensively in the future in welding all types of ferrous and nonferrous metals in both gauge and plate thicknesses.

Resistance welding is primarily a production welding process. It is especially designed for the mass production of domestic goods, automobile bodies, electrical equipment, hardware, etc. Probably the outstanding characteristic of this type of welding is its adaptability to rapid fusion of seams.

Occupational Opportunities in Welding

The wide spread use of welding in American industry provides a constant source of employment for welders. According to the U.S. Department of Labor there are approximately 573,000 persons employed as welders. Over half of these work in industries that manufacture durable goods, such as transportation equipment, machinery, and household products. Most of the others work for construction firms and repair shops.

Employment outlook. Employment of welders is projected to increase at least as fast as the average for all occupations, with growth and replacement needs averaging 105,000 to 120,000 each year. This is particularly true in ship building, tank and boiler fabrication, rail, automotive, and aircraft manufacture, building construction, piping, and many other metalworking industries. Although there is no uniform wage rate for welders, they are considered to be in one of the higher classifications of job wages.

Training. Learning the essential skills needed to fulfill the many welding job requirements varies from a few months of on-the-job training to several years of formal training. Most employers prefer applicants who have a high school education or vocational school training in welding. Courses in mathematics, mechanical drawing, blueprint reading and general metals are very helpful. See Figure 1-11.

Young people planning careers as welders need manual dexterity, good eyesight and good eye-hand coordination. They should be able to concentrate on detailed work for long periods and must be free of any physical disabilities that would prevent them from bending, stooping or working in awkward positions. Welders must be able to lift 50 pounds regularly and 100 pounds occasionally.

Job classification. Welding jobs differ in the degree of skill required. Welding machine operators can learn the required procedures in several hours, while skilled manual welders need years of on-the-job training to master their craft. A beginner usually starts on simple production jobs and gradually works up to higher levels of skill as his or her experience increases and ability improves. Before being assigned to work where the quality and strength of the weld are critical, a welder will generally have to pass a certification test given by the employer, government agency or some other inspection authority.

Skilled manual welders have a wide range of technical knowledge involving properties of metals, effects of heat on welded structures, control of expansion and contraction forces, reading welding symbols, and recognizing welding defects. The skilled welder may be proficient in several welding areas encompassing oxyacetylene or one of the arc welding processes. As a rule the skilled welder is always certified for the particular welding job required.

The semi-skilled welders usually do repetitive work, that is, production work which generally does not involve critical safety and strength requirements. They

Figure 1-11. To attain the required welding skills, a person usually must complete a formal course of instruction under a competent instructor.

primarily weld surfaces in only one position and may or may not have to be certified.

The following are some of the principal job titles of welders:

Welding engineer
Arc welder—shielded metal arc
              —gas metal arc
              —gas tungsten arc
Welding cutter
  Submerged-arc welder
  Welder—resistance, spot, automatic
  Pipe welder
  Boilermaker welder
  Structural welder
  Maintenance welder
  Welding layout and set-up person
  Welding inspector
  Welding tester
  Welding supervisor

Skilled welders may, by promotion, become inspectors or supervisors. Actually there are unlimited opportunities for those who become thoroughly acquainted with the techniques, materials, designs, and new applications of welding processes.

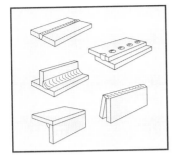

WELDING SAFETY

Introduction to Welding

Have you ever heard the saying "some people are accident-prone"? The implication is that accidents just seem to follow some individuals no matter what they do. They just seem plagued with bad luck. Actually, there is no such thing as being accident-prone. People have accidents simply because they are careless, or indifferent to safety regulations.

Each year thousands of people suffer the pain of injury because they have failed to use good judgment, Figure 2-1. In many ways, safety can be considered a habit, a kind of behavior. A habit is acquired; you are not born with it. It is the result of repetition—doing something over and over again until it becomes part of you. Thus if you consistently follow good safety practices you subconsciously build within yourself a safety awareness that usually keeps you from making foolish mistakes. Taking a chance is another way of saying "I don't know any better." Thus safety simply means using a little common sense and in so doing avoiding many serious accidents.

Finally safety is not something you read about or practice only on occasion. It has to be observed constantly. Industry places a high premium on safety—ask anyone in industry and they will tell you that a tremendous amount of time and effort is given to safety. So never take chances; you will enjoy your work more if you learn to become a safe worker.

The primary purpose of this chapter is simply to alert you to some of the general precautions which should be followed while performing various welding operations. More definite safety practices are included throughout the text where they deal with specific welding situations.

Accident Reporting

Always report an accident regardless of how slight it may be. Even a little scratch might lead to an infection, or a minute particle could result in a serious eye injury. Prompt attention to any injury usually will minimize what may become serious if neglected.

Generally, where any physical work is performed, either in a learning situation or on an actual payroll job in industry, a definite accident reporting procedure is established. Since this reporting procedure

is for the best interest of the individual, it is foolhardy to ignore it or go around it. Consequently, make it a practice to become fully informed about what should be done and then take immediate action if an accident occurs.

Work Behavior

On some occasion you may be tempted to engage in what might appear to be a harmless prank. Any form of horseplay in a shop is dangerous and can lead to an accident. There are many recorded incidents where foolish play ended in a serious injury. Most work areas are reasonably safe if proper work precautions are taken, but no one is safe if good work attitudes are ignored.

Welding Equipment Familiarization

No welding equipment of any kind should ever

Figure 2-1. Follow safety procedures to avoid serious accidents.

be used until exact instructions on how to operate it have been received. Manufacturer's recommended methods are very important and should be followed at all times. Attempting to operate a piece of equipment without instruction not only may damage the equipment but could result in a serious injury. Welding equipment of all kinds is safe to operate, providing it is used in the proper manner.

WARNING: When operating welding equipment, never try to remedy a malfunction without first consulting supervising personnel. This applies to a whole range of items from a leaking gas hose to a loose cable on a welding generator. The instructor or supervisor is knowledgeable of what should be done, and it is that person who should decide on the course of action to be taken.

Ventilation

All welding should be done in well-ventilated areas. There must be sufficient movement of air to prevent accumulation of toxic fumes or possible oxygen deficiency. Adequate ventilation becomes extremely critical in confined spaces where dangerous fumes, smoke, and dust are likely to collect, Figure 2-2.

Figure 2-2. A ventilation system is required to remove toxic fumes, smoke, and dust caused from welding. (*Nederman, Inc.*)

Where considerable welding is to be done, an exhaust system is necessary to keep toxic gases below the prescribed health limits. An adequate exhaust system is especially necessary when welding or cutting zinc, brass, bronze, lead, cadmium, or beryllium bearing metals. This includes galvanized steel and metal painted with lead-bearing paint. Fumes from these materials are toxic, and they are very hazardous to your health.

WARNING: Even with proper ventilation a respirator should be used when welding metals giving off toxic fumes. See Figure 2-3.

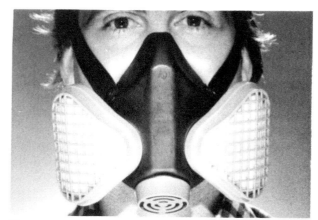

Figure 2-3. A respirator should be worn when welding metals that produce toxic fumes. (*American Optical Corporation*)

Body Protection

In any welding or cutting operation sparks, dangerous ultraviolet and infrared rays are generated. Consequently, suitable clothing and proper eye protection are necessary. See Figure 2-4. Sparks may lead to serious burns and rays are extremely dangerous to the eyes. More specific instructions concerning correct apparel and eye shields will be found in other sections dealing with various phases of welding and cutting. At the moment it is sufficient to point out that a welder must be aware of possible body dangers during any welding or cutting operation and learn the safe practices for his or her personal welfare.

Welding and Cutting Containers[1]

WARNING: Explosions, fires, and health hazards may result if welding, cutting, or other hot work is performed on containers that are not free of hazardous substances such as combustible or toxic sol-

---

1. AWS F4.1

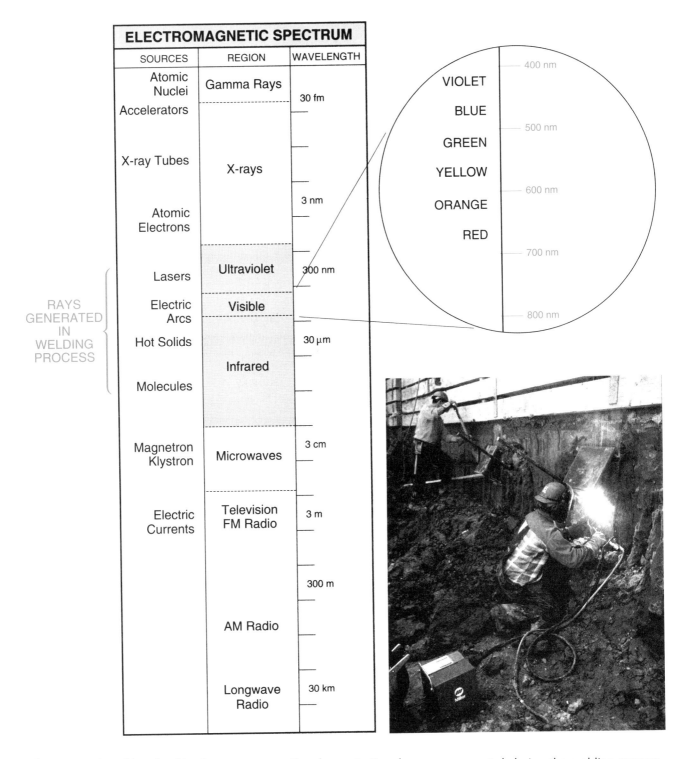

Figure 2-4. A welder should take every precaution for protection from rays generated during the welding process. *(Miller Electric Manufacturing Company)*

ids, liquids, vapors, dusts, and gases. No such container should be presumed to be clean or safe; but these containers can be made safe for such work provided the necessary steps and precautions are followed.

Hazardous substances include those that are combustible, toxic, or corrosive. They may be present in a container having previously held one of the following:

1. A volatile liquid that releases potentially hazardous, flammable, and/or toxic vapors at atmospheric conditions.

2. An acid or alkaline material that reacts with metals to produce hydrogen.

3. A nonvolatile liquid or solid that at ordinary temperatures will not release potentially hazardous vapors, but will do so if the container is heated.

4. A dust cloud of finely divided airborne particles that may still be present in an explosive concentration.

5. A flammable or toxic gas.

WARNING: Cleaning and welding a container that has held unknown substances should never be performed as this involves unknown risk.

Before any container is cleaned, the hazardous characteristics of the substance previously held by the container must be determined. Information about the substance, and safety precautions are contained in the material safety data sheet. A Material Safety Data Sheet (MSDS) is printed material used to relay chemical hazard information. Chemical manufacturers, distributors, and importers must develop an MSDS for each hazardous chemical. If an MSDS is not provided, the employer must write to the manufacturer, distributor, or importer to obtain the missing MSDS.

An MSDS has no prescribed format but must contain certain information related to identification, physical and chemical characteristics, fire hazards, reactivity and health hazard data, handling precautions, and control measures. MSDS files must be kept up-to-date and well-organized to allow quick access to information in an emergency situation.

WARNING: In all container cleaning operations, suitable eye protection and personal protective equipment must be used and adequate ventilation provided.

For maximum safety, only qualified personnel shall designate the container cleaning method. The cleaning method used depends upon the substance previously held in the container. The water method of cleaning is used when the substance is known to be readily soluble in water. The residue can be removed by completely filling the container with water and draining several times. When the substance originally held in the container is not readily soluble in water, the container should be treated by one of the following methods:

a. The *hot chemical solution method* uses trisodium phosphate (a strong washing powder) or a commercial caustic cleaning compound dissolved in hot water.

b. The *steam method of cleaning* uses low-pressure steam and a hot soda or soda ash.

c. The *mechanical cleaning method* is generally used when scaly, dry, or insoluble residues are left on the surface. Mechanical cleaning may be performed by scraping, sand or grit blasting, high-pressure water washing, brushing, filling the container one-quarter full of clean dry sand and rolling it on the floor, or any method in which the contaminant can safely be dislodged. During mechanical cleaning the container should be grounded to minimize the possibility of static charge buildup and spark charges.

d. The *chemical cleaning method* is generally used when the container contains insoluble deposits or when it cannot be mechanically cleaned. Care must be used in selecting a chemical solvent; some solvents may be as hazardous as the deposits they are intended to remove. When selecting chemical solvents consult the manufacturer of the material to be removed.

Occasionally, combinations of all the methods of cleaning prior to welding or cutting must be utilized. Care must be exercised to protect personnel and prevent hazardous reactions when combining some of the cleaning methods.

As a final precaution after cleaning, a container should be vented and filled with water before welding or cutting is undertaken. The container should be arranged so that water can be kept filled within a few inches of the point where the welding or cutting is to take place. See Figure 2-5.

Figure 2-5. Before welding or cutting a container, partially fill the container with water.

WARNING: Be sure that there is a vent or opening to allow for the release of air pressure or steam.

Here are some other safety precautions to follow when welding or cutting containers.
WARNINGS:
1. Never use oxygen to ventilate a container as it may start a fire or cause an explosion.

2. Never rely on your nose or eyes to determine if it is safe to weld or cut a closed container. A small amount of residual flammable liquid or gas may cause a serious explosion.

3. Always use a spark-resistive tool when heavy sludge or scale must be removed by scraping or hammering.

4. Never weld or cut drums, barrels, or tanks until you know there is no danger of fire or explosion.

SAFETY IN CUTTING

Fires often occur in a cutting operation simply because proper precautions were not taken. Too often a worker forgets that sparks and falling slag can travel as much as 35 feet and can pass through cracks out of sight of the goggled operator. Persons responsible for supervising or performing cutting of any kind should observe the following:
WARNINGS:
1. Never use a cutting torch where sparks will be a hazard, such as near rooms containing flammable

materials, especially dipping or spraying rooms.

2. If cutting is to be over a wooden floor, sweep the floor clean and wet it down before starting the cutting. Provide a bucket or pan containing water or sand to catch the dripping slag.

3. Keep a fire extinguisher nearby whenever any cutting is done. See Figure 2-6.

4. Whenever possible perform the cutting operation in wide open areas so sparks and slag will not become lodged in crevices or cracks.

5. If cutting is to be done near flammable materials, and the flammable materials cannot be moved, suitable fire-resisting guards, partitions, or screens must be used.

6. In plants where a greasy, dirty or gassy atmosphere exists, extra precaution should be taken to avoid explosions resulting from electric sparks or open fire during a cutting or welding operation.

7. Keep flame and sparks away from oxygen cylinders and hose.

8. Never do any cutting near ventilators.

9. Move combustible materials at least 40 feet away from any cutting or welding operation.

10. Where the risk of fire is great, always have standby watchers with fire extinguishers.

11. Never use oxygen to dust off clothing or work.

12. Never use oxygen as a substitute for compressed air.

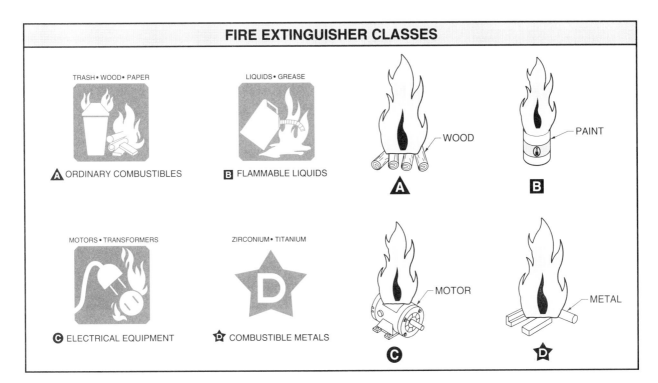

Figure 2-6. Fire extinguishers are classified as A, B, C, or D.

SAFETY IN GAS WELDING

Specific instructions dealing with safety in gas welding are listed in the units involving oxyacetylene welding. These precautions cover proper handling of cylinders, operation of regulators, use of oxygen and acetylene, welding hose, testing for leaks and lighting a torch. All of these safety regulations are extremely important and should be followed with the utmost care and regularity.

In addition to the normal precautions to be observed in gas welding, a very significant safety procedure involves the piping of gas. All piping and fittings used to convey gases from a central supply system to work stations must withstand a minimum pressure of 150 psi. Oxygen piping can be of black steel, wrought iron, brass, or copper. Only oil-free compounds should be used on oxygen threaded connections. Piping for acetylene must be of wrought iron. After assembly all piping must be blown out with air or nitrogen to remove foreign materials.

WARNING: Copper piping must never be used with acetylene. Acetylene and copper form copper acetylide. Copper acetylide collects in low points in the piping system and will explode violently if shocked or moved suddenly.

There are five basic rules which contribute to the safe handling of oxyacetylene equipment. These are:
1. Keep oxyacetylene equipment clean, free of oil, and in good condition.
2. Avoid oxygen and acetylene leaks.
3. Open cylinder valves slowly.
4. Purge oxygen and acetylene lines before lighting torch.
5. Keep heat, flame and sparks away from combustibles.

SAFETY IN ARC WELDING

Arc welding includes shielded metal arc, inert gas shielded arc and resistance welding. Only general safety measures can be indicated for these areas because arc welding equipment varies considerably in size and type. Equipment may range from a small portable shielded metal-arc welder to highly mechanized production spot or gas shielded-arc welders. In each instance specific manufacturers' recommendations should be followed.

Safety practices which are generally common to all types of arc welding operations are as follows.
CAUTIONS:
1. Install welding equipment according to provisions of the National Electrical Code®.

2. Be sure a welding machine is equipped with a power disconnect switch which is conveniently located at or near the machine so the power can be shut off quickly. See Figure 2-7.

3. Don't operate the range switch under load. The range switch which provides the current setting should be operated only while the machine is idling and the current is open. Switching the current while the machine is under a load will cause an arc to form between the contact surfaces.

WARNINGS:

1. Don't make repairs to welding equipment unless the power to the machine is shut OFF. The high voltage used for arc welding machines can inflict severe and fatal injuries.

2. Don't use welding machines without proper frame grounding connection in accordance with the National Electrical Code. Stray current may develop which can cause severe shock when ungrounded parts are touched. Make sure the ground to your

Figure 2-7. Be sure that there is a power disconnect switch close at hand.

work is securely attached. See Figure 2-8. Do not ground to pipelines carrying gases or flammable liquids.

3. Don't use electrode holders with loose cable connections. Keep connections tight at all times. Avoid using electrode holders with defective jaws or poor insulation.

4. Don't change the polarity switch when the machine is under a load. Wait until the machine idles and the circuit is open. Otherwise, the contact surface of the switch may be burned and the person throwing the switch may receive a severe burn from the arcing.

been welded or heated. Figure 2-10.

9. Always wear protective safety glasses. A face shield may be necessary when chipping or grinding. A small particle of slag or metal may cause a severe eye injury. See Figure 2-11. Contact lenses should not be worn when welding. Lenses rest on the fluid of eyes and there is a possibility that particles, fumes, and gases could become lodged between the eye and the lens during the welding process.

10. Don't weld on hollow (cored) castings unless they have been properly vented; otherwise an explosion may occur. See Figure 2-12.

11. Be sure press-type welding machines are ef-

Figure 2-8. Before welding, attach the ground securely to your work.

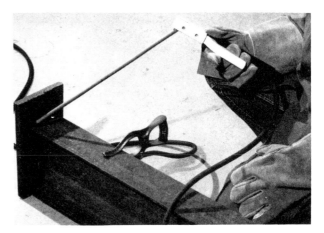

Figure 2-9. Operating with currents above the rated cable capacity can cause overheating of the welding cables.

5. Don't overload welding cables or operate a machine with poor connections. Operating with currents beyond the rated cable capacity causes overheating. Neatly arrange the welding cables and secure proper connections. Poor connections may cause the cable to arc when it touches metal grounded in the welding circuit. See Figure 2-9.

6. Don't weld in damp areas and keep hands and clothing dry at all times. Dampness on the body may cause an electric shock. Never stand or lie in puddles of water, on damp ground, or against grounded metal when welding without suitable insulation. Use a dry board or rubber mat to stand on.

7. Don't strike an arc if someone without proper eye protection is nearby. Arc rays are harmful to the eyes and skin. If other persons must work nearby, the welding area should be partitioned off with a fire-retardant canvas curtain to protect them from the arc welding flash.

8. Never pick up pieces of metal which have just

Figure 2-10. Assume all welded objects are hot before touching.

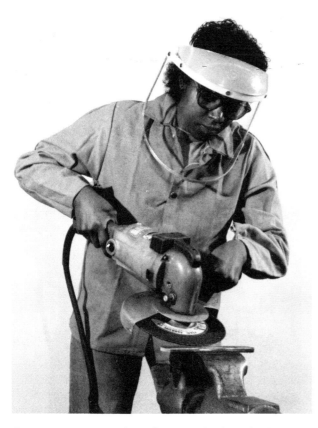

Figure 2-11. Wear safety glasses and a face shield when chipping and grinding.

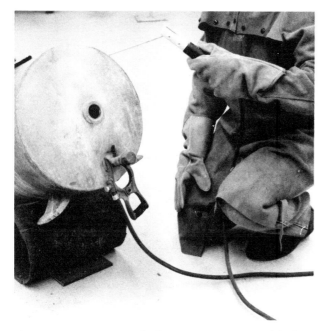

Figure 2-12. Be sure a hollow casting is vented before welding. (*Airco*)

fectively guarded.

12. Be sure suitable spark shields are used around equipment in flash welding.

13. When welding is completed, turn OFF the machine, pull the power disconnect switch and hang the electrode holder in its designated place.

14. Inspect cables for cuts, nicks, or abrasion.

Final Precaution

Remember, accidents do not just happen. Invariably they occur because of indifference to regulations, lack of information, or just plain carelessness.

Injury of any kind is painful and very often can incapacitate a person, or even produce a permanent deformity. If more thought were given to the consequences of injuries there would be less tendency to ignore safety precautions and thus, fewer accidents.

QUESTIONS FOR STUDY AND DISCUSSION

1 Why is there very little basis for saying some people are accident-prone?

2 What are some of the main causes of accidents?

3 Why should all accidents be reported immediately?

4 How is it possible to become involved in an accident when playing around in the shop?

5 What may happen if you attempt to use welding equipment without proper instruction?

6 What should be done if some malfunction occurs in any welding equipment?

7 What general practice should be followed regarding ventilation during the performance of any welding operation?

8 Why should strict attention be given to proper clothing and eye protection?

9 Why should used containers be thoroughly cleaned and safety processed before any welding or cutting is done on them?

10 Why do fires often occur during a cutting operation?

11 What are some of the precautions that should be taken when using a cutting torch?

12 Why is it dangerous for acetylene to come in contact with high alloyed copper piping?

13 What is the significance of having each welding machine equipped with a power disconnect switch?

14 Why should a welding machine never be overloaded?

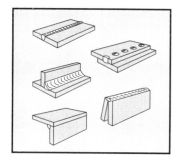

WELDING METALLURGY

Introduction to Welding

In preparing to become a skillful welder you should become familiar with the effects of heat on the structure of metal and with what happens to metal when certain alloying elements are added to it.

You will also need to know what safeguards must be followed in welding metals because application of heat during the welding process may destroy the very elements which were originally added to improved the structure of the metal. For example, metals expand and contract, thereby setting up great stresses which often result in severe distortions. Improper welding of stainless steel may result in a complete loss of its corrosion-resistant qualities, and welding high-carbon steel in the same manner as low-carbon steel may produce such a brittle weld as to make the welded piece unusable.

This chapter deals with the metallurgy of welding; that is, the formation of impurities and the effects of heat on the chemical, physical, and mechanical properties of metals.

PROPERTIES OF MATERIALS

Chemical, physical, and mechanical properties have a very significant influence in any welding operation. This will become more apparent in later chapters dealing with specific welding techniques. These properties can be defined as follows:

Chemical properties. Chemical properties are those which involve corrosion, oxidation, and reduction. *Corrosion* is a wasting away of metal due to various atmospheric elements. *Oxidation* is the formation of metal oxides which occur when oxygen combines with a metal. *Reduction* refers to the removal of oxygen from the surrounding molten puddle to reduce the effects of atmospheric contamination.

In any welding situation, it is important to remember that oxygen is a highly reactive element. When it comes in contact with metal, especially at high temperatures, undesirable oxides and gases are formed, thereby complicating the welding process. Hence, the success of any welding operation depends on how well oxygen can be prevented from contaminating the molten metal.

Physical properties. Physical properties are those which affect metals when they are subject to heat generated by welding such as *melting point, thermal conductivity*, and *grain structure*. Solid metals change into a liquid state at different temperatures. When cooling from a liquid state the atoms will form various crystal patterns (lattices). The strength of a weld often depends on how these lattices are controlled and how much heat is necessary to produce proper fusion of metal. Equally important is being aware that some metals have a high rate of heat conductivity. Also a welder needs to understand how heat will affect the grain structure of metals since the grain size of the crystalline structure has a direct bearing on the strength of the welded joint.

Mechanical properties. Mechanical properties are those which determine the behavior of metals under applied loads. These include a wide range of properties such as *tensile strength, ductility, toughness, brittleness* and others, all of which are extremely important in their relationship to welding.

STRUCTURE OF METALS

When you examine a polished piece of metal under a microscope, you will see small grains. Each of these grains is made up of smaller particles, called atoms, of which all matter is composed.

The grains, or crystals as they are often called, vary in shape and size. The arrangement of the atoms determines the shape of the crystalline structure. In general, the crystals of the more common types of metals arrange themselves in three different patterns. These are known as *space-lattices*.

A space-lattice is a visual representation of the orderly geometric pattern into which the atoms of all metals arrange themselves upon cooling from a liquid to a solid state.

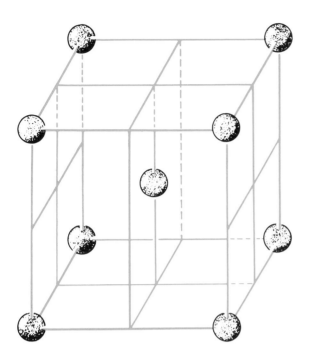

Figure 3-1. In a body-centered cubic crystal, there is one atom at each corner and one in the center. Iron, molybdenum, chromium, columbium, tungsten, and vanadium have this structure.

The first type of space-lattice, illustrated in Figure 3-1, is the *body-centered cube*. Here you will find nine atoms—one at each corner of the cube and one in the center. This crystal pattern is found in such metals as iron, molybdenum, chromium, columbium, tungsten, and vanadium.

The second crystal pattern in the *face-centered cube*. Notice in Figure 3-2 how the atoms are arranged. Metals having this space-lattice pattern are aluminum, nickel, copper, lead, platinum, gold, and silver.

The third space-lattice is called the *close packed hexagonal form*. See Figure 3-3. Among the metals having this type of crystalline structure are cadmium, bismuth, cobalt, magnesium, titanium, and zinc.

Metals with the face-centered lattice are generally *ductile*; that is, plastic and workable. Metals with close-packed hexagonal lattice lack plasticity and cannot be cold-worked, with the exception of zirconium and titanium. Metals with body-centered

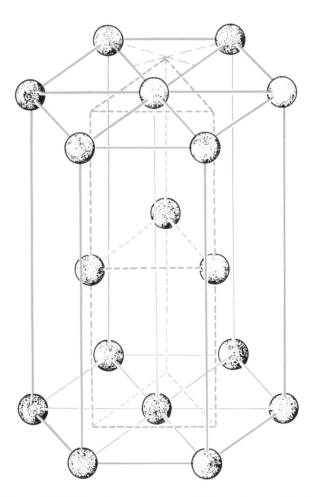

Figure 3-2. This is the face-centered cubic crystal, found in such metals as aluminum, nickel, copper, lead, platinum, gold, and silver.

Figure 3-3. The atoms are arranged in a close-packed hexagonal crystal in cadmium, bismuth, cobalt, magnesium, titanium, and zinc.

crystals have higher strength but lower cold working properties than those with the face-centered pattern.

Crystallization of Metals

All metals solidify in the form of crystals. Each metal has its own characteristic geometric pattern. Some metals may even change from one crystal structure to another crystal structure at various temperature levels. For example, iron when heated changes completely to a face-centered cubic structure at a temperature of 1670°F [910°C].

As liquid metal is cooled it loses thermal energy (heat) to the air and walls of the container. At the *solidification temperature* the atoms of the metal assume their characteristic crystal structure. Crystals begin growing at random in the melt at points of lowest energy. If the rate of cooling is fast, more crystals will form instantaneously than at slow rates of cooling. The more crystals that are growing simultaneously the finer will be the grain size of the metal.

Grain size is important since fine-grained steels have far superior mechanical properties than coarse-grained steels. Hence, it is important for a welder to preserve the grain size of the parent metal. The use of excessive heat leads to a slow rate of cooling, thus producing coarse grains and brittleness in a weldment.

Heating Effect on Grain Structure of Steel

When steel, which is carbon and iron, is heated from room temperature to above 1333°F [835°C], the pearlite grains change from a body-centered lattice to a face-centered structure. Such an arrangement of iron atoms is known as *gamma iron*.

What has happened is that while the steel went through its *critical temperature* (temperature above which steel must be heated so it will harden when quenched), the iron carbide separated into carbon and iron, with the carbon distributing itself evenly in the iron. The material is now called *austenite*.

If the heating is continued beyond the critical point, the grains grow larger or coarser until the melting point is reached. When the steel melts, the crystal structure is completely broken and the atoms float about without any definite relationship to one another.

Cooling Effect on Grain Structure of Steel

If you cool a metal from a molten state to room temperature, the change that takes place, under proper conditions, is exactly the opposite of what occurs while the metal is heating.

As the metal begins to cool, the crystals of pure iron start to solidify. This is followed by a crystallization of austenitic grains, and eventually the entire mass becomes solid.

During the range of temperatures at which various stages of solidification takes place, the metal passes from a mushy condition to a solid solution. While in a mushy stage the metal can be shaped easily. After it has reached a solid state, even though the alloy is still hot, it can be formed only by applying heavy pressure or hammering (forging).

With continued cooling of the solid metal, the austenite contracts evenly as the temperature falls. When it reaches its *transformation temperature*, the temperature drop stops for a time. At this point there occurs a rearrangement of *gamma iron* to *alpha iron* as well as a separation of iron carbide and pure iron into *pearlite* grains.

The transformation of the metal from a liquid to a solid is important because the proper rearrangement of the atoms depends on the rate of cooling. If, for example, a piece of 0.83 percent carbon steel is cooled rapidly after its critical temperature is reached, certain actions are arrested before the pearlitic structure can be formed. The result is a metal that is hard, but very brittle, known as *martensite*. See Figure 3-4. Martensite is the constituent found in fully hardened steel which is hard and brittle. On the other hand, if the rate of quenching (cooling) is somewhat slower, the structure will be much more ductile.

Figure 3-4. Martensite is a very hard but brittle metal produced through rapid cooling of carbon steel from its critical temperature.

IMPORTANCE OF CARBON IN STEEL

Carbon is the principal element controlling the structure and properties that might be expected from any carbon steel. The influence that carbon has in strengthening and hardening steel is dependent upon the amount of carbon present and upon its microstructure. Slowly cooled carbon steels have a relatively soft iron pearlitic microstructure; whereas rapidly quenched carbon steels have a strong, hard, brittle, martensitic microstructure.

In carbon steel, at normal room temperature, the atoms are arranged in a body-centered lattice. This is known as *alpha iron*. Each grain of the structure is made up of layers of pure iron (ferrite) and a combination of iron and carbon. The compound of iron and carbon, or iron carbide, is called *cementite*. The cementite is very hard and has practically no ductility.

In a steel with 0.83 percent carbon, the grains are *pearlitic*, meaning that all the carbon is combined with iron to form iron carbide. This is known as a *eutectoid mixture* of carbon and iron. See Figure 3-5.

If there is less than 0.83 percent carbon, the mixture of pearlite and ferrite is referred to as *hypoeutectoid*. An examination of such a mixture would show grains of pure iron and grains of pearlite as shown in Figure 3-6.

When the metal contains more than 0.83 percent carbon, the mixture consists of pearlite and iron carbide and is called *hypereutectoid*. Notice in Figure 3-7 how the grains of pearlite are surrounded by iron carbide. In general, the greatest percentage of steel used is of the hypoeutectoid type, that which has less than 0.83 percent carbon.

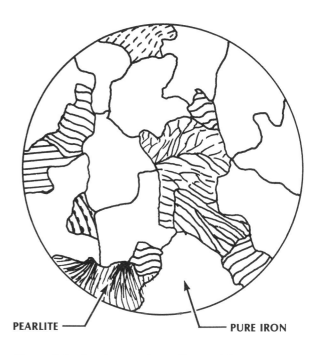

PEARLITE ——————| |—————— PURE IRON

Figure 3-6. A mixture of pearlite and ferrite grains in steel with less than 0.83 percent carbon is called hypoeutectoid.

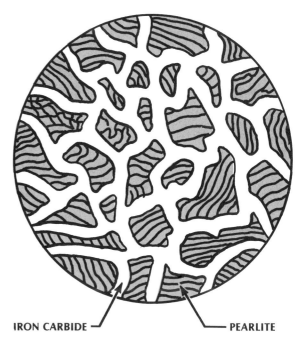

IRON CARBIDE ——————| |—————— PEARLITE

Figure 3-5. Pearlite grains arrange themselves in a eutectoid mixture.

Figure 3-7. In steel that contains more than 0.83 percent carbon, iron carbide surrounds the grains of pearlite. This mixture is called hypereutectoid.

Other Factors Altering Strength and Structure

When a metal is *cold-worked* (that is; hammered, rolled or drawn through a die) the ferrite and pearlite grains are made smaller and the metal becomes stronger and harder. If, after cold working, the metal is heated and allowed to cool, the grain size is again increased and the metal softened.

The grain size of some metals is reduced and the strength improved through a heating and quenching process. Thus, if a high-carbon steel is heated to a prescribed temperature and then immediately quenched in oil or water, followed by a tempering process, the grain size remains fine. But if you allow the same metal to heat for a long time or if you subject it to temperatures beyond the critical range, then the grain size increases and the metal is weakened. This point is particularly important to remember in welding various steel alloys. The problem of structural change is not too serious in welding mild steel. On the other hand, alloy steels are greatly dependent on space-lattice formation and grain size for their strength. Therefore, you must take extreme care during welding to avoid seriously altering a metal's space-lattice pattern through excessive application of heat or improper treatment of the weld during its cooling stages in order to avoid this problem.

Effects of Heat of the Welding Process

In welding you must realize, too, that one edge of the metal may cool rapidly, thereby resulting in the formation of hard spots which cause cracks or failure in the weld. Also, there will be conditions where the metal is in a molten state at one point while the surrounding areas may have a temperature ranging from near the molten point down to room temperature. This means that in some areas the crystal structure is completely broken down while elsewhere recrystallization is taking place.

Keep in mind that when hardenable steels are being fused, and you make no effort to control the structural changes either through preheating or by slowing down the cooling rate, the completed weld will be too brittle to be of any value. If a piece of steel, such as an automobile spring, is welded, the heat will remove the springiness from the metal. Moreover, you must remember that if a weld is made on a hardened structure, the act of welding will usually soften the steel and lower its strength. Such metals must then be heat treated to restore their original properties. It is evident then, that in welding any alloy steel, an understanding of the effects of heating and cooling is important.

Heat Treating Metals

Heat treatment is used to soften metal and relieve internal stresses (annealing), harden metal, and temper metal (to toughen certain parts). An understanding of these processes is important to a welder because often he must be aware of how welding heat will affect the structure which he is welding.

Annealing is a softening process which allows metal to be more readily machined and also eliminates stresses in metal after it has been welded. The steel is heated to a certain temperature and held at this temperature to allow the carbon to become evenly distributed throughout the steel. The degree of annealing temperature varies with different kinds of steel. After the metal has been heated for a sufficient period, it is allowed to cool slowly either in the furnace or by burying it in ashes, lime, or in some other insulating material.

For some metals, the *normalizing treatment* is used. It differs from standard annealing in that the steel is heated to a higher temperature for shorter periods and then air cooled.

Stress relieving is a means of removing the internal stresses which develop during the welding operation. The process consists of heating the structure to a temperature below the critical range (approximately 1100°F [594°C] and allowing it to cool slowly. Another method of relieving stresses is *peening* (hammering). However, peening must be undertaken with considerable care because there is always danger of cracking the metal.

Stress relieving is done only if there is a possibility that the structure will crack upon cooling and no other means can be used to eliminate expansion and contraction forces.

Hardening increases the strength of pieces after they are fabricated. It is accomplished by heating the steel to some temperature above the critical point and then cooling it rapidly in air, oil, water, or brine. Only medium, high, and very-high-carbon steels can be hardened by this method. The temperature at which the steel must be heated varies with the steel used.

The tendency of a steel to harden may or may not be desirable depending upon how it is going to be processed. For example, if it is to be welded, a strong tendency to harden will make a steel brittle and susceptible to cracking during the welding process. Special precautions such as preheating and a very careful control of heat input and cooling will be necessary to minimize this condition. During welding, an extremely high localized temperature difference exists between the molten metal of the weld and the metal being welded. The cold parent metal acts as a quench to the weld metal and the metal nearby which has been heated above the upper *critical temperature* (the metal's temperature of transformation). The resulting structure of these areas is hard, brittle martensite. The greater the hardenability of a steel, the less severe the rate of heat extraction necessary to cause it to harden. This is one of the reasons that alloy and high-carbon steels have to be welded with greater care than ordinary low-carbon steels.

Case Hardening

Case hardening is a process of hardening low-carbon or mild steels by adding carbon, nitrogen, or a combination of carbon and nitrogen to the outer surface, forming a hard, thin outer shell. The three principal case hardening techniques are known as carburizing, cyaniding, and nitriding.

Carburizing consists of heating low-carbon steel in a furnace containing a gas atmosphere with the desired amount of carbon monoxide. An alternate method is to heat the steel in contact with a carbon material such as charcoal, coal, nuts, beans, bone, leather or a combination of these. However, modern methods of carburizing use gas atmospheres almost exclusively.

The piece is heated to a temperature between 1650° and 1700°F [899° to 927°C] where steel in the austenitic condition readily absorbs carbon on its surface. The length of the heating period depends on the thickness of the hardened case desired. After heating, the steel is quenched, which produces a material with a hard surface and a relatively tough inner core.

Cyaniding involves heating a low-carbon steel in sodium cyanide or potassium cyanide. The cyanide is heated until it reaches a temperature of 1500°F [815°C] and then the steel is placed in the liquid bath. This produces a very thin outer case which is harder than that obtained by the carburizing process.

Nitriding is a case hardening method which produces the hardest surface of any hardening process. Hardness is obtained by the formation of hard, wear-resistant nitrogen compounds in certain alloy steels where distortion must be kept to a minimum. The alloy is heated to about 900° to 1000°F [482°C to 538°C] in an atmosphere of dissociated ammonia gas.

MECHANICAL PROPERTIES OF METALS

Mechanical properties are measures of how materials behave under applied loads. Another way of saying this is how strong is a metal when it comes in contact with one or more forces. If you know the strength properties of a metal, you can build a structure that is safe and sound. Likewise, when a welder knows the strength of his weld as compared with the base metal, he can produce a weldment that is strong enough to do the job. Hence strength is the ability of a metal to withstand loads (forces) without breaking down.

Some of the basic terms that are associated with mechanical properties of metals are included in the paragraphs that follow. A welder should become familiar with them because they are often directly related to his ability to produce sound welds.

Stress is the internal resistance a material offers to being deformed and is measured in terms of the applied load over the area. See Figure 3-8A.

Strain is the deformation that results from a stress and is expressed in terms of the amount of deformation per inch. See Figure 3-8B.

STRESS

Figure 3-8A. Stress, measured in terms of the applied load over the area, is the internal resistance a material offers to being deformed.

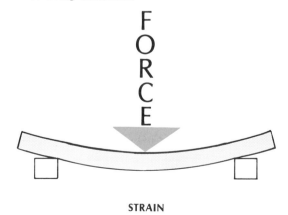

STRAIN

Figure 3-8B. Strain, expressed in terms of amount per inch, is the deformation resulting from a stress.

Elasticity is the ability of a metal to return to its original shape after being elongated or distorted, when the forces are released. See Figure 3-9. A rubber band is a good example of what is meant by elasticity. If the rubber is stretched, it will return to its original shape after you let it go. However, if the rubber is pulled beyond a certain point, it will break. Metals with elastic properties react in the same way.

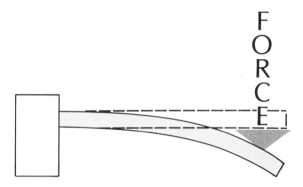

Figure 3-9. A metal having elastic properties returns to its original shape after the load is removed.

Elastic limit is the last point at which a material may be stretched and still return to its undeformed condition upon release of the stress.

Modulus of elasticity is the ratio of stress to strain within the elastic limit. The less a material deforms under a given stress the higher the modulus of elasticity. By checking the modulus of elasticity the comparative stiffness of different materials can readily be ascertained. Rigidity or stiffness is very important for many machine and structural applications.

Tensile strength is that property which resists forces acting to pull the metal apart. See Figure 3-10. It is one of the more important factors in the evaluation of a metal.

Compressive strength is the ability of a material to resist being crushed. See Figure 3-11. Compression is the opposite of tension with respect to the direction of the applied load. Most metals have high tensile strength and high compressive strength. However, brittle materials such as cast iron have high compressive strength but only moderate tensile strength.

Bending strength is that quality which resists forces from causing a member to bend or deflect in the direction in which the load is applied. Actually a bending stress is a combination of tensile and compressive stresses. See Figure 3-12A to grasp the idea.

Torsional strength is the ability of a metal to withstand forces that cause a member to twist. See Figure 3-12B.

Shear strength refers to how well a member can withstand two equal forces acting in opposite directions. See Figure 3-12C.

Fatigue strength is the property of a material to resist various kinds of rapidly alternating stresses. For example, a piston rod or an axle undergoes complete reversal of stresses from tension to compression.

Impact strength is the ability of a metal to resist loads that are applied suddenly and often at high velocity. The higher the impact strength of a metal the greater the energy required to break it. Impact strength may be seriously affected by welding since it is one of the most structure sensitive properties.

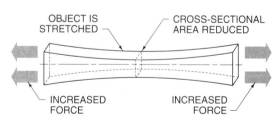
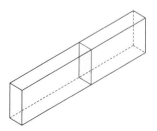

Figure 3-10. A metal with tensile strength resists pulling forces.

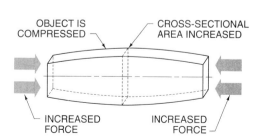

Figure 3-11. Compressive strength refers to a metal's ability to resist crushing forces.

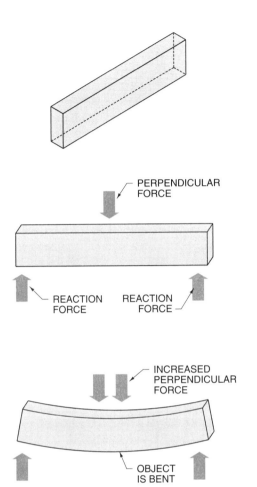

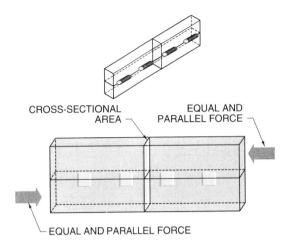

Figure 3-12C. Two equal forces acting in opposite directions test a metal's shear strength.

Ductility refers to the ability of a metal to stretch, bend, or twist without breaking or cracking. See Figure 3-13. A metal having high ductility, such as copper or soft iron, will fail or break gradually as the load on it is increased. A metal of low ductility, such as cast iron, fails suddenly by cracking when subjected to a heavy load.

Hardness is that property in steel which resists indentation or penetration. See Figure 3-14. Hardness is usually expressed in terms of the area of an indentation made by a special ball under a standard load, or the depth of a special indenter under a specific load.

Brittleness is a condition whereby a metal will easily fracture under low stress. It is a property which often develops because of improper welding techniques. Brittleness is a complete lack of ductility.

Toughness may be considered as strength, together with ductility. A tough material or weld is one which may absorb large amounts of energy without breaking. It is found in metals which exhibit a high elastic limit and good ductility. Welding materials of this kind must be done with a great deal of care. For example, improper application of heat may change the grain size and carbon distribution in the metal so its inherent toughness will be completely destroyed.

Malleability is the ability of a metal to be deformed by compression forces without developing defects, such as encountered in rolling, pressing, or forging.

Figure 3-12A. A bending stress is a combination of tensile and compressive stresses.

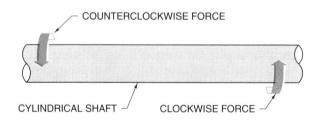

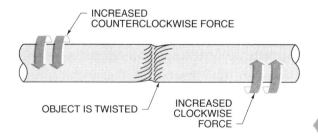

Figure 3-12B. The ability of a metal to withstand twisting forces is called torsional strength.

Figure 3-13. A metal with high ductility can stretch, bend, or twist without breaking or cracking.

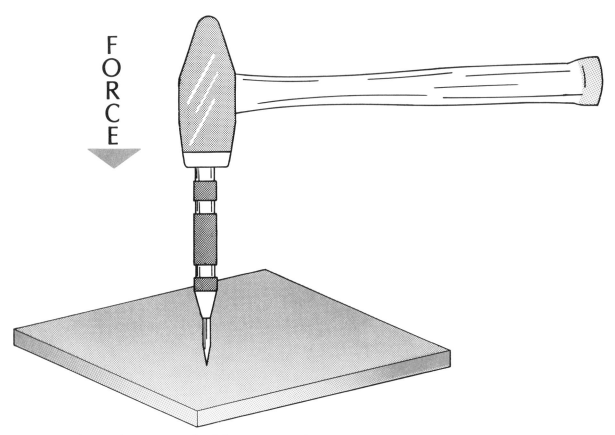

Figure 3-14. Hardness refers to a metal's ability to resist indentation or penetration.

Creep is a slow but progressively increasing strain, usually at high temperatures, causing the metal to fail.

Cryogenic properties of metals represent behavior characteristics under stress in environments of very low temperatures. In addition to being sensitive to crystal structure and processing conditions, metals are also sensitive to low and high temperatures. Some alloys which perform satisfactorily at room temperatures may fail completely at low or high temperatures. The change from ductile to brittle failure occurs rather suddenly at low temperatures.

Coefficient of expansion is the amount of expansion in one inch or one foot produced by a temperature rise of 1°F. The expansion rate of metals is always an important factor in welding.

CLASSIFICATION OF CARBON STEELS

A plain carbon steel is one in which carbon is the only alloying element. The amount of carbon in the steel controls its hardness, strength, and ductility. The higher the carbon content, the harder the steel. Conversely, the less the carbon the greater the ductility of the steel.

Carbon steels are classified according to the percentage of carbon they contain. They are referred to as low, medium, high, and very-high-carbon steels.

Low-carbon steels. Steels with a carbon range of 0.05 to 0.30 percent are called low-carbon steels. Steels in this class are tough, ductile, and easily machined, formed, and welded. Most of them do not respond to any heat treating process except case hardening. Low-carbon steel, when subjected to the spark test, will throw off long, white-colored streamers with very little or no sparklers. See Figure 3-15.

Medium-carbon steels. These steels have a carbon range from 0.30 to 0.45 percent. They are strong and hard but cannot be worked or welded as easily as low-carbon steels. Because of their higher carbon content, they can be heat treated. Successful welding of these steels often requires special electrodes, but even then greater care must be taken to prevent formation of cracks around the weld area.

The spark test will show more numerous sparklers, beginning closer to the wheel, with the streamers much lighter in color.

High and very-high-carbon steels. Steels with a carbon range of 0.45 to 0.75 percent are classified as high-carbon and those with 0.75 to 1.7 percent carbon as very-high-carbon steels. Both of these steels respond well to heat treatment. As a rule, steels up to 0.65 percent carbon can be welded with special electrodes, although preheating and stress relieving techniques must often be used after the welding is completed. Usually it is not practical to weld steels in the very-high-carbon range.

The spark test for high-carbon steels can easily be recognized by the numerous explosions or sparklers given off, which are practically white in color. See Figure 3-15.

ALLOY STEELS

An alloy steel is a steel to which one or more of such elements as nickel, chromium, manganese, molybdenum, titanium, cobalt, tungsten, or vanadium have been added. The addition of these elements gives steel greater toughness, strength, resistance to wear, and resistance to corrosion.

Alloy steels are called by the predominating element which has been added. Most of them can be welded, provided special electrodes are used. The more common elements added to steel are:

Chromium. When quantities of chromium are added to steel the resulting product is a metal having extreme hardness and resistance to wear without making it brittle. Chromium also tends to refine the grain structure of steel, thereby increasing its toughness. It is used either alone in carbon steel or in combination with other elements such as nickel, vanadium, molybdenum, or tungsten.

Manganese. The addition of manganese to steel produces a fine grain structure which has greater toughness and ductility.

Molybdenum. This element produces the greatest hardening effect of any element except carbon and at the same time it reduces the enlargement of the grain structure. The result is a strong, tough steel. Although molybdenum is used alone in some alloys, often it is supplemented by other elements, particularly nickel or chromium or both.

Nickel. The addition of nickel increases the ductility of steel while allowing it to maintain its strength. When large quantities of nickel are added (25 to 35 percent), the steels not only become tough but develop high resistance to corrosion and shock.

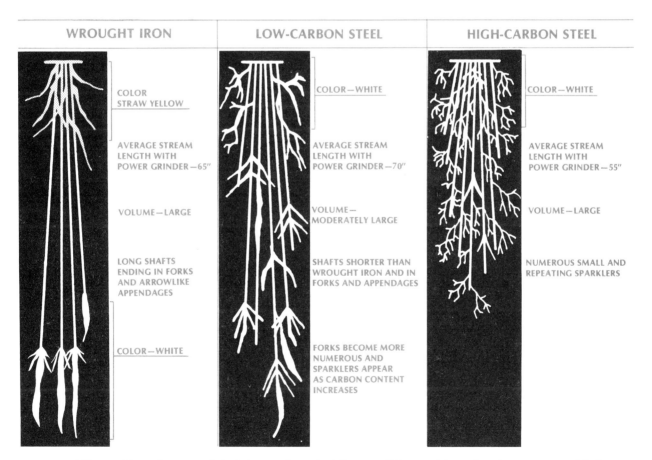

Figure 3-15. When subjected to a spark test, low-carbon steel throws off long, white colored streamers and little or no sparklers. High-carbon steel, in contrast, gives off numerous short sparklers which are practically white in color.

Vanadium. Addition of this element to steel promotes fine grain structure when the steel is heated above its critical range for heat treatment. It also imparts toughness and strength to the metal.

Tungsten. This element is used mostly in steels designed for metal cutting tools. Tungsten steels are tough, hard, and very resistant to wear.

Cobalt. The chief function of cobalt is to strengthen the ferrite. It is used in combination with tungsten to develop red hardness; that is, the ability to remain hard when red hot.

STEEL CODE CLASSIFYING SYSTEMS

A uniform steel classification system has been adopted by the Society of Automotive Engineers (SAE) and the American Iron and Steel Institute (AISI). Identification is based on a four or five digit code. The first digit indicates the type of steel; thus 1 is a carbon steel, 2 a nickel steel, 3 a nickel-chromium steel, and so on. In the case of simple alloy steels the second number of the series indicates the approximate amount of the predominating alloying element. The last two or three digits refer to the carbon content and are expressed in hundredths of 1 percent. For example, a *2335* steel indicates a nickel steel of about 3 percent nickel and 0.35 percent carbon.

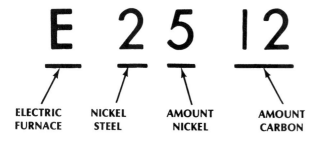

ELECTRIC FURNACE | NICKEL STEEL | AMOUNT NICKEL | AMOUNT CARBON

The following are the basic classification numerals for various steels:

Type of Steel	Series Designation
Carbon steels	1XXX
Plain carbon	10XX
Free machining, resulfurized (screw stock)	11XX
Free machining, resulfurized, rephosphorized	12XX
Manganese steels	13XX
High-manganese carburizing steels	15XX
Nickel steels	2XXX
3.50 percent nickel	23XX
5.00 percent nickel	25XX
Nickel-chromium steels	3XXX
1.25 percent nickel, 0.60 percent chromium	31XX
1.75 percent nickel, 1.00 percent chromium	32XX
3.50 percent nickel, 1.50 percent chromium	33XX
Corrosion and heat resisting steels	30XXX
Molybdenum steels	4XXX
Carbon-molybdenum	40XX
Chromium-molybdenum	41XX
Chromium-nickel-molybdenum	43XX
Nickel-molybdenum	46XX and 48XX
Chromium steels	5XXX
Low chromium	51XX
Medium chromium	52XXX
Corrosion and heat resisting	51XXX
Chromium-vanadium steels	6XXX
Chromium 1.0 percent	61XX
Nickel-chromium-molybdenum	86XX and 87XX
Manganese-silicon	92XX
Nickel-chromium-molybdenum	93XX
Manganese-nickel-chromium-molybdenum	94XX
Nickel-chromium-molybdenum	97XX
Nickel-chromium-molybdenum	98XX
Boron (0.0005% boron minimum)	XXBXXX

AISI also uses a prefix to indicate the steel-making process. These prefixes are:
- A—Open-hearth alloy steel
- B—Acid Bessemer carbon steel
- C—Basic open-hearth carbon steel
- D—Acid open-hearth carbon steel
- E—Electric furnace steel of both carbon and alloy steels

Examples:

C1078—Basic open-hearth carbon steel; carbon 0.72 to 0.85 percent

E50100—Electric furnace chromium steel 0.40 to 0.60 percent; chromium, 0.95 to 1.10 percent carbon.

E2512—Electric furnace nickel steel, 4.75 to 5.25 percent nickel; 0.09 to 0.14 percent carbon

WELDING DEFECTS

In the process of welding various materials, precautions must be taken to prevent the development of certain defects in the weld metal otherwise these defects will severely weaken the weld. The following are some of the principal defects that are significant in any welding or brazing process.

Grain growth. A wide temperature differential will exist between the molten metal of the actual weld and the edges of the heat-affected zone of the base metal. This temperature may range from a point far above the critical temperature down to an area unaffected by the heat. Thus the grain size can be expected to be large at the molten zone of the weld puddle and gradually reducing in size until recrystallization is reached. Grain growth can be kept to a minimum by effective control of preheating and postheating.

Where heavy sections require successive passes, it is possible to use the heat of each successive pass to refine the grain of the previous pass. This can be done only if the metal is allowed to cool below the lower critical temperature between each pass. High-carbon and alloy steels are especially vulnerable to coarse growth if cooled rapidly. These metals usually require a certain amount of preheating before welding and then allowed to cool slowly after the weld is completed.

Porosity. Porosity is the formation of small holes or cavities caused by gas entrapment during the solidification of the weld metal. Gases are formed by chemical reactions in the weld area and are released in the weld as the weld metal cools. The main causes of porosity are overheating and underheating, excessive current, and excessive arc length. Overheating increases the amount of gas present in the weld metal. Underheating prevents the weld metal from staying in a molten state to release trapped gases.

Porosity can be avoided by proper electrode or torch manipulation, correct current, and arc length. Proper manipulation maintains the molten pool of weld metal long enough to allow trapped gas, slag, and other foreign matter to escape. Consistent welding speeds help maintain a uniform temperature, which allows the weld metal to solidify slowly. Porosity is most likely to occur at the beginning and end of welds.

The mechanical properties of the weld can be adversely affected by excessive porosity. Welding codes and standards may specify the maximum allowable amount of porosity in a weld.

Inclusions. Inclusions are impurities or foreign substances which are forced in a molten puddle during the welding process. Any inclusion tends to weaken a weld because it has the same effects as a crack. A typical example of an inclusion is slag which normally forms over a deposited weld. If the electrode is not manipulated correctly, the force of the arc causes some of the slag particles to be blown into the molten pool. When the molten metal freezes before these inclusions can float to the top, they become lodged in the metal, producing a defective weld.

Inclusions are more likely to occur in overhead welding, since the tendency is not to keep the molten pool too long to prevent it from dripping off the seam. However, if the electrode is manipulated correctly and the right electrodes are used with proper current settings, inclusion can be avoided, or at least kept to a minimum.

Segregation. Segregation is a condition where some regions of the metal are enriched with an alloy ingredient while surrounding areas are actually impoverished. For example, when metal begins to solidify, tiny crystals form along grain boundaries. These so-called crystals or dendrites tend to exclude alloying elements. As other crystals form, they become progressively richer in alloying elements leaving other regions without the benefits of the alloying ingredients. Segregation can be remedied by proper heat treating or slow cooling.

RESIDUAL STRESSES

The strength of a welded joint depends a great deal on the way you control the expansion and contraction of the metal during the welding operation. Whenever heat is applied to a piece of metal, expansion forces are created which tend to change the dimensions of the piece. Upon cooling, the metal undergoes a change again as it attempts to resume its original shape.

No serious consideration is given these factors when there are no restricting forces to prevent the free movements of the expansion and contraction forces or when welding ductile metal, because the flow of metal will usually relieve the stresses. When free movement is restricted there is likely to occur a warping or distortion if the metal is malleable or ductile, and a fracture if the metal is brittle, as with cast iron.

To better understand the effects of expansion and contraction, assume that the bar shown in Figure 3-16 is thoroughly and uniformly heated. Since the bar is not restricted in its movements, expansion is free to take place in all directions. Consequently, the overall size of the bar is increased. If the bar is allowed to cool without restraint of any kind, it will contract to its original shape.

Suppose now that a similar bar is clamped in a vise, as shown in Figure 3-17, and heated. Because the ends of the bar cannot move, expansion must take place in another direction. In this case the expansion occurs at the sides.

If heat is applied to one section only, the expansion becomes uneven. The surrounding cold metal prevents free expansion and the displacement of metal takes place only in the heated area. When this

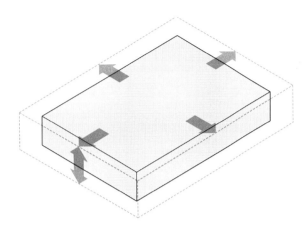

Figure 3-16. Metal that is not restricted in any way will expand in all directions when heated thoroughly and uniformly.

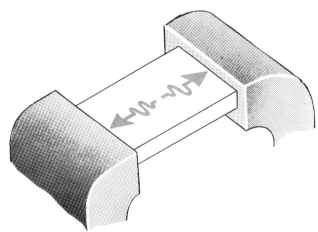

Figure 3-17. The expansion forces are hindered when the bar has restricting forces applied.

area starts to cool, contraction will also be uneven and some of the original displaced metal will become permanently distorted as illustrated in Figure 3-18.

To show just how the expansion and contraction forces affect metal, study the results of welding two different pieces. In the first case, assume a break has occurred in the middle of a bar, as in Figure 3-19. Upon welding the break, the heat naturally will cause the metal to expand. Since there are no obstructions on the ends of the bar the metal is permitted to move to whatever limits it desires. When the piece begins to cool, there are still no forces to prevent the metal from assuming its original shape.

Suppose the break was in a center section as shown in Figure 3-20. Note that in this case the ends of the bar are rigidly fastened to a solid frame. If the same procedure is used to weld the fracture as in the first case, something is bound to happen to the casting if no provisions are made for expansion and contraction. Since the vertical and horizontal sections (outside) of the frame will prevent expanding the ends of the center piece, there is only one direction in which this movement can go while the metal is being heated. That is at the point where fusion takes place. Now consider what will happen when the section begins to cool. The frame around the center section has not moved and, when contraction sets in, the center piece will be shortened. When the rigid frame resists this pull, a fracture or deformation at the line of weld or in some other place is bound to occur.

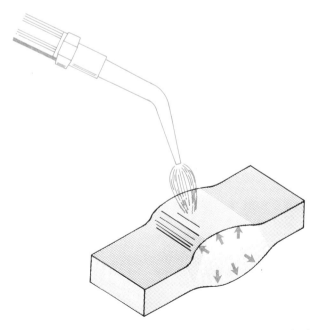

Figure 3-18. Expansion occurs where the heat is applied when forces are restricted.

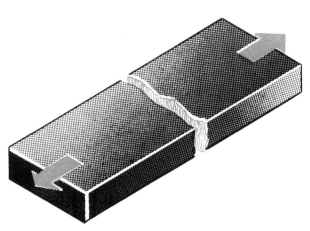

Figure 3-19. In this bar, where a break has occurred in the middle, expansion forces are unrestricted when heat from welding is applied.

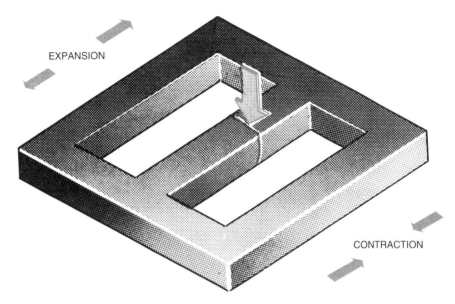

Figure 3-20. In this solid frame, with a break in the center section, expansion forces are restricted as the weld area is heated. Unless provisions are made, a fracture or deformation is bound to occur.

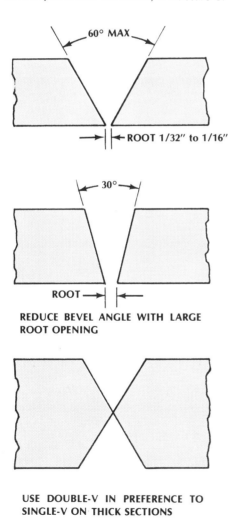

60° MAX

ROOT 1/32" to 1/16"

30°

ROOT

REDUCE BEVEL ANGLE WITH LARGE ROOT OPENING

USE DOUBLE-V IN PREFERENCE TO SINGLE-V ON THICK SECTIONS

Figure 3-21. Proper edge preparation minimizes distortion and insures good weld penetration.

Controlling Residual Stresses

The following are a few simple procedures which will help control the forces caused by expansion and contraction:

Proper edge preparation and fit-up. Make certain that the edges are correctly beveled. Proper edge beveling will not only restrict the effects of distortion but will insure good weld penetration. See Figure 3-21. Although sometimes the bevel angle can be reduced, care must be taken to insure that there is sufficient room in the joint to permit proper manipulation of the electrode when doing the weld.

Less distortion will occur if the welds are balanced around the center of gravity which is designated as the *neutral axis.* See Figure 3-22 top, left. Furthermore, distortion is reduced if the joint nearest to the neutral axis is welded first, followed by welding the unit that is farthest from the neutral axis, Figure 3-22 top, right, and bottom.

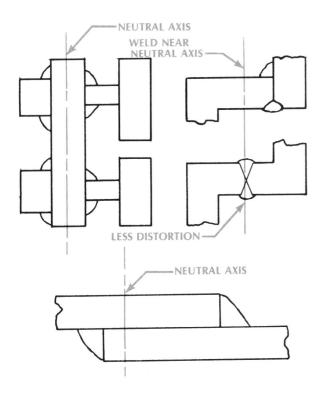

Figure 3-22. Welding near the neutral axis helps reduce distortion.

On long seams, especially on thin sections, the practice is to allow about ⅛″ [0.125″] at the end for each foot in length of the weld for expansion. See Figure 3-23 for example.

Tack welds are also used to control expansion on long seams as shown in Figure 3-24. Tack welds are spaced about 12″ [305 mm] apart and run approximately twice as long as the thickness of the weld. When tack welds are used, progressive spacing is not necessary. The plates are simply spaced an equal amount throughout the seam. Also, a long longitudinal (end-ways) seam is welded before a short transverse (side-ways) seam. See Figure 3-25.

Minimizing heat input. Controlling the amount of heat input is somewhat more difficult for the be-

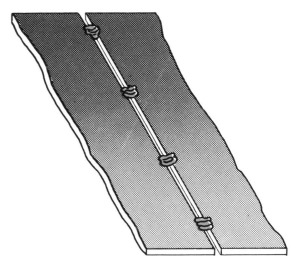

Figure 3-24. Tack welds hold the plates in position and help control expansion forces on long seams.

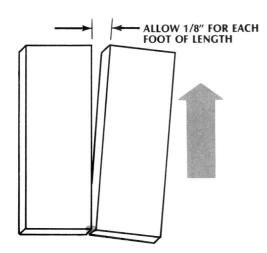

Figure 3-23. On long seams (especially on thin sections), provide a space between the edges to be welded.

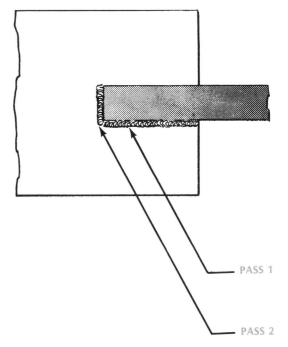

Figure 3-25. To minimize distortion, weld a long longitudinal seam before a short transverse seam.

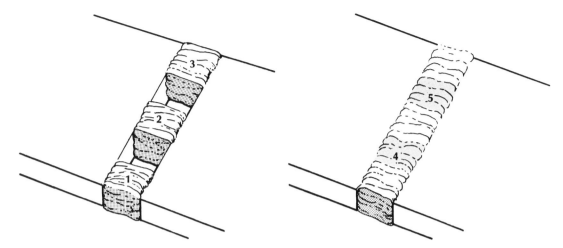

Figure 3-26. The intermittent weld, sometimes called a skip weld, can prevent distortion by minimizing heat input.

ginner. An experienced welder is able to join a seam with the minimum amount of heat by rapid welding.

A technique often employed to minimize the heat input is the *intermittent*, or *skip weld*. Instead of making one continuous weld, a short weld is made at the beginning of the joint. Next a few inches is welded at the center of the seam, and then a short length is welded at the end of the joint. Finally you return to where the first weld ended and proceed in the same manner, repeating the cycle until the weld is completed. See Figure 3-26.

The use of the *back-step*, or *step-back*, welding method also minimizes distortion. With this technique, instead of laying a continuous bead from left to right, you deposit short sections of the beads from right to left as illustrated in Figure 3-27, along the entire seam.

Preheating and postheating. The temperature difference between the localized heat caused by the welding heat source and the metal being welded causes varied rates of contraction and expansion within the welded structure. To control the rate of expansion and contraction particularly on alloy steels and cast iron, the entire structure is preheated before welding. Welding at low ambient temperatures or on steel that has been stored in a cold environment also increases the need for preheating. Torches using natural gas, acetylene, Mapp, or other gases are commonly used to preheat to the necessary temperature. See Figure 3-28.

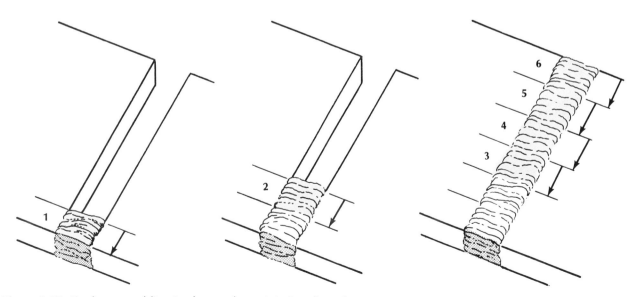

Figure 3-27. Back-step welding is also used to minimize distortion.

Figure 3-28. A torch is commonly used to preheat an entire structure before welding.

Factors affecting the correct preheat temperature include the composition of the metal being welded, material thickness, and the welding procedure used. To measure the preheat temperature of the weld structure, temperature indicating crayons may be used. See Figure 3-29.

Preheating must be uniform throughout the welding operation. It may be necessary to postheat the structure. Postheating is the application of heat to the structure after the weld is completed. This relieves internal stresses caused by the concentration of heat in the weld area. Postheating uses the same equipment as preheating. A second person may be required for each operation.

Peening. To help a welded joint stretch as it cools, a common practice is to peen it lightly with the round end of a ball peen hammer. However, peening should be done with care because too much hammering will add stresses to the weld or cause the weld to work-harden and become brittle. See Figure 3-30.

Figure 3-30. Peening a weld helps relieve internal stresses.

Figure 3-29. A temperature indicating crayon may be used to determine the temperature of the weld structure. *(Tempil Division, Big Three Industries)*

Stress relieving. A common stress relieving method is heat treating. The welded component is placed in a furnace capable of uniform heating and temperature control. The metal must be kept in a soaking temperature until it is heated throughout. Correct temperatures are important to prevent injury to the metal being treated. For example, mild steels require temperatures of 1100° to 1200°F (595° to 650°C) while other alloy steels must be heated to temperatures of 1600°F (870°C) or more.

After the proper soaking period the heat must be reduced gradually to nearly atmospheric temperature.

Jigs and fixtures. The use of jigs and fixtures will help prevent distortion, since holding the metal in a fixed position prevents excessive movements. A jig or a fixture is any device that holds the metal rigidly in position during the welding operation. Figure 3-31 illustrates a simple way to hold pieces firmly in a flat position. These heavy plates not only prevent distortion but they also serve as heat sinks to avoid excessive heat building up in the work. Special chill blocks made of copper or other metal having good conductivity are particularly effective in dissipating heat away from the weld area. See Figure 3-32.

Jigs and fixtures are used extensively in production welding since they permit greater welding speed while reducing to a minimum any form of distortion. By and large, industrial jigs and fixtures are designed to accommodate the specific production work being done. Figure 3-33 shows such a device.

Number of passes. Distortion can be kept to a minimum by using the number of passes necessary to meet the requirements of the job with a minimum amount of heat generated in the weld. The amount

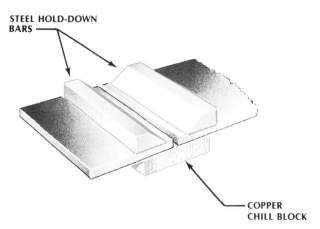

Figure 3-32. Copper chill blocks help to reduce heat and warpage in the weld area. (*LTV Steel*)

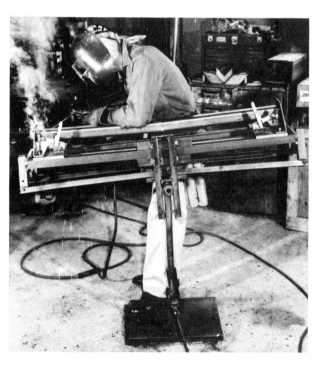

Figure 3-33. An industrial welding jig is usually designed for a specific production task. (*The Lincoln Electric Company*)

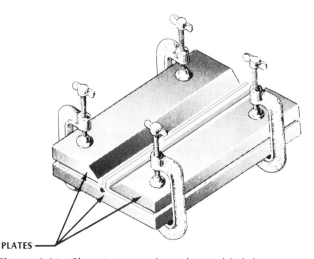

Figure 3-31. Clamping metal to be welded between heavy plates prevents distortion. (*LTV Steel*)

of distortion, the position, and the type of welding process used will determine the number of passes required on a given weld joint. See Figure 3-34.

Parts out of position. When a single-V butt joint is welded, the greater amount of hot metal at the top than at the root of the V will cause more contraction across the top of the welded joint. The result is a distortion of the plate as shown in Figure 3-35.

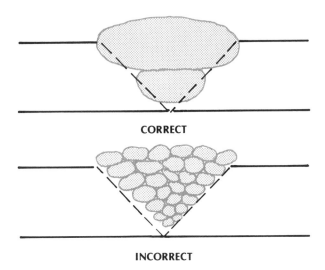

Figure 3-34. The number of passes necessary depends upon the requirements of the weld.

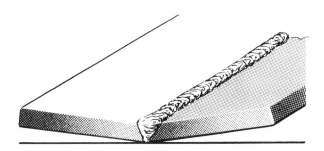

Figure 3-35. A butt joint will distort from contraction of the weld metal if the plates are not clamped when welding.

Figure 3-36. A T-joint is likely to be distorted in this manner.

In a T-joint, the weld along the seam will bend both the upright and flat piece. See Figure 3-36.

To minimize these distortions, the simplest thing to do is to angle the pieces slightly in the opposite direction in which contraction is to take place. Then, upon cooling, the contraction forces will pull the pieces back into position. Thus, the distortion shown in Figs. 3-35 and 3-36 can be prevented by tack welding the pieces out of alignment before welding as illustrated in Figure 3-37.

Final note.   A skillful welder is one who possesses a considerable amount of technical information. Merely being able to lay a good bead is not enough, because in the process of making a weld he may, from lack of understanding, jeopardize the strength of the welded structure. Consequently, such factors as properties of metals, expansion and contraction, grain growth, effects of heat, and others should definitely be considered essential knowledge for any welder.

As a final consideration a welder should always remember that the higher the carbon content of a metal the more extensive should be the planning, selection, and execution of the welding process. Additional details on the application of this kind of knowledge in specific welding situations will be covered in subsequent chapters.

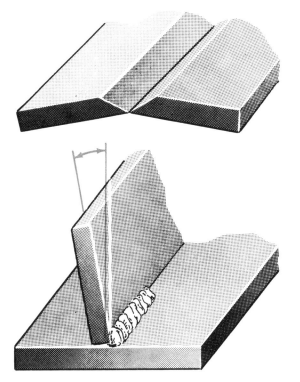

Figure 3-37. To minimize distortion, set the pieces slightly out of alignment before welding.

Points to Remember

1 Be sure you know the kind of metal you are welding and the effects heat may have on the welded structure.

2 When welding alloy steels make certain that the piece is not subjected to prolonged periods of heat that are beyond the critical point.

3 Be sure provisions are made for expansion and contraction forces in any welding job.

4 For most butt welds, allow about ⅛″ space at the end of the joint for each foot in length of the weld.

5 Keep the heat as low as possible on a piece being welded.

6 Heat can be controlled on a weld by using intermittent or back-step welding techniques.

7 Always guard against porosity and inclusions in a weld.

8 Select the type of joint that best meets the load requirements of the welded structure.

QUESTIONS FOR STUDY AND DISCUSSION

1 What is the difference between a stress and a strain?

2 Why is elasticity an important property in metals?

3 How does tensile strength differ from compressive strength?

4 What is meant by torsional strength?

5 What is shear strength?

6 Why is fatigue strength in some structures very important?

7 Impact strength refers to what particular quality in metals?

8 What do cryogenic properties refer to?

9 Why is ductility important in some metals?

10 How are carbon steels classified?

11 What is meant by an alloy steel?

12 What are some of the alloying elements that are added to steel?

13 Identify the elements present in a piece of steel labeled C1024.

14 Why are space-lattice formations important in steels?

15 Ductile metals have what kind of space-lattice pattern?

16 How does grain size affect the strength of steel?

17 What is alpha iron?

18 What is meant by cementite?

19 What is meant by pearlite?

20 What is gamma iron?

21 What effects does heating beyond the critical range have on the grain size?

22 What is martensite?

23 What may cause brittleness in a weld area?

24 What are the principal functions of the annealing process?

25 What is stress relieving?

26 What is the difference between hardening and tempering?

27 Why are some metals case hardened?

28 What are the principal case hardening processes?

29 How can grain growth in a weld be controlled?

30 What is porosity?

31 What cause inclusions in a weld?

32 How can porosity in a weld be prevented?

33 Why must a welder take into account expansion and contraction forces when preparing to make a weld?

34 What means can be used to minimize distortion?

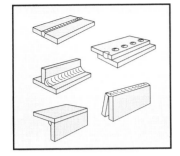

Introduction to Welding

Joint design is greatly influenced by the cost of preparing the joint, the accessibility of the weld, its adaptability for the product being designed or welded, and the type of loading the weld is required to withstand.

The five basic joints used in welding are butt, T, lap, edge, and corner. See Figure 4-1. Each has certain advantages and limitations. A welder should be especially concerned with their limitations since the effectiveness of a weld is often contingent on the type of joint that is used as well as the skill used in welding.

Weld Types

The various joint configurations are used with the following types of welds: *surfacing, fillet, groove, plug* and *slot*. See Figure 4-2.

Surfacing weld. A type of weld composed of one or more stringer or weave beads deposited on an unbroken surface to obtain desired properties or dimensions.

Fillet weld. A fillet weld is approximately a triangle in cross-section, joining two surfaces at right angles to each other in a lap, T or corner joint.

Groove weld. A groove weld is a weld made in the groove between two members to be joined. The weld is adaptable for a variety of joints as discussed in the section on butt joints later in the chapter.

Plug and slot-weld. These welds are used to join two overlapping pieces of metal by welding through circular holes or slots. Such welds are often used instead of rivets.

Joint Selection

Just what type of joint is best suited for a particular job depends on many factors. Although the designer or engineer is primarily responsible for determining the kind of joint that is to be used, nevertheless if a welder knows something about joint design a weld will be produced that will better meet the established specifications for the job. In general there are five basic considerations in the selection of any

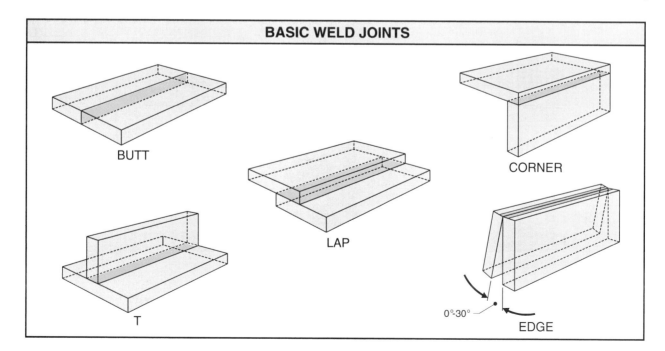

BASIC WELD JOINTS

BUTT

CORNER

LAP

T

0°-30°

EDGE

Figure 4-1. The five basic types of joints used in welding are the butt, T, lap, edge, and corner.

welding joint. These five considerations are:

1. Whether a load is in tension or in compression and if bending, fatigue, or impact stresses will be encountered.

2. How a load is applied, that is, whether the load is steady, sudden, or variable. See Figure 4-3.

3. Direction of the load as applied to the joint.

4. Thickness of the load as applied to the joint.

5. Cost of preparing the joint.

Joint Geometry[1]

Proper joint geometry is based on the following principles:

1. *Fit-up must be consistent for the entire joint.* Sheet metal and most fillet and lap joints should be clamped tight for their entire length. Gaps or bevels must be accurately controlled over the entire joint. Any variation in a given joint will force the operator to adjust the welding speed to avoid burnthrough and to handle the different electrode manipulation required by the fit-up variation.

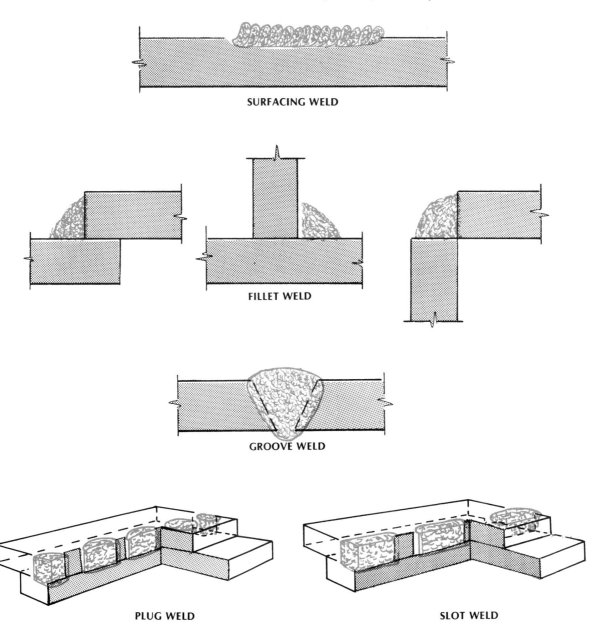

SURFACING WELD

FILLET WELD

GROOVE WELD

PLUG WELD

SLOT WELD

Figure 4-2. Weld types used in various joint configurations include the surfacing, fillet, groove, plug, and slot welds.

1. The Lincoln Electric Co.

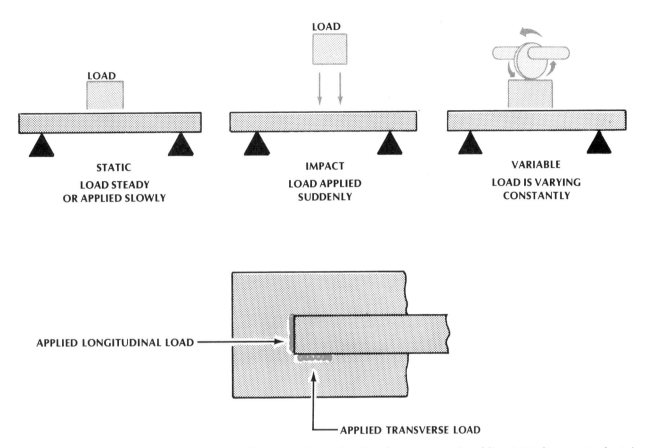

Figure 4-3. How a load is applied is one consideration when selecting the best type of welding joint for a particular job.

2. *Correct groove angle is required for good bead shape and penetration.* See Figure 4-4. Insufficient bevel prevents getting the electrode into the joint. A deep narrow bead may lack penetration and has a strong tendency to crack. For complete penetration a wider bevel is used in pipe welding. See Figure 4-5.

3. *Excess bevel wastes weld metal.* See Figure 4-6. Since filler metal in the form of electrodes and wire is expensive, any variation from the recommended groove angle size simply contributes to the cost of making a weld both in terms of material and time.

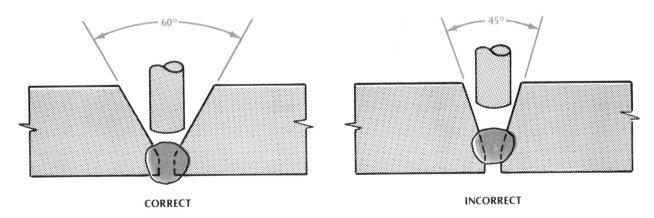

Figure 4-4. A correct groove bevel is essential for a good weld.

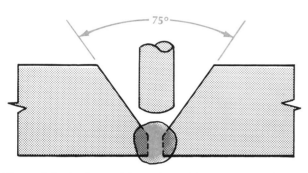

Figure 4-5. A pipe weld requires a 75° groove angle.

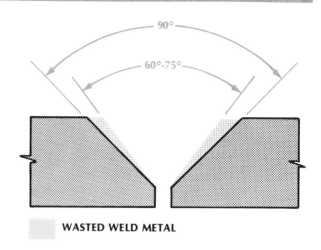

WASTED WELD METAL

Figure 4-6. A groove angle that is too wide wastes metal and time, resulting in greater welding costs.

4. *Sufficient gap is needed for full penetration.* See Figure 4-7. Unless there is adequate penetration, a welded joint will not withstand the loads imposed on it. Although proper penetration depends to some extent on electrode manipulation, the first essential is providing a correct root opening.

5. *Either a $^1/_8$" [0.125"] land (root face) or a back-up strip is required for fast welding and good quality.* See Figure 4-8 and Figure 4-17. Feather-edge preparations require a slow costly seal (root) bead. However, double-V butt joints without a land are practical when the root bead is offset by easier edge preparations and the gap can be limited to about $^3/_{32}$".

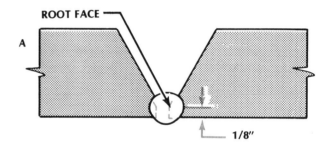

ROOT FACE

A

1/8"

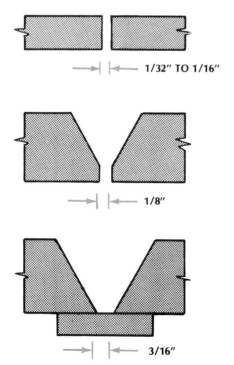

1/32" TO 1/16"

1/8"

3/16"

Figure 4-7. A correct root opening is essential to making a sound weld.

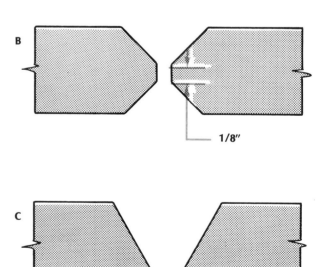

B

1/8"

C

BACKING STRIP

Figure 4-8. Provide a proper root face for quality welds.

Butt Joints

In a butt joint the weld is made between the edge surfaces of the two sections to be fused. The joint may be either of the square or grooved type. See Figure 4-9.

Square butt joint. The square butt joint is intended primarily for materials that are ³⁄₁₆″ or lighter in thickness, with adequate root opening, and require full and complete fusion for optimum strength. For submerged-arc welding, materials up to ³⁄₈″ with a minimum gap of ⅛″ can be welded. The joint is reasonably strong in static tension but is not recommended when it is subjected to fatigue or impact loads, especially at low temperatures. The preparation of the joint is relatively simple since it requires only matching the edges of the plates, consequently the cost of making the joint is low. See Figure 4-9A for this type of joint.

Single bevel butt joint. The single bevel butt joint is a partial penetrating single bevel groove weld. It is welded from one side and generally is used on metals having a maximum thickness of ½″. See Figure 4-9B.

Single-V butt joint. This joint is used on plate ³⁄₈″ or greater in thickness, but not to exceed ¾″. Preparation is more costly because a special beveling operation and more filler material are required. The joint is strong in static loading but as in the square joint it is not particularly suitable when subject to bending at the weld root. See Figure 4-9C.

Double-V butt joint. The double-V butt joint is best for all load conditions. It is often specified for stock that is heavier than metal used for a single-V (up to ¾″). For maximum strength the penetration must be complete on both sides. The cost of preparing the joint is higher than the single-V, but often less filler material is required because a narrower included angle can be used. To keep the joint symmetrical and warpage to a minimum the weld bead must be alternated, welding first on one side and then the other. See Figure 4-9D.

Single-U butt joint. A joint of this type readily meets all ordinary load conditions and is used for work requiring high quality. It has greatest applications for joining plates ½″ to ¾″. See Figure 4-9E. The joint needs less filler metal than the single or double-V groove, and less warpage is likely to occur.

Double-U butt joint. The double-U butt joint is intended for metals ¾″ or more in thickness where welding can readily be accomplished on both sides. The joint meets all regular load conditions although the preparation cost is higher than the single-U butt joint. See Figure 4-9F.

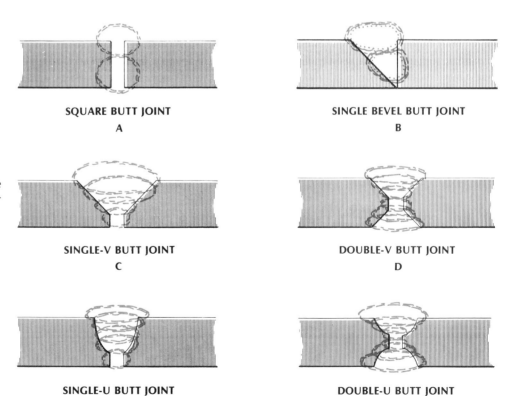

Figure 4-9. Butt joints are classified by the preparation of their edges.

SQUARE BUTT JOINT
A

SINGLE BEVEL BUTT JOINT
B

SINGLE-V BUTT JOINT
C

DOUBLE-V BUTT JOINT
D

SINGLE-U BUTT JOINT
E

DOUBLE-U BUTT JOINT
F

T-Joints

A T-joint is made by placing the edge of one piece of metal on the surface of the other piece at approximately a 90° angle and is used for all ordinary plate thicknesses. The basic T-joints are classified as square, single bevel, double bevel, single-J and double-J.

Square T-joint. The square T-joint requires a fillet weld which can be made on one or both sides. It can be used for light or reasonably thick materials where loads subject the weld to longitudinal shear. Since the stress distribution of the joint may not be uniform this factor should be considered where severe impact or heavy transverse loads are encountered. For maximum strength, considerable weld metal is required. See Figure 4-10A.

Single bevel T-joint. This joint will withstand more severe loadings than the square T-joint due to better distribution of stresses. It is generally confined to plates ½" or less in thickness where welding can be done from one side only. See Figure 4-10B for example.

Double bevel T-joint. The double bevel T-joint is intended for use where heavy loads are applied in both longitudinal and transverse shear and where welding can be done on both sides. See Figure 4-10C.

Single-J T-joint. The single-J T-joint is used on plates one inch or more in thickness where welding is limited to one side. It is especially suitable where severe loads are encountered. See Figure 4-10D

Double-J T-joint. A joint of this kind is particularly suitable for heavy plates 1½" or more in thickness where unusually severe loads must be absorbed. Joint location should permit welding on both sides. See Figure 4-10E.

Lap Joints

A lap joint as the name implies is made by lapping one piece of metal over another. It is one of the strongest joints, despite the lower unit strength of

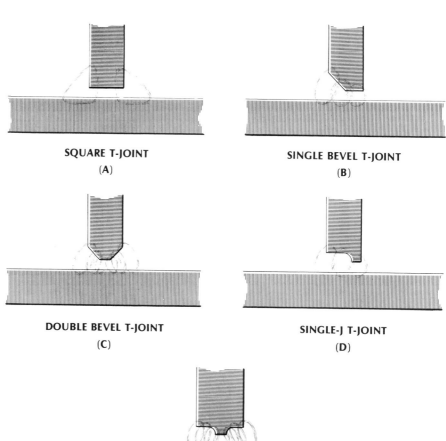

Figure 4-10. The basic T-joints are the square, single bevel, double bevel, single-J and double-J.

SQUARE T-JOINT
(A)

SINGLE BEVEL T-JOINT
(B)

DOUBLE BEVEL T-JOINT
(C)

SINGLE-J T-JOINT
(D)

DOUBLE-J T-JOINT
(E)

the filler metal. For joint efficiency, an overlap greater than three times the thickness of the thinnest member is recommended. The two basic lap joints are known as single fillet and double fillet lap joints.

Single fillet lap joint. The single fillet lap joint is very easy to weld. Filler metal is simply deposited along the seam. Actually the strength of this weld depends on the size of the fillet. Metal up to ½″ in thickness can be welded satisfactorily with a single fillet if the loading is not too severe. See Figure 4-11A.

Double fillet lap joint. This joint can withstand greater loads than the single fillet lap joint. It is one of the more widely used joints in welding. As a rule if the weld is properly made its strength is very comparable to that of the parent metal. See Figure 4-11B.

Corner Joints

Corner joints have wide applications in joining sheet and plate metal sections where generally severe loads are not encountered. The common corner joints are classified as flush, half open and full open.

Flush corner joint. This joint is designed primarily for welding sheet 12 gauge and lighter. It is restricted to lighter materials because deep penetration is sometimes difficult and it supports only moderate loads. See Figure 4-12A.

Half-open corner joints. The half-open corner joint is usually more adaptable for materials heavier than 12 gauge. It is suitable for loads where fatigue or impact are not too severe and where the welding can be done from only one side. Since the two edges

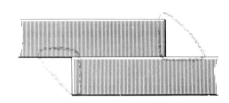

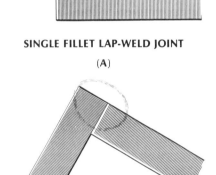

Figure 4-11. The two basic lap joints are the single fillet and double fillet. A lap joint, in which one piece of metal laps over the other, is one of the strongest joints.

SINGLE FILLET LAP-WELD JOINT

(A)

DOUBLE FILLET LAP-WELD JOINT

(B)

FLUSH CORNER JOINT

(A)

HALF-OPEN CORNER JOINT

(B)

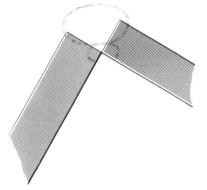

Figure 4-12. Corner joints are classified as flush, half open, and full open. They are generally used only where severe loads are not encountered. The edge joint can sustain only light loads.

FULL-OPEN CORNER JOINT

(C)

EDGE JOINT

(D)

of the pieces are shouldered together there is less tendency to burn through the plates at the corner. See Figure 4-12B for example.

Full-open corner joint. Since this joint permits welding on both sides it produces a strong joint capable of carrying heavy loads. Plates of all thicknesses can be welded. It is recommended for fatigue and impact applications because of good stress distribution. See Figure 4-12C.

Edge Joint

The edge joint is suitable for plate ¼″ or less in thickness and can sustain only light loads. See Figure 4-12D.

Basic Welding Terms

Before proceeding with various welding operations, an understanding of the following terms is important:

Welding positions. The four positions assumed in welding are flat, horizontal, vertical, and overhead. See Figure 4-13. The flat position is the most widely used because welding can be done faster and easier.

Horizontal welding is difficult because the molten puddle has a tendency to sag. Vertical welding is done in a vertical line from the bottom to the top or top to bottom of the plates. On thin material a down-hand or downhill welding technique is usually more applicable. Overhead welding is more difficult because the molten metal will sag and considerable skill is required to secure a uniform bead with proper penetration.

Base metal or parent metal. Metal to be welded. See Figure 4-14.

Electrode. Wire that conducts electrical current and melts to form the weld metal. Shielded metal arc welding electrodes are covered with a flux coating.

Bead. Narrow layer or layers of metal, as shown in Figure 4-14, deposited on the base metal as the electrode melts.

Ripple. Shape within deposited bead that is caused by movement of the welding heat source is shown in Figure 4-14.

Pass. Each layer of beads deposited on the base metal as in Figure 4-14.

Crater. Depression in the base metal, shown in Figure 4-15, made by the welding heat source.

Penetration. Depth of fusion with the base metal shown in Figure 4-16.

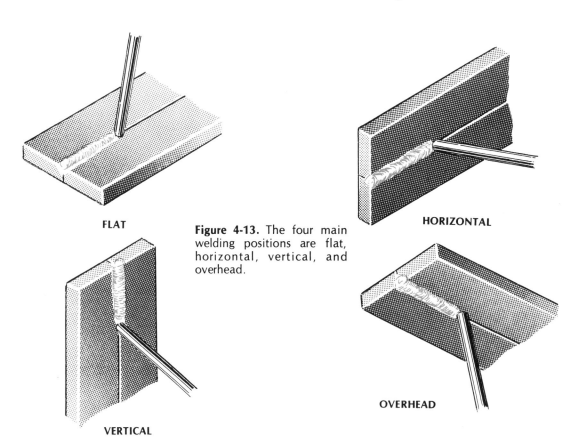

FLAT

Figure 4-13. The four main welding positions are flat, horizontal, vertical, and overhead.

HORIZONTAL

VERTICAL

OVERHEAD

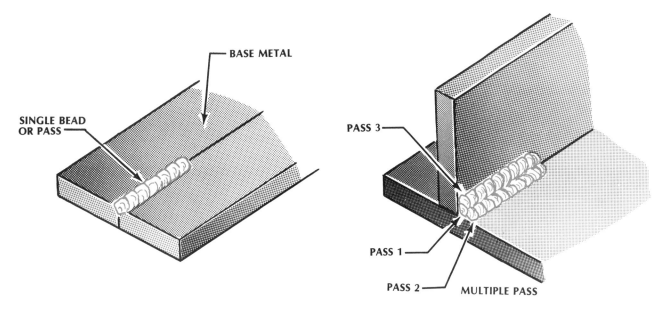

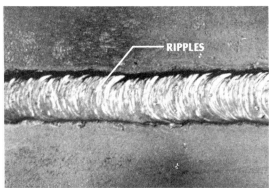

Figure 4-14. Ripples in the deposited bead are caused by movement of the welding heat source.

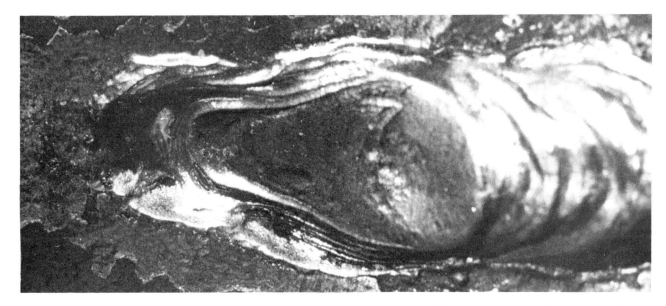

Figure 4-15. The crater is the depression in the base metal made by the welding heat source. (*The Lincoln Electric Company*)

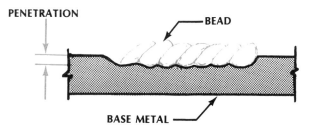

Figure 4-16. Penetration is the distance from the surface on the base metal to the bottom of the bead.

Reinforcement. Refers to the amount of weld metal that is piled up above the surface of the pieces being joined. It is particularly applicable to butt welds. See Figure 4-17.

Root face. The bottom lip near the slanted surface of the groove. See Figure 4-17.

Root opening. The distance between the root faces of the pieces being joined.

Weld width. Distance from toe to toe across the face of the weld. See Figure 4-17.

Toes. The points where the base metal and weld metal meet. See Figure 4-18.

Face. The exposed surface of the weld bounded by the toes of the weld. The face may be either concave or convex. See Figure 4-18.

Root. The point of the weld triangle opposite the face of the weld. See Figure 4-18.

Throat. The distance through the center of the weld from the face to the root. See Figure 4-18.

Weld legs. Size of fillet welds made in lap or T-joints. See Figure 4-18.

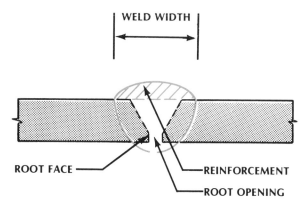

Figure 4-17. This groove weld shows the application of various welding terms.

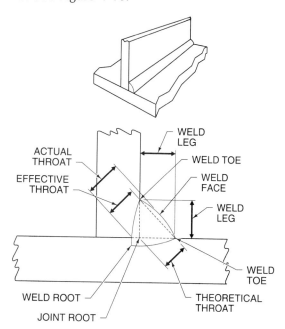

Figure 4-18. This T-joint identifies the various parts of a fillet weld.

QUESTIONS FOR STUDY AND DISCUSSION

1 What are some of the factors that must be considered in the type of joint that should be used in welding any structural unit?

2 When is a surfacing weld used?

3 What is a fillet weld?

4 In what type of joints are groove welds made?

5 What is a plug weld?

6 Why are grooved butt joints usually superior to square butt joints for welding thick plates?

7 What are the basic types of T-joints?

8 How would you describe a double fillet lap joint?

9 Of the various types of corner joints, which is the strongest?

10 Why is the flat position the most practical for welding?

11 Why is overhead welding more difficult to perform?

12 What is meant by downhill welding? By uphill welding?

13 What does it mean when a weld requirement calls for three passes?

14 What do we mean when we speak of the toes of a weld?

15 What is meant by the root of a weld?

16 What are some of the basic principles which contribute to good joint geometry?

17 When are double bevel T-joints normally used?

18 What are the different types of loads that can be applied to weld parts?

19 Which butt joint requires the least amount of preparation before welding?

20 What is meant by reinforcement of the weld?

21 What causes the formation of ripples in the welding process?

22 How is the root opening size determined?

23 Why is the proper groove angle required?

24 How is the size of a weld leg determined?

25 How is the base metal affected in the welding process?

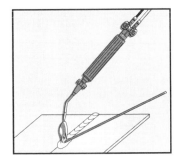

EQUIPMENT

Oxyacetylene Welding—OAW

The oxyacetylene welding process has to a large extent been superseded by other types of welding for industrial production purposes. However, oxyacetylene welding still has its place in the welding family. Undoubtedly its greatest use is in general maintenance, auto body shops, and in repairing small parts where other welding processes would be too expensive to use both in terms of material and equipment set-up costs.

The process of oxyacetylene welding is possible because when acetylene is mixed with oxygen in correct proportions and ignited, the resulting flame is one of the hottest known. This flame, which reaches a temperature of 5700°F to 6300°F (3150°C to 3482°C), melts all commercial metals so completely that metals to be joined actually flow together to form a complete bond without application of any mechanical pressure or hammering. In most instances, some extra metal in the form of a wire rod is added to the molten metal in order to build up the joint slightly for greater strength. On very thin material the edges are usually flanged and just melted together. In either case, if the weld is performed correctly, the section where the bond is made will be as strong as the base metal itself.

With the oxyacetylene flame, such metals as iron, steel, cast iron, copper, brass, aluminum, bronze, and other alloys may be welded. In many instances dissimilar metals can be joined, such as steel and cast iron, brass and steel, copper and iron, brass and cast iron.

The oxyacetylene flame is also employed for a variety of other purposes, notably for cutting metal, case hardening, and annealing. It is used for several types of metallic spray guns which spray fine particles of molten metal on worn surfaces that need refacing or building up. The oxyacetylene flame can be used in practically any situation which involves joining metal parts. See Figure 5-1 for flame hardening application.

Oxygen for Welding

The atmosphere which we commonly refer to as air is composed of approximately 20 percent oxygen, the rest being nitrogen and a small percentage of

Figure 5-1. Flame hardening with oxyacetylene flame permits hardening just those surfaces of the gear teeth that are subjected to more wear in service. (*Wall Colmonoy Company*)

rare gases such as helium, neon, and argon. To obtain the oxygen in a state that makes it usable for welding, it is necessary to separate it from the other gases.

The two general methods that may be used to produce oxygen are the electrolytic and the liquid-air methods. in the electrolytic method, the fact is utilized that water is a chemical compound consisting of oxygen and hydrogen. By sending a current through a solution of water containing caustic soda, oxygen is given off at one terminal plate, and hydrogen at the other, thereby making possible the separation of the two gases. Because of the greater cost in manufacturing oxygen in this manner, the liquid-air method is more commonly used to produce commercial oxygen.

In a plant where oxygen is made by the liquid-air method, the air is drawn from the outside into huge containers known as washing towers, where the air is washed and purified of carbon dioxide. A solution of caustic soda is used to wash the air, this solution being run from a nearby tank and circulated through the towers by means of centrifugal pumps.

As the air leaves the washing towers, it is compressed and passed through *oil-purging cylinders* in which oil particles and water vapor are removed. From this point the air goes into drying cylinders. These cylinders contain dry, caustic potash which dries the air and removes any remaining carbon dioxide and water vapor. On the top of each cylinder, special cotton filters are provided to prevent any particles of foreign matter from being carried into the high-pressure lines.

The dry, clean, compressed air then goes into rectifying or liquification columns where the air is cooled and then expanded to approximately atmospheric pressure. The process of changing the extremely cold air under high pressure to the lower atmospheric pressure causes the air to liquefy.

The separation of the nitrogen from the oxygen becomes possible at this stage because of the difference in the boiling point between the nitrogen ($-320°F$ or $-195.5°C$) and the oxygen ($-296°F$ or $-182°C$). The nitrogen, having the lower boiling point, evaporates first, leaving the liquid oxygen in the bottom of the condenser. The liquid oxygen next passes through a heated coil which changes the liquid into a gaseous form. From here the gas goes into a storage tank, flowing through a gas meter which registers the amount of gas entering the storage tank. The gas is then drawn from this tank and compressed into receiving cylinders.

The oxygen cylinder. Oxygen cylinders are made from seamless, drawn steel and tested with a water pressure of 3360 psi. The cylinders are equipped with a high-pressure valve which can be opened by turning the hand wheel on top of the cylinder. See Figure 5-2. *This valve should always be opened by hand and not with a wrench. The hand wheel must be turned slowly to permit a gradual pressure load on the regulator, and then opened as far as the valve will turn, to full gas pressure.* Unless the valve is wide open the high oxygen pressure may cause the oxygen to leak around the valve, resulting in considerable waste.

There are three common sizes of oxygen cylinders. The large size which is the one popularly used by industrial plants and shops consuming large quantities of gas, is filled with 244 cubic feet (cu ft) of oxygen. The medium sized cylinder contains 122 cu ft, and the small cylinder 80 cu ft at standard conditions.

Cylinders are charged with oxygen at a pressure of 2200 psi at a temperature of 70°F (21°C). Gases

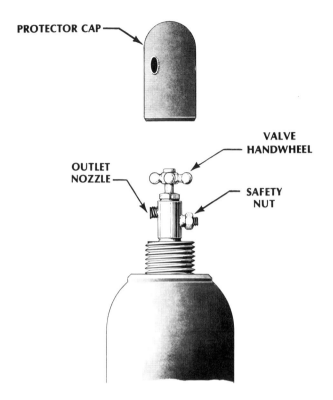

Figure 5-2. A protector cap protects the high-pressure valve of the oxygen cylinder when the cylinder is not in use. (*Linde Company*)

expand when heated and contract when cooled, so the oxygen pressure will increase or decrease as the temperature changes. For example, if a full cylinder of oxygen is allowed to stand outdoors in near freezing temperature, the pressure of the oxygen will register less than 2200 psi. However, this does not mean that any of the oxygen has been lost; cooling has only reduced the pressure of the oxygen.

Since the pressure of gas will vary with the temperature, all oxygen cylinders are equipped with a safety nut that permits the oxygen to drain slowly in the event the temperature increases the pressure beyond the safety load of the cylinder. Thus if a cylinder were exposed to a hot flame, the safety nut would relieve the pressure before the cylinder would reach the point where it would explode.

A *protector cap* which screws onto the neck ring of the cylinder is furnished to protect the valve from any damage. This cap must always be in place when the cylinder is not in use.

Acetylene for Welding

Acetylene is a gas formed by the mixture of calcium carbide and water. The commercial generator in which the gas is made consists of a huge tank containing water. A specified quantity of carbide is dumped into a hopper and raised to the top of the generator. The carbide is then allowed to fall into the water, and upon coming in contact with the water, bubbles of gas are given off. This gas is collected, purified, cooled, and slowly compressed into cylinders.

Acetylene is a colorless gas, with a very distinctive nauseating odor. It is highly combustible when mixed with oxygen or air. Although it is very stable under low pressures, it becomes very unstable if compressed to more than 15 psi.

WARNING: acetylene becomes dangerous if used beyond a 15 pound pressure!

The acetylene cylinder. To insure safety in storing acetylene, the cylinder is packed with a porous material. This material is saturated with acetone, which is a chemical liquid that dissolves or absorbs large quantities of acetylene under pressures greater than 15 psi without changing the nature of the gas. See Figure 5-3.

The acetylene cylinder is equipped with a fusible plug to relieve any excess pressure if the cylinder

should be subjected to undue heat, or any other mechanical pressure.

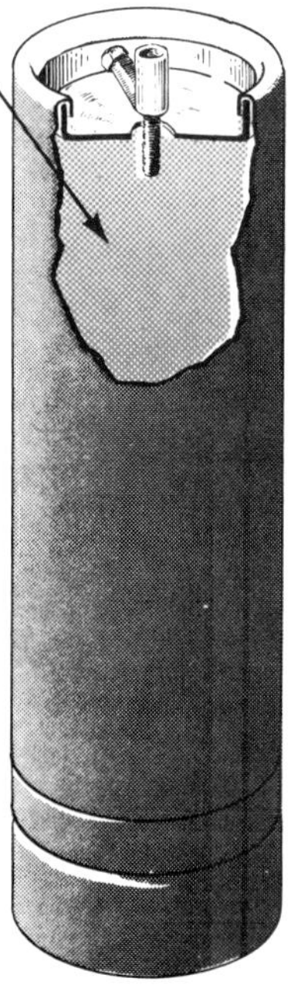

The cylinder valve is operated by means of a wrench. *This valve should never be opened more than one and one-half turns.* A slight opening is advisable since it permits closing the valve in a hurry in case of an emergency. See Figure 5-4.

Figure 5-4. The acetylene cylinder valve is operated with a wrench. (*Linde Company*)

When considerable welding is done, such as in industry or in a school welding shop, the acetylene cylinders are frequently connected to a manifold with pipe lines carrying the gas to the welding stations as shown in Figure 5-5. A flash arrestor prevents a flashback from the torch from reaching the acetylene cylinders.

Safety in Handling Cylinders

To move a cylinder, rotate it on its bottom edge. Place the palm of one hand over the protector cap and tilt the cylinder toward you. Start the tank rolling by pushing it with the other hand as shown in Figure 5-6.

Precautions must be followed when handling oxygen and acetylene:

1. Never use the valve-protector caps for lifting cylinders.

2. Do not allow cylinders to lie in a horizontal position.

3. Never permit grease or oil to come in contact with cylinder valves. Although oxygen is in itself non-

POROUS MATERIAL

Figure 5-3. The acetylene cylinder is packed with a porous material that is saturated with acetone. Acetone absorbs large quantities of acetylene without changing the nature of the gas for safe storage. (*Linde Company*)

Figure 5-5. A flash arrestor is connected in line to prevent a flashback reaching the acetylene cylinders in a manifold system.

Figure 5-6. To safely move a cylinder, the operator tilts it backwards with the palm of one hand over the protector cap, and pushes with the other hand to start the cylinder rolling on its bottom edge.

flammable, if it is allowed to come in contact with any flammable material it will quickly aid combustion.

4. Avoid exposing cylinders to furnace heat, radiators, open fire, or sparks from the torch.

5. Never transport a cylinder by dragging, sliding, or rolling it on its side. Avoid striking it against any object that might create a spark. There may be just enough gas escaping to cause an explosion.

6. If cylinders have to be moved, be sure that the cylinder valves are shut off.

7. Never tamper with or attempt to repair the cylinder valves. If valves do not function properly or if they leak, notify the supplier immediately.

8. Keep valves closed on empty cylinders.

9. If cylinder valves cannot be opened by hand, do not use a hammer or wrench—notify the supplier.

10. When not in use, keep cylinders covered with valve protector caps.

WELDING APPARATUS

A set of welding apparatus consists of a torch with an assortment of different-sized tips: two lengths of hose, one red for the acetylene and the other green for the oxygen; two pressure regulators; two cylinders, one containing acetylene and the other oxygen; a pair of goggles; and a welding sparklighter. See Figure 5-7.

As a rule, the cylinders are chained to a two-wheel truck to permit moving the equipment to any desired place. If the cylinders are positioned near the work bench, they should be chained to some fixed object.

CAUTION: Securing cylinders is important; otherwise they may tip over and cause an explosion or ruin the regulators. See Figure 5-8.

The Welding Torch

The torch, or blowpipe as it is sometimes called, is a tool which mixes acetylene and oxygen in the

Figure 5-8. Always chain cylinders to prevent them from tipping over.

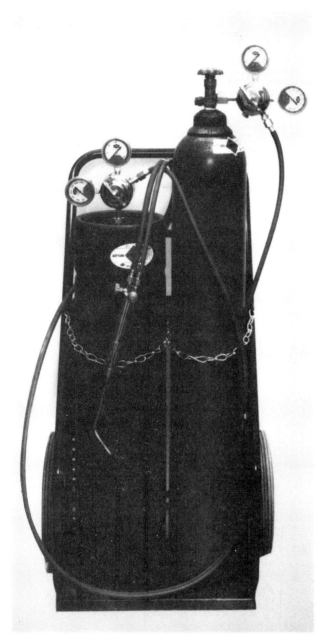

Figure 5-7. A welding unit on a hand truck permits transporting the unit to where the welding job is to be done. (*Linde Company*)

correct proportions and permits the mixture to flow to the end of a tip where it is burned. Although torches vary to some extent in design, basically they are all made to provide complete control of the flame during the welding operation.

The two main types of torches are the *injector* and the *medium (equal) pressure*. The injector torch is designed to use acetylene at very low pressures, from 1 psi to zero. The equal pressure type requires acet-

ylene pressures of 1 to 15 psi. Both will operate when acetylene is supplied from cylinders or medium-pressure generators.

With the injector blowpipe, the oxygen, as it passes through a small opening in the injector nozzle, draws acetylene into the oxygen stream. One advantage of this blowpipe is that small fluctuations in the oxygen supplied to it will produce a corresponding change in the amount of acetylene drawn, thereby making the proportions of the two gases constant while the torch is in operation. In the equal-pressure type the acetylene and oxygen are fed independently to a mixing chamber, after which they flow out through the tip. See Figure 5-9. The equal pressure type of blowpipe is the one most generally used.

Both kinds of torches are equipped with two needle valves; one regulates the flow of oxygen and the other the acetylene. On the rear end of the torch there are two fittings for connecting the two hoses. In order to eliminate any danger of interchanging the hoses, the oxygen fitting is made with a right-hand thread and the acetylene with a left-hand thread.

Care of the torch. When the welding is completed, always suspend the torch securely so it will not fall. The needle valves are especially delicate,

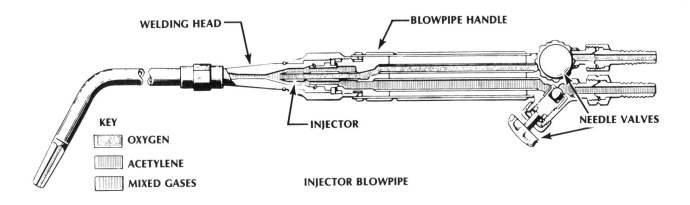

WELDING HEAD

BLOWPIPE HANDLE

INJECTOR

NEEDLE VALVES

KEY
- OXYGEN
- ACETYLENE
- MIXED GASES

INJECTOR BLOWPIPE

KEY
- OXYGEN
- ACETYLENE
- MIXED GASES

MIXER

EQUAL PRESSURE BLOWPIPE

Figure 5-9. The two main types of oxyacetylene torches are the injector and equal pressure. The equal pressure type is the one most commonly used. (*Linde Company*)

and care must be taken never to drop the torch so they will strike some hard object. Occasionally the needle valves will turn too freely, making it difficult to secure and keep the adjustment for the proper mixture. When this occurs, give the packing nuts on the stem of the needle valves a slight turn with a correct fitting wrench as shown in Figure 5-10.

Welding Tips

To make possible the welding of different thicknesses of metal, torches are equipped with an assortment of different size heads, or tips, as shown in Figure 5-11. The size of the tip is governed by the diameter of its opening which is marked on the tip.

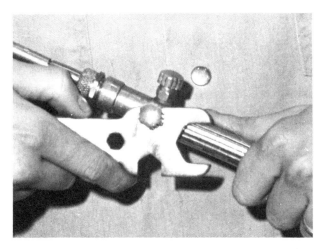

Figure 5-10. Tighten the packing nuts on the stem of the needle valves with a correct fitting wrench.

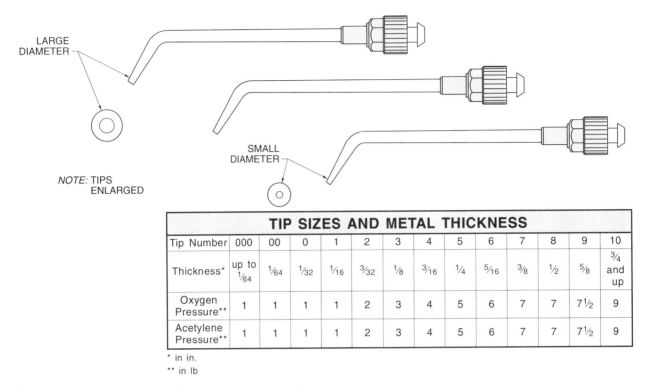

LARGE DIAMETER

SMALL DIAMETER

NOTE: TIPS ENLARGED

TIP SIZES AND METAL THICKNESS													
Tip Number	000	00	0	1	2	3	4	5	6	7	8	9	10
Thickness*	up to 1/64	1/64	1/32	1/16	3/32	1/8	3/16	1/4	5/16	3/8	1/2	5/8	3/4 and up
Oxygen Pressure**	1	1	1	1	2	3	4	5	6	7	7	7½	9
Acetylene Pressure**	1	1	1	1	2	3	4	5	6	7	7	7½	9

* in in.
** in lb

Figure 5-11. The size of the welding tip is governed by the diameter of the opening.

The system of identifying tip sizes depends largely on the manufacturer of the welding equipment. The most common system consists of numbers which range from 000 to 15. With this system, the higher the number the larger the tip diameter.

Care of welding tips. A welding tip is designed to be mounted and removed by hand.

CAUTION: Under no conditions should pliers be used to remove a welding tip. Pliers quickly ruin the nut on the tip, rendering it nearly useless.

CAUTION: Never mount a tip while the torch is hot. Heat expands the threads, causing the tip to freeze onto the torch after it has cooled. Attempting to force a tip loose from the torch later may result in breaking off the threaded section, thereby ruining the tip. Moreover you also have the problem of removing the broken section from the opening of the torch. Keep the tip clean at all times to produce satisfactory welds.

Frequent use of the torch will cause a formation of carbon in the passage of the tip. Remove this carbon by inserting a tip cleaner straight in and pulling it straight out. See Figure 5-12. Brushing the end of the tip with a fine grade of sandpaper is also advisable at times.

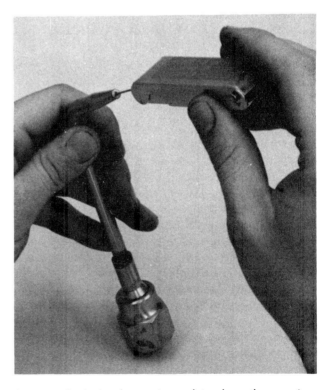

Figure 5-12. A tip cleaner is used to clean the opening of the welding tip.

Regulators

The oxygen and acetylene pressure regulators perform two functions. They reduce the cylinder pressure to the required working pressure and produce a steady flow of gas under varying cylinder pressures. To illustrate, assume that the oxygen in the cylinder is under a pressure of 1800 psi and a pressure of 6 pounds is needed at the torch. The regulator must maintain a constant pressure of 6 pounds even if the cylinder pressure drops to 500 pounds.

There are two types of regulators, the *two-stage* and the *single-stage*. With the two-stage regulator, the reduction of the cylinder pressure to that required at the torch is accomplished in two stages. In the first stage, the gas flows from the cylinder into a high-pressure chamber. A spring and diaphragm keeps a predetermined gas pressure in this chamber. For oxygen such a pressure is usually set at 200 psi and for acetylene 50 psi. From the high-pressure chamber the gas then passes into a second reducing chamber. Control of the pressure in the reducing chamber is governed by an adjusting screw.

The single-stage regulator is not as expensive as the two-stage type. With this regulator, there is no intermediate chamber through which the gas passes before it enters the low pressure chamber. The gas from the cylinder flows into the regulator and is controlled entirely by the adjusting screw.

Both types of regulators have two gauges. One gauge indicates the actual pressure in the cylinder and the other shows the working or line pressure used at the torch. The oxygen high pressure gauge is usually graduated to 4000 pounds. As a rule this gauge also has a second scale which is calibrated to register the content of the cylinder in cubic feet. The acetylene high pressure gauge is graduated to 350 to 400 pounds as shown in Figure 5-13. The acetylene gauge usually is marked with a warning color above 15 psi on the working pressure gauge.

The oxygen working pressure gauge is graduated in divisions between 0 to 60 pounds and the working pressure gauge on the acetylene regulator in divisions of 0 to 30 pounds.

The most important thing to remember when using a regulator is to *make absolutely certain that the adjusting screw is released (turned out) before the cylinder valve is opened.* If the adjusting screw is not released and the cylinder valve opened, the tremendous pressure of the gas in the cylinder, forced on to the working pressure gauge may result in damage to the regulator.

Care of regulators. Regulators are sensitive instruments and must at all times be regarded as such. It takes only a slight jar to put a regulator out of

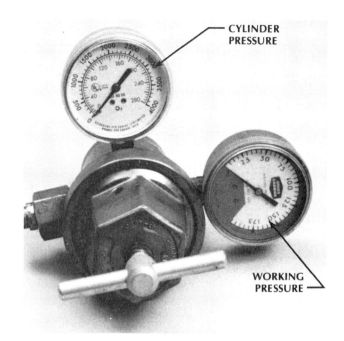

OXYGEN

ACETYLENE

Figure 5-13. The oxygen and acetylene regulators control the flow of gas to be used for welding. (*Linde Company*)

commission. Be extremely careful in handling the regulator while removing it from the cylinder. *Never allow a regulator to remain on a bench top or floor for any length of time,* as someone may come along and carelessly move it, which may result in damaging the regulator. Here are a few more rules that should be followed:

1. Always check the adjusting screw before the cylinder valve is turned on, and release this screw when the welding has been completed.

2. Never use oil on a regulator. *Use only soap or glycerine to lubricate the adjusting screw.*

3. Do not try to interchange the oxygen and acetylene regulators.

4. If a regulator does not function properly, shut off the supply of gas and have a qualified repairman check it.

5. If a regulator creeps (does not remain at set pressure), have it repaired immediately. Creeping will be noticed on the working pressure gauge after the needle-valves on the torch are closed. A creeping regulator usually requires a change of valve seat or stem.

6. If the gauge pointer fails to go back to the pin when the pressure is released, this condition should be repaired. The trouble is probably due to a sprung mechanism brought about by allowing the pressure to enter a gauge suddenly.

7. Always keep a tight connection between the regulator and the cylinder. If the connection leaks after a reasonable force has been used to tighten the nut, close the cylinder valve and remove the regulator. Clean both the inside of the cylinder valve seat and the regulator inlet-nipple seat. If the leak persists, the seat and threads are probably marred, in which case the regulator will have to be returned to the manufacturer for repair.

Oxygen and Acetylene Hose

A special nonporous hose is used for welding. To prevent mistakes in connecting them, *the oxygen hose is green and the acetylene hose red.* If oxygen were to pass through an old acetylene hose, a dangerous combustible mixture might result.

A standard connection is used to attach the hose to the regulator and torch. This connection consists of a nipple which is forced into the hose and a nut that connects the nipple to the regulator and torch. The acetylene nut may be distinguished from the oxygen by the notch that runs around the center, indicating a left-hand thread as illustrated in Figure 5-14. A clamp of some type is used to squeeze the hose around the nipple to prevent it from working loose.

Care of welding hose. The acetylene and oxygen hose is an important part of the welding equipment. First of all, as previously mentioned, all the

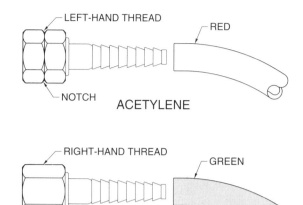

Figure 5-14. The nut on the acetylene connection has a notch that runs around the center, distinguishing it from the nut on the oxygen connection.

connections on the hose must be perfectly tight. These connections should be tightened with close-fitting wrenches to prevent damaging the nuts.

Avoid dragging the hose around on a greasy floor, since grease or oil eventually will soak into it. Be careful in pulling the hose around sharp objects and especially over hot metal. Prevent anyone from stepping or dropping anything on the hose. When the welding has been completed, roll up the hose and suspend it in such a manner that it will not drop to the floor.

Here are a few additional precautions:

1. All new hose is dusted with talcum powder inside. This powder should be blown out before using the hose.

2. Do not use ordinary wire to bind hose to a connection. Use regular hose binders or clamps.

3. When splicing hose, use standard brass splicing nipples—never use copper tubing.

4. Long lengths of hose tend to kink.

5. Do not repair leaking hose with tape. Splice in a new piece or discard.

The Sparklighter

The sparklighter, sometimes called a striker, is the tool used for igniting the torch. See Figure 5-15. Form the habit of always employing a sparklighter to light a torch. *Never use matches.* The use of matches for this purpose is very dangerous because the puff of the flame produced by the ignition of the acetylene flowing from the tip is likely to burn your hand.

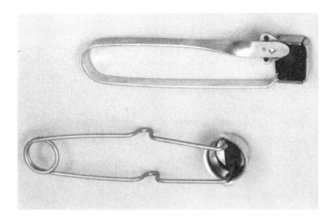

Figure 5-15. A sparklighter is used for lighting the welding torch.

Goggles

An oxyacetylene flame produces light rays of great intensity and also heat rays which, if meeting the naked eye, may eventually prove destructive to the eye tissues. Consequently, always wear goggles having suitable, approved, colored glass. The density of the colored lenses should be such that damaging light and heat rays are not allowed to pass through to the welder.

For most gas welding, goggles with shade numbers of 4, 5 and 6 are recommended. See Figure 5-16. These also protect the eyes from flying sparks and pieces of molten metal that may splatter around safety glasses. If in doubt about which lens to use, refer to American Welding Society standards for eye protection.

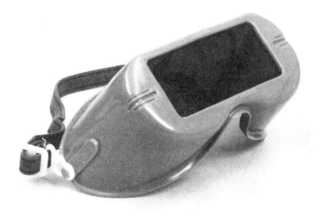

Figure 5-16. Always wear welding goggles with the recommended shade number.

Protective Clothing

It is a good idea to wear an apron, shop coat, or coveralls while welding with oxyacetylene equipment. Sparks will invariably shoot away from the molten metal, and, without some suitable covering, numerous holes will be burned in your clothes. *Under no circumstances should a sweater or other flammable garments be worn.* A small spark falling on these garments may burst into a rapid-spreading flame that could produce dire consequences. Some type of welding cap is also desirable to prevent any hot metal particles from falling on the hair.

The beginner should wear a pair of lightweight gloves to avoid possible burns. Occasionally the hot end of a rod or a piece of metal which has been momentarily laid aside to cool is picked up by mistake, and without gloves serious burns are apt to result. However, oxyacetylene welders frequently do not wear gloves.

Special Gas Welding Processes

Although oxyacetylene is very common for certain types of welding processes, other gases also may be used. The most common of these gases are natural gas, propane, hydrogen, and methylacetylene propadiene stabilized (Mapp). The principal difference between these and oxyacetylene is in the type of gas employed in the burning mixture; otherwise the welding technique is the same.

Oxy-Mapp welding. Mapp gas is a Dow Chemical Company product having many of the physical properties of acetylene, but it lacks the shock sensitivity of acetylene. The gas is the result of rearranging the molecular structure of acetylene and propane. Although propane itself is very stable, its limiting factor for welding is its low flame temperature. Acetylene on the other hand produces a very high flame temperature but is very unstable. By combining the two gases and changing their molecular structure we have a very stable fuel with a flame temperature nearly comparable to acetylene, without the dangers of acetylene.

Generally, with Mapp gas a slightly larger tip is required because of its greater gas density and slower flame propagation rate. The only significant difference is in the flame appearance. A neutral flame for welding will have a longer inner cone than with oxyacetylene gas.

Since Mapp gas is not sensitive to shock it can be stored and shipped in lighter cylinders. Acetylene to be kept safe must be stored in cylinders filled with a porous filler and acetone. Where empty acetylene cylinders weigh around 220 pounds, Mapp cylinders weigh only 50 pounds. Normally a filled cylinder of acetylene weighs 240 pounds while a filled cylinder of Mapp gas weighs 120 pounds. See Figure 5-17.

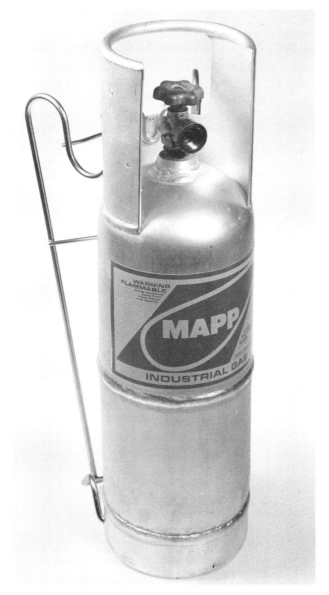

Figure 5-17. Since Mapp gas is not as sensitive to shock as acetylene is, it can be stored and shipped in lighter cylinders. (Airco).

Oxy-hydrogen welding. This combination generates a low temperature flame used primarily for welding thin sections of metal where low temperatures are required. One of the unusual characteristics of oxy-hydrogen is that the flame is practically non-luminous. Consequently, difficulty is often experienced in adjusting for a neutral flame. To avoid welding with what might be an oxidizing flame, the practice is to adjust the regulator so it produces a greater flow of hydrogen at first.

Air-acetylene welding. The air-acetylene flame is generated by burning a mixture of acetylene with air. As the acetylene flows to the torch under pressure it draws the right amount of air for proper combustion. The temperature of the air-acetylene flame is even lower than that of the oxy-hydrogen and therefore is used mostly for soldering and brazing very light metals. It has wide applications in the plumbing industry for joining copper tubing.

Points to Remember (*WARNINGS*)

1 *Handle oxygen and acetylene cylinders with care. Never expose them to excessive heat and prevent oil and grease from coming in contact with them.*

2 *Be sure the adjusting screw on a regulator is fully released before opening a cylinder valve.*

3 *Always hang up a torch when not in use to prevent it from dropping to the floor and being bent or damaged.*

4 *Never remove tips with pliers. If a tip has to be cleaned, use a tip cleaner.*

5 *Do not lubricate the adjusting screw on a regulator with oil. Use soap or glycerine.*

6 *Never interchange the hoses. Avoid dragging them over greasy floors.*

7 *Always wear proper goggles when welding, as well as suitable protective clothing.*

8 *Never light a torch with a match.*

9 *Never use oxygen or acetylene from the tank to blow dirt and dust from clothing.*

QUESTIONS FOR STUDY AND DISCUSSION

1 What safety devices are used to prevent cylinders from exploding when subjected to intense pressure?

2 What is the purpose of the protector cap on a cylinder?

3 How much should the cylinder valve be opened on an acetylene cylinder? On the oxygen cylinder?

4 Why is it dangerous to allow grease or oil to come in contact with the oxygen cylinder valve?

5 What is the function of the needle valves on a welding torch?

6 Why are the oxygen and acetylene hose fittings made with different screw threads?

7 How are sizes of welding tips indicated?

8 Why should a close-fitting wrench be used in removing the various welding fittings?

9 Why is it incorrect to use pliers for removing welding tips?

10 What is a tip cleaner, and when and why should it be used?

11 What is meant by a two-stage pressure regulator?

12 Why should the adjusting screw on a regulator be fully released before opening a cylinder valve?

13 What precautions should be observed in handling the pressure regulator?

14 Why is it a poor practice to light a torch with a match?

15 Why is Mapp gas sometimes used instead of acetylene?

16 Oxy-hydrogen and air-acetylene are often used for what operations?

17 What welding goggle shade numbers are commonly used for most oxyacetylene welding?

18 What type of protective clothing is commonly worn when oxyacetylene welding?

19 What are three ways of identifying oxygen hoses from acetylene hoses?

20 Who is responsible for repairing a damaged regulator?

21 How are oxygen and acetylene cylinders moved safely?

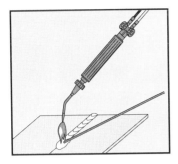

Oxyacetylene Welding–OAW

One of the first things you should learn in starting your oxyacetylene welding operations is to assemble a welding outfit. A certain sequence must be followed if the equipment is to be properly and safely connected. Once you have learned to do this then you need to know how to light the torch and adjust the flame. This chapter explains how these operations may be performed.

ASSEMBLING THE WELDING OUTFIT

To assemble a welding outfit follow these directions:

Chain cylinders. Fasten the cylinders securely to a truck or to some other fixed object where they are to be located. Remove the protector cap from each cylinder and examine the outlet nozzles closely. Make sure the connection seat or screw threads are not damaged. A damaged screw thread may ruin the regulator nut, while a poor connection seat will cause the gas to leak.

Crack cylinder valves. Particles of dirt frequently collect in the outlet nozzle of the cylinder valve. If not cleaned out, this dirt will work into the regulator when the pressure is turned on. To avoid any possibility of dirt clogging up any passage, open the valves of each cylinder for just an instant. Then with a clean cloth, wipe out the connection seat. See Figure 6-1.

Attach regulators. Connect the oxygen pressure regulator to the oxygen cylinder and the acetylene regulator to the acetylene cylinder as shown in Figure 6-2. Use a close-fitting wrench to tighten the nuts, and avoid stripping the threads. Always use a proper wrench to tighten the nuts, as a loose-fitting wrench will eventually ruin the corners of the regulator nuts.

Figure 6-1. Crack cylinder valves to blow out dirt which may be lodged in the outlet nozzles. *(Linde)*

Connect hoses to regulators. Connect the green hose to the oxygen regulator and the red hose to the acetylene regulator as illustrated in Figure 6-3. Check the adjusting screw on each regulator to make certain that it is released, and open the cylinder valves. Then blow out any dirt that may be lodged in the hoses by opening the regulator adjusting screws. This also will purge the hoses of any residual gases. Promptly close the regulator adjusting screws. See Figure 6-4.

Figure 6-2. Attach the regulator to the cylinder with a proper fitting wrench. *(Linde)*

Figure 6-3. Connect the green hose to the oxygen regulator and the red hose to the acetylene regulator. *(Linde)*

Connect check valves and hoses to the torch. To prevent the reverse flow of gases that would result in a combustible mixture in the welding hose, check valves are mounted to the welding torch. Under normal conditions the gas flows toward the welding torch. Any condition that might cause a reverse flow of gas will close the valve. Check valves should be left in place on the torch when the hose is detached.

Connect the hoses to the check valves mounted on the torch. The red hose is connected to the acetylene check valve mounted on the needle valve fitting marked *AC*. The green hose is connected to the check valve mounted on the needle valve fitting marked *OX*. See Figure 6-5. *Remember, acetylene hose connections always have left-hand threads as indicated by the notched nut and oxygen hose connections have right-hand threads.*

Figure 6-4. Purge each hose by opening the regulator adjusting screw. *(Linde)*

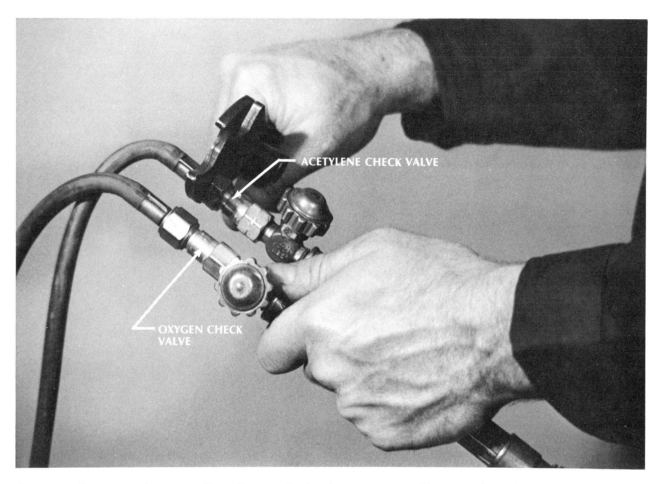

Figure 6-5. To prevent the reverse flow of gases, check valves are mounted between the welding torch and hose.

Testing for leaks. All new welding apparatus needs to be tested for leaks before being operated. Thereafter it is advisable to periodically apply this same test in order to insure that no leakage has developed. A leaky apparatus is very dangerous, since a fire may develop. Furthermore, leaks mean wasted gas.

To test for leaks, open the oxygen and acetylene cylinder valves and with the needle valves on the torch closed, adjust the regulators to give about normal working pressure. Apply soapy water with a brush on the following points as shown in Figure 6-6:

    A—Oxygen cylinder valve
    B—Acetylene cylinder valve
    C—Oxygen regulator inlet connection
    D—Acetylene regulator inlet connection
    E—Hose connections at the regulators and torch
    F—Oxygen and acetylene needle valves

Inspect each point carefully. Note any noise or bubbles at each point as an indication of leakage. If a leak is detected by a connection, use a wrench to properly tighten the fitting. If tightening does not

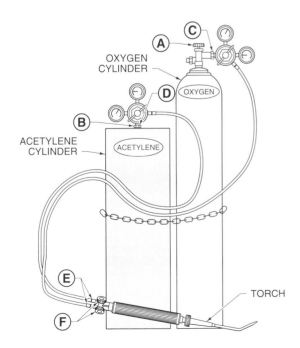

Figure 6-6. Check for leaks at these points. (*Linde Company*)

remedy the leak, shut off the gas pressure, open the connections, and examine the screw threads.

Under no circumstances should any other test for leakage be performed. If there is a leak in the cylinder or cylinder valve, move the cylinder out of doors away from possible sources of ignition and notify the supplier immediately.

To check for leakage in the welding hose adjust the regulators to working pressure. Submerge the hose in clean water. Check for any bubbles indicating a leak. On the sections of the welding hose that cannot be submerged, brush on soapy water and check for bubble formation. Welding hoses should be routinely inspected for cuts and worn areas that could be sites of future leakage.

Selecting the Proper Welding Tip

The size of the tip will depend upon the thickness of the metal to be welded. If very light sheet metal is to be welded, a tip with a small opening is used, while a large-sized tip is needed for thick metal. Table 6-1 lists the sizes of tips for various thicknesses of metals and the approximate working pressures required.

It is very important to use the correct tip, with the proper working pressure. If too small a tip is employed, the heat will not be sufficient to fuse the metal to the proper depth. When the tip is too large, the heat is too great, thereby burning holes in the metal. A good weld must have the right penetration and smooth, even, overlapping ripples. Unless the conditons are just right, it is impossible for the torch to function the way it should and, consequently, the weld will be poor.

TABLE 6-1. TIP SIZES VS METAL THICKNESS.

TIP NUMBER	THICKNESS (inches)	OXYGEN PRESSURE (pounds)	ACETYLENE PRESSURE (pounds)
000	up to 1/64	1	1
00	1/64	1	1
0	1/32	1	1
1	1/16	1	1
2	3/32	2	2
3	1/8	3	3
4	3/16	4	4
5	1/4	5	5
6	5/16	6	6
7	3/8	7	7
8	1/2	7	7
9	5/8	7 1/2	7 1/2
10	3/4 & up	9	9

Lighting the Torch

1. Select a tip for welding ⅛" or ⅟₁₆" metal and mount it on the torch.

2. Stand aside and open the oxygen and acetylene cylinder valves slowly. Set the working pressure to correspond to the size of tip that is to be used. *Do not face the regulator when opening the cylinder valve.* Stand to one side of the regulator. See Figure 6-7. A defect in the regulator may cause the gas to blow through, shattering the glass and blowing it into your face. Remember, oxygen and acetylene are charged in the tanks under a high pressure and if the gas is permitted to come against the regulator suddenly, it may cause some damage to the equipment. Open the acetylene cylinder valve approximately one complete turn and the oxygen all the way. Next turn the oxygen and acetylene regulator adjusting valves to the required working pressures.

Figure 6-7. Stand to one side of the regulator when opening a cylinder valve.

To light the torch turn the acetylene needle valve on the torch approximately one half of a turn. With the sparklighter held about one inch away from the end of the tip, ignite the acetylene as it leaves the tip as shown in Figure 6-8. Do this as rapidly as possible to avoid unnecessary wasting of gas. If not enough acetylene is turned on, the flame will produce considerable smoke; therefore, quickly turn on more acetylene until the flame has a slight tendency to jump away from the tip.

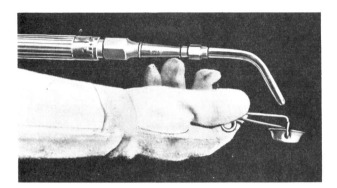

Figure 6-8. Hold the sparklighter approximately one inch away from the tip when lighting the torch.

WARNING: Never use a match to light a torch. This procedure brings your fingers too close to the tip and the sudden ignition of the acetylene is very apt to burn them.

When igniting a torch, keep the tip facing downward. Lighting the torch while it is facing outward or upward may result in burning someone nearby as the ignited flame spurts out.

WARNING: If other people are welding around the same area, never reach over another person for a light. Such an action will not only disturb their work, but might cause a serious accident.

WARNING: Make no attempt to relight a torch from the hot metal when welding. In an enclosed box, tank, drum, or other small cavity, there may be just enough unburned gas in this confined space to cause an explosion as the acetylene from the tip comes in contact with the hot metal. Instead, *move the torch into the open,* relight it in the usual manner, and make the necessary adjustments before resuming the weld.

Adjusting the Flame

With the acetylene burning, gradually open the oxygen needle valve until a well-defined white cone appears near the tip, surrounded by a second, bluish cone that is faintly luminous. This is known as a *neutral flame* because there is an approximate one-to-one mixture of acetylene and oxygen resulting in a flame which is chemically neutral. The brilliant white cone should be from 1/16″ to 3/4″ long, depending on the tip size. See Figure 6-9. The neutral flame is used for most welding operations.

Any variation from the one-to-one mixture will alter the characteristics of the flame. When an excess amount of oxygen is forced into the mixture, the resulting flame is said to be *oxidizing.* This flame resembles the neutral flame slightly, but has a shorter

and more pointed inner cone with an almost purple color rather than brilliant white. It is sometimes used for brazing. See Figure 6-10.

If the mixture consists of a slight excess of acetylene, the flame is *carburizing,* or *reducing.* This flame can easily be identified by the existence of three flame zones instead of the usual two found in the neutral flame. The end of the brilliant white cone is no longer as well defined, and it is also surrounded by an intermediate white cone, which has a feathery edge in addition to the usual bluish outer envelope. See Figure 6-10 for this feathery edge.

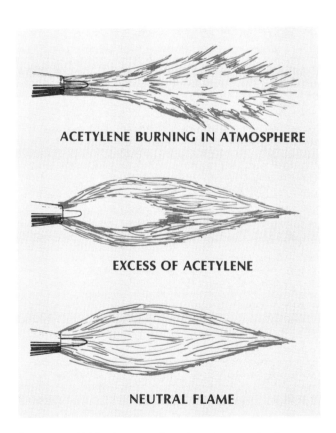

Figure 6-9. With the acetylene burning, gradually open the oxygen needle valve to obtain a neutral flame. (*Linde Company*)

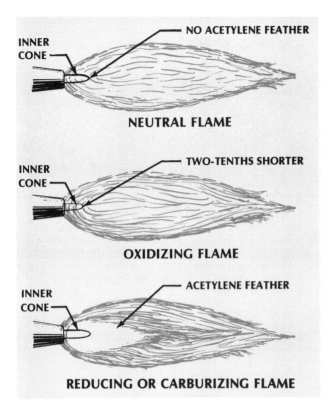

NEUTRAL FLAME

INNER CONE — / — NO ACETYLENE FEATHER

OXIDIZING FLAME

INNER CONE — / — TWO-TENTHS SHORTER

REDUCING OR CARBURIZING FLAME

INNER CONE — / — ACETYLENE FEATHER

Figure 6-10. An oxidizing flame is the result of an excess of oxygen in the mixture. A slight excess of acetylene produces a reducing flame. (*Linde Company*)

Testing the Flames   **Exercise 1**

You will probably have a little difficulty at first in making the correct flame adjustment, but if the characteristics of the carburizing and oxidizing flames are understood, the adjustment for a neutral flame will be relatively easy. As an aid in becoming familiar with the effects of these various flames, complete the following exercise:

1   *Carburizing flame.* Obtain a piece of scrap metal. Light the acetylene and turn on the oxygen until a white cone appears on the end of the tip enveloped by another fan-shaped cone which has a feathered edge.

Put on a pair of goggles and apply this flame to the metal, holding the point of the white cone close to the metal. You will notice that as the metal melts it has a tendency to boil. This is an indication that carbon is entering the molten metal. You will also discover, after the metal has cooled, that the surface is pitted and very brittle.

2   *Oxidizing flame.* Now, open the oxygen needle valve. The white cone becomes short and the color changes to a purplish hue. The flame burns with a decided roar.

Apply this flame to a piece of metal, allowing the cone to come in contact with the surface. You will find that as the metal melts there are numerous sparks given off and a white foam or scum forms on the surface. After this piece cools, the metal will be shiny.

3   *Neutral flame.* Now, adjust the needle valve until the flame is balanced. Apply this neutral flame to a piece of metal. The molten metal flows smoothly like syrup, with very few sparks, clean and clear.

Flame Characteristics

A flame may be harsh or quiet. The harsh type is induced by forcing too much pressure of both gases to the tip. This flame is undesirable, since it has a tendency to depress the molten surface and cause the metal to splatter around the edges of the puddle. Such a flame is noisy, and its use makes it extremely difficult to bring about perfect fusion, with smooth, uniform ripples.

The quiet flame is just the opposite of the harsh flame and is achieved by the correct pressure of gases flowing to the tip. The flame is not a forcing, noisy flame but one that permits a continuous flow of the molten puddle without any undue amount of splatter.

To secure a soft neutral flame, see that the tip is absolutely clean and the mixture is correct. Even if the proportion of acetylene and oxygen is right, a good weld is difficult to achieve unless the opening in the tip permits a free flow of gases. Any foreign matter in the tip will simply restrict the source of heat necessary to melt the metal.

Flame Control

Once the flame has been properly set, it does not mean that further adjustments are unnecessary. From time to time, as the welding progresses, it is necessary to observe the flame cone to be certain that the mixture has not altered. Changes in the flame will occur as a result of some slight fluctuation in the flow of the gases from the regulators. A slight turn of one needle valve or the other will quickly readjust the flame.

In the course of welding, a torch may occasionally start popping. This noise is an indication that there is an insufficient amount of gases flowing to the tip. Such popping can be stopped by opening both the oxygen and acetylene needle valves on the torch to a greater extent. Another reason for popping is the overheating of the molten pool by lingering or keeping the flame too long in one position and not melting enough rod into the pool.

Backfire and Flashback

When the flame goes out with a loud pop, it is called a *backfire.* A backfire may be caused by (1) operating the torch at lower pressures than required for the tip used, (2) by touching the tip against the work, (3) by overheating the tip, or (4) by an obstruction in the tip. If a backfire should occur, shut the needle valves and after remedying the cause, relight the torch.

A *flashback* is a condition that results when the flame flashes back into the torch and burns inside with a shrill hissing or squealing noise. If this should happen, close the needle valves immediately. A flashback generally is an indication that something is wrong. Perhaps it is a clogged tip, or the improper functioning of the needle valves, or even an incorrect acetylene or oxygen pressure. In any case, investigate the cause before relighting the torch.

Shutting Off the Torch

Following is the correct sequence of steps for shutting OFF a torch:

1. Close the oxygen needle valve.
2. Close the acetylene needle valve.
3. If the entire welding unit is to be shut down, shut off both the acetylene and the oxygen cylinder valves.
4. Remove the pressure on the working gauges by opening the needle valves until the lines are drained. Then promptly close the needle valves.
5. Release the adjusting screws on the pressure regulators.

Points to Remember

1 Be sure to fasten the cylinders securely so they will not fall over.

2 Crack the cylinder valves before attaching regulators.

3 Periodically test the welding outfit for leaks. Use soapy water only.

4 Use the correct size tip for the thickness of metal to be welded.

5 Always stand to one side of the regulator when opening cylinder valves and be sure the regulator adjusting valves are fully released.

6 Never use a match to light a torch; use a regulation sparklighter.

7 Adjust the torch to a soft, neutral flame for welding unless the nature of the metal calls for a different type of flame.

8 Keep the passage in the welding tip clean.

9 Avoid conditions that may cause a backfire or flashback.

QUESTIONS FOR STUDY AND DISCUSSION

1 Why should the cylinders be securely fastened before being used?

2 Why should the cylinder outlet nozzles be examined closely?

3 What is meant by the term *cracking the cylinder?* Why is this done?

4 What is the proper order for setting up the welding apparatus?

5 Why should a close-fitting wrench be used in fastening all connections?

6 Why are check valves used?

7 What is the proper method of testing for gas leaks?

8 What governs the size of the tip which should be used?

9 Why must the working pressure be correct for the size of tip that is to be used?

10 How far should the acetylene needle valve be opened when lighting the torch?

11 What is meant by an *oxidizing flame?*

12 What is a *carburizing flame?*

13 What is the difference between a neutral flame and a carburizing flame?

14 What are the characteristics of a *neutral flame?*

15 What is the difference between a harsh and soft flame?

16 What are some of the conditions that may cause a *backfire?*

17 What is meant by a *flashback* when one is using an oxyacetylene torch?

18 What causes excessive smoke when lighting the torch?

19 How is a carburizing flame changed to a neutral flame?

20 Why are hoses purged after being connected to the regulators?

21 Why should the welder stand to one side when opening cylinder valves?

22 What is the last step completed to the regulator when shutting off the torch?

23 What kind of mixture of acetylene and oxygen is required to achieve a neutral flame?

24 What happens when an oxidizing flame is used to melt the metal?

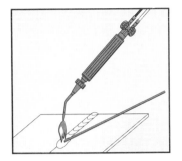

Oxyacetylene Welding—OAW

To master the skill of welding with an oxyacetylene torch, you will have to practice a series of operations in a definite order. These operations involve carrying a puddle without a rod, laying beads with a filler rod, and welding various types of joints. All of this welding should be done with the metal lying in a flat position.

Carrying a Puddle without a Filler Rod

Holding the Torch. A torch may be held in either one of two ways, depending on which is the more comfortable for you. When welding light-gauge metal, most operators prefer to grasp the handle of the torch with the hose over the outside of the wrist, in the same manner a pencil ordinarily is held. See Figure 7-1. In the other grip, the torch is held like a hammer, with the fingers lightly curled underneath as shown in Figure 7-2. In either case the torch should balance easily in the hand to avoid fatigue.

Position and motion of the torch. Hold the torch so the flame points in the direction you are going to weld and at an angle of about 45° with the completed part of the weld. See Figure 7-3. If you are right handed, start the weld at the right edge of the metal and bring the inner cone of the neutral flame to within ⅛" of the surface of the plate. The left-handed person reverses this direction. Hold the torch still until a pool of molten metal forms. Then move the puddle across the plate. As the puddle travels forward, rotate the torch to form a series of overlapping ovals as shown in Figure 7-3.

Do not move the torch ahead of the puddle, but slowly work forward, giving the heat a chance to melt the metal. If the flame is moved forward too rapidly, the heat fails to penetrate far enough and the metal does not melt properly. If the torch is kept in one position too long, the flame will burn a hole through the metal. See Exercise 1.

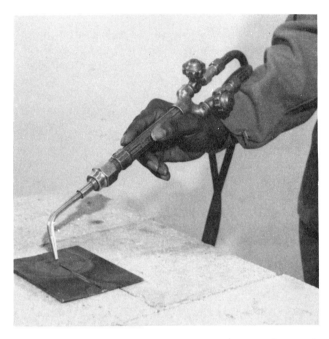

Figure 7-1. When welding in the flat position, the torch may be held like a pencil.

Figure 7-2. Another way to hold the torch is to grip it like a hammer, with fingers curled underneath.

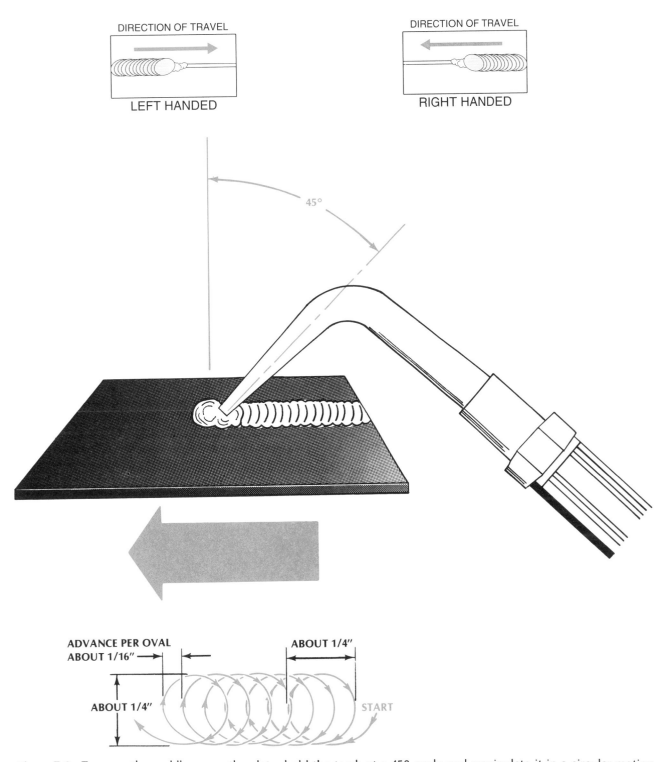

DIRECTION OF TRAVEL

LEFT HANDED

DIRECTION OF TRAVEL

RIGHT HANDED

45°

ADVANCE PER OVAL
ABOUT 1/16″

ABOUT 1/4″

ABOUT 1/4″

START

Figure 7-3. To move the puddle across the plate, hold the torch at a 45° angle and manipulate it in a circular motion.

Carrying a Puddle without a Filler Rod   **Exercise 1**

1 Obtain a piece of metal ¹⁄₁₆″ to ⅛″ in thickness and approximately 3″ in width and 5″ in length.

2 Be sure the surface is free of oil, dirt, and scale.

3 Light the torch and adjust it for a neutral flame.

4 Holding the inner cone of the flame approximately ⅛″ from the work, angle the torch to a 45° angle to the piece. Using a circular manipulation, move the torch from the right side of the plate to the left. Reverse the direction of travel if you are left handed.

5 Maintain a consistent travel speed to prevent burn-throughs in the plate.

6 Practice running beads without filler rod until properly formed beads are produced consistently.

BURN-THROUGH

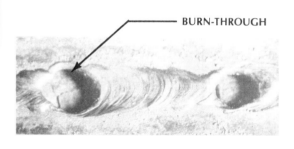

Laying Beads with a Filler Rod

On some types of joints it is possible to weld the two pieces of metal without adding a filler rod. In most instances the use of a filler rod is advisable because it builds up the weld, thereby adding strength to the joint. The strength of a weld depends largely on the skill with which the rod is blended, or interfused, with the edges of the base metal.

The use of a filler rod requires coordination of the two hands. One hand must manipulate the torch to carry a puddle across the plate, while the other hand must add the correct amount of filler rod.

Selecting the filler rod. A welded joint should always possess as much strength as the base metal itself. If this is to be accomplished it is necessary to employ a welding rod that has the same properties as the base metal. It is a mistake to attempt to use just any kind of wire, because an inferior rod contains so many impurities that it is difficult to use and makes a weld that is weak and brittle. A good welding rod will flow smoothly and readily unite with the base metal without any excessive amount of sparking. A rod of poor weldability will spark profusely, flow irregularly, and leave a rough surface filled with punctures like pinholes.

Filler rods come in a variety of sizes ranging from ¹⁄₁₆″ to ⅜″ in diameter. The size of rod to use will depend largely upon the thickness of the metal. *The general rule is to use a rod with a diameter equal to the thickness of the base metal.* In other words, if a ¹⁄₁₆″ metal is to be welded, a ¹⁄₁₆″ diameter rod should be used.

A great many different kinds of rods are available for welding a variety of metals. For example, a mild-steel rod for cast iron, a nickel rod for nickel steel, a bronze rod for bronzing malleable cast iron and other dissimilar metals, an aluminum rod for aluminum welding, a copper rod for copper products, etc.

Manipulating the filler rod. Hold the rod at approximately the same angle as the torch but slant it away from the torch. See Figure 7-4.

Figure 7-4. Hold the torch and filler rod at the same angle when adding filler rod to the weld.

Melt a small pool of the base metal and then insert the tip of the rod in this pool. Remember, to secure proper fusion, the correct diameter rod is important. If the rod is too large the heat of the pool will be insufficient to melt the rod, and if the rod is too small the heat cannot be absorbed by the rod, with the result that a hole is burned in the plate.

As the rod melts in the pool, advance the torch forward. Concentrate the flame on the base metal and not on the rod. Do not hold the rod above the pool. If you do this the molten metal will fall through the air to the puddle. When this happens it combines with the oxygen of the air and part of it burns up, causing a weak, porous weld. Always dip the rod in the center of the molten pool.

Laying beads. Rotate the torch to form overlapping ovals and keep raising and lowering the rod as the molten puddle is moved forward. *An alternate torch movement* is the semicircular motion as shown in Figure 7-5. When the rod is not in the puddle, keep the tip just inside the outer envelope of the flame. A beginner often experiences difficulty in holding the welding rod steady which can cause the rod to stick to the base metal. Instead of inserting the rod in the middle of the puddle where the heat is sufficient to melt it readily, the beginner may insert it near the edge of the pool where the temperature is lower. The heat at the edge is not enough to melt the rod. When this happens, do not try to jerk it loose, since such an action will simply interrupt the welding. To loosen the rod, play the flame directly on the tip of the rod and the rod will be freed immediately. In all probability, while the rod is being freed, the puddle will have solidified; therefore be sure to reform the puddle before moving forward.

Travel speed. To secure beads of uniform width and height you must keep the forward movement of the torch just right. If the puddle is carried forward too slowly, it becomes too large and you may even burn through the metal. If it is moved too rapidly, the rod is not actually fused with the base metal but merely stuck on the surface. Furthermore, it will be impossible to form ripples evenly.

When it appears that the puddle is getting too large, withdraw the flame slightly so only the outer envelope of the flame is touching the molten puddle. *Do not move the flame off to one side, since such a movement allows the air to strike the hot metal and oxidize it.* See Exercise 2.

Laying Beads with a Filler Rod **Exercise 2**

1 Obtain a piece of metal ¹⁄₁₆″ to ⅛″ in thickness and approximately 3″ in width and 5″ in length.

2 Be sure the surface is free of oil, dirt, and scale.

3 Light the torch and adjust it for a neutral flame.

4 While manipulating the torch and the filler rod at the correct angles, practice running consistent straight beads.

5 As the torch is withdrawn at the end of the pass, fill the crater by adding filler rod.

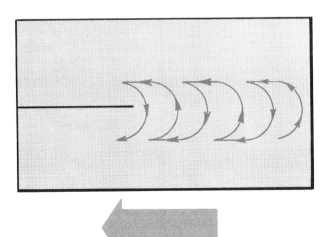

Figure 7-5. A semicircular torch motion can be used to maintain an adequate puddle.

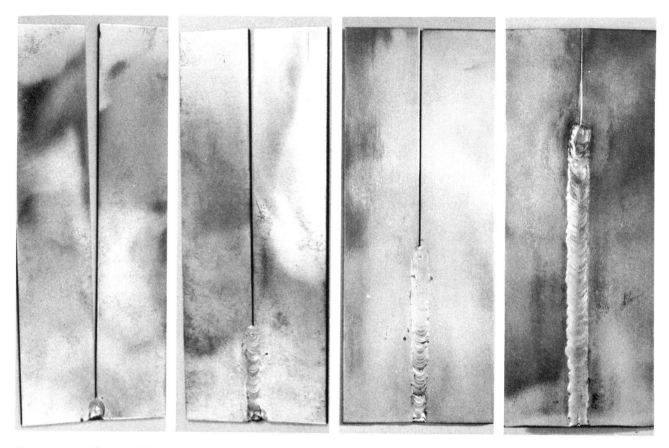

Figure 7-6. When welding two plates in a butt joint, put more space between the plates at the end of the weld to allow for expansion and contraction of the metal.

Welding Butt Joints

After you have mastered carrying a puddle across the surface of a plate while adding filler rod, your next task is to fuse two pieces together in a butt joint.

Space plates. Allow a gap of about 1/16″ at the starting end of the joint and approximately 1/8″ at the other. This is known as progressive spacing. The purpose of the space is to allow for expansion of the metal; otherwise the edges will overlap before the weld is completed as illustrated in Figure 7-6. Furthermore, this space permits the flame to melt the edges all the way to the bottom of the plates.

Tack plates. If progressive spacing between the edges of a seam is not used, then the plates must be tacked at various intervals to restrict expansion forces. See Figure 7-7.

To make a tack weld, simply apply the flame to the metal unit until it melts and then add a little filler rod. See Exercise 3.

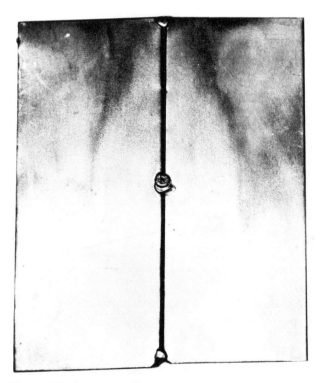

Figure 7-7. In cases where progressive spacing is not used, the plates must be tacked before welding to restrict expansion forces.

Welding Butt Joints in the Flat Position — **Exercise 3**

1 Obtain two pieces of metal ¹⁄₁₆″ to ⅛″ in thickness and approximately 1½″ in width and 5″ in length.

2 Space for progressive spacing or tack the two plates together on two firebricks.

3 Begin welding at the right end (or the left end if you are left handed), using the same torch and rod motion as when running beads with a rod.

4 Work the torch slowly to give the heat a chance to penetrate the joint, and add sufficient filler rod to build up the weld about ¹⁄₁₆″ above the surface. Be sure the puddle is large enough and the metal is flowing freely before you dip the rod.

5 Maintain a molten puddle approximately ¼″ to ⅜″ in width.

6 Advance the puddle about ¹⁄₁₆″ with each complete motion of the torch while maintaining a uniform bead width.

7 Uniform torch motion will produce smooth even ripples.

Check for Defects. It is only natural to assume that your first few welds will break easily. Continue trying until you can make a straight, smooth weld that will not open when bent.

The following are some of the common defects that you may expect to find in your first few attempts to weld:

1. Uneven weld bead, caused by moving the torch too slowly or too rapidly.

2. Holes in the joint, caused by holding the flame too long in one spot.

3. A brittle weld, due to improper flame adjustment during welding.

4. Excessive metal hanging underneath the weld, showing too much penetration. See Figure 7-8 for example.

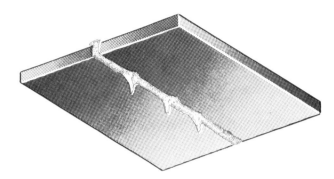

Figure 7-8. Too much penetration causes excessive metal hanging underneath the weld.

5. Insufficient penetration, caused by moving the torch forward too rapidly. When the penetration is correct, the underside of the seam should show that fusion has taken place clear to the bottom of the joint as shown in Figure 7-9.

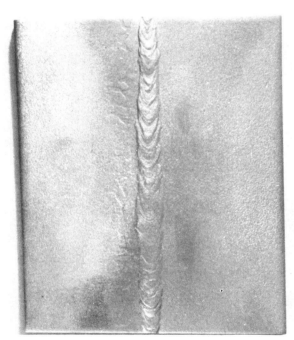

Figure 7-9. When penetration is correct, complete fusion is evident on the underside of the joint.

6. Hole in the end of the joint, caused by not lifting the torch when the end of the weld has been reached.

7. Very often a joint appears to have correct penetration and still cracks open when tested. This may be caused by a number of reasons, such as:

a. Improper space allowances between the edges of the plates.

b. Filling the space between the metal plates with molten rod without sufficiently melting the edges of the plates to insure a good bond between the parent metal and rod.

c. Holding the torch too flat, causing the molten puddle to lap over an area that has not been properly melted to merge with it.

Welding Other Joints

When the ability to weld a correct butt joint is mastered, other joints may be welded employing techniques similar to those used on the butt joint. A *flange joint* is used a great deal in sheetmetal work, particularly on material that is 20 gauge or less. The flange should extend above the surface of the sheet a distance equal to about the thickness of the sheet. See Exercise 4.

The *corner joint* is used extensively in fabricating numerous products such as tanks and vessels as well as in repair work. The edges are fused without a filler rod as in welding a flange joint. See Exercise 5.

A *lap joint* is formed when one piece of metal is laid on top of another. Careful control of the direction of heat is needed in this weld. See Exercise 6.

A T-joint is made by standing one piece of metal at a right angle to the second plate. This joint necessitates a greater amount of filler metal added; thus, correct rod usage is critical. See Exercise 7.

Welding a Flange Joint in the Flat Position — Exercise 4

1. Place two pieces with the flange edges touching and tack weld them.

2. Hold the torch on the end until a puddle is formed.

3. Carefully manipulate the torch to maintain the puddle as the puddle is carried across the entire joint.

4. Withdraw the torch at the end of the joint to prevent burning a hole in the joint.

Welding a Corner Joint in the Flat Position — Exercise 5

1. Tack two pieces to form a corner joint.

2. Hold the torch on the end of the joint until a puddle is formed.

3. Using a technique similar to that employed on the flange joint, manipulate the torch to maintain a puddle across the entire joint.

4. Withdraw the torch at the end of the joint to prevent burning a hole in the joint.

5. If additional build-up is required, filler rod may be added as the puddle is carried across the joint.

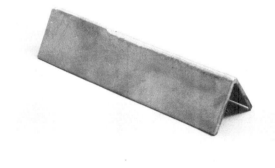

Welding a Lap Joint in the Flat Position

Exercise 6

1 Obtain two pieces of metal approximately 1/16″ to 1/8″ in thickness, 1½″ in width, and 4″ to 5″ in length.

2 Lay one plate on top of the other and tack them in place to form a lap joint.

3 Weld the two plates using a semicircular motion of the torch.

4 While manipulating the torch, direct more of the heat to the bottom plate. This may be accomplished by increasing the duration of the torch motion on the bottom plate. The top plate requires less heat and may overheat if too much heat is applied.

5 Weld one side of the plate and then practice on the reverse side.

Welding a T-Joint in the Flat Position

Exercise 7

1 Obtain two pieces of metal approximately 1/16″ to 1/8″ in thickness, 1½″ in width, and 4″ to 5″ in length.

2 Stand one plate on top of the other and tack in place to form a T-joint.

3 Tilt the tacked pieces 45° to the work surface, and place a firebrick under one side to support the pieces.

4 Hold the torch so the tip forms an angle of about 45° to the bottom plate.

Exercise 7, cont.

5 Using the same technique employed when welding a butt joint, keep the inner cone of the flame about ⅛″ away from the deepest part of the weld.

6 Manipulate the torch constantly while adding filler rod to produce a consistent weld free of any undercutting.

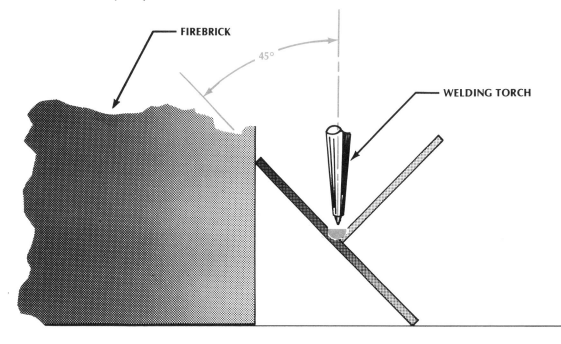

FIREBRICK

45°

WELDING TORCH

Points to Remember

1 Move the torch just fast enough to keep the puddle active and flowing forward.

2 When reaching the end of a joint, raise the flame away momentarily to give the puddle a chance to solidify partly.

3 Use a filler rod that has the same composition as the base metal.

4 Use a rod with a diameter equal to the thickness of the base metal.

5 Do not hold the rod too high above the pool so the molten metal falls drop by drop onto the puddle.

6 When welding with a filler rod, move the torch in a semicircular or circular motion.

7 Allow a space between plates to compensate for expansion forces.

QUESTIONS FOR STUDY AND DISCUSSION

1 What governs the rate at which the flame should be moved forward?

2 What happens if the torch is moved forward too slowly?

3 Why should the torch be raised when it nears the end of the weld?

4 What are some of the common defects of a beginner's weld?

5 How should the top and bottom surfaces appear when proper fusion has taken place?

6 If the metal does not melt readily, what is the probable cause for it?

7 Why is a filler rod used in welding?

8 What are some signs of correct filler rod selection?

9 What determines the size of the rod that should be used?

10 What happens if the rod is too large for the size of metal that is being welded? If it is too small?

11 How should the torch be manipulated when using a filler rod on a butt weld?

12 Why is the length of the flame prolonged on the bottom plate of a lap joint?

13 What precautions should be observed in welding a T-joint?

14 If the metal does not melt freely, what steps should be taken?

15 What should be the position of the torch for welding a T-joint?

16 What are causes of uneven bead width across the length of the weld joint?

17 Where is the filler rod inserted when laying beads with a filler rod?

18 What are the causes of excessive penetration when welding a butt joint?

19 How are tack welds used when welding butt joints?

20 How are smooth, even ripples formed in the weld bead?

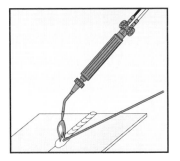

Welding with an oxyacetylene torch cannot always be done with the work in a flat position. On some occasions the location of the piece will be such as to require horizontal, vertical, or overhead welding. Undoubtedly welding in a flat position is easier and somewhat faster; nevertheless after a little practice, welds in other positions can be performed without too much difficulty.

In all three positions — horizontal, vertical, and overhead — the main obstacle to obtaining a sound weld is the gravitational pull downward on the molten metal. The overhead weld is a little more difficult to perform because of the unusual position in which you are required to work and the skill needed to keep the molten puddle from dropping off the plates.

Overhead welding is possible because of the fact that molten metal has cohesive (sticky) qualities as long as the puddle is not permitted to get too large. In other words, molten metal will not fall if the puddle is not allowed to form in complete drops. The amount of heat directed on the seam must be very carefully regulated, since excessive heat will increase the flow of the molten metal. With the correct flame, proper torch manipulation, and practice, the overhead weld can be mastered quickly.

Horizontal Welding a Butt Joint   Exercise 1

1 Tack two pieces of ⅟₁₆″ or ⅛″ plates to form a butt joint. Allow a space between the edges for expansion.

2 Clamp the joint in a jig so the surfaces are in a vertical position with the line of weld running horizontally.

3 Start welding from the right edge (or left, if you are left handed) using a semicircular torch motion. As the welding progresses, you will find that the metal has a tendency to build up much more on the edge of the lower plate. To overcome this, direct the flame longer on the lower edge of the plate without allowing the molten puddle to drop. Keep the tip of the filler rod nearer the upper plate.

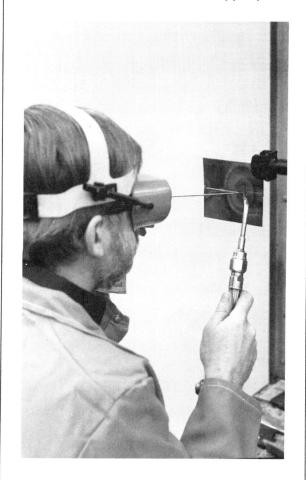

Horizontal Welding a T-Joint

1 Tack two plates of ¹⁄₁₆″ to ⅛″ thickness at a 90° angle to form a T-joint.

2 Start the welding at the right end if you are right handed, at the left end if you are left handed. Proceed to the opposite end using a semicircular torch movement.

3 Hold the torch so the tip forms an angle of 45° to the flat plate and the same angle to the line of the weld.

4 Point the rod toward the welding tip at an angle of approximately 30°·to the line of the weld and 15° to 20° to the horizontal plate.

5 Direct the flame evenly over both plates. To prevent undercutting, add filler rod nearer the vertical plate.

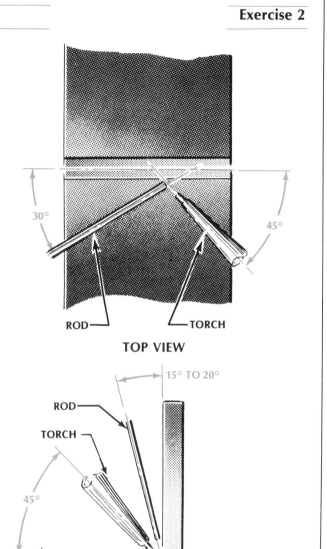

Vertical Welding a Butt Joint

1 Obtain two pieces of ¹⁄₁₆″ or ⅛″ metal and tack them to form a butt joint. Then mount the plates in a jig.

2 Hold the torch and rod at about the same angle as in flat welding. As the welding progresses you may have to vary the angle slightly to control the puddle.

3 Start the weld at the bottom and work upward, using a semicircular torch motion. Do not allow the puddle to become too large. If the puddle gets too big or too fluid, it will get out of control and run down the face of the weld. When you find that the weld is getting too hot, pull the flame away slightly so that it does not play directly on the puddle.

Exercise 3, cont.

4 To prevent the puddle from getting too fluid, direct more of the flame on the filler rod. Continue to practice this weld until you can obtain beads that appear good and have correct penetration.

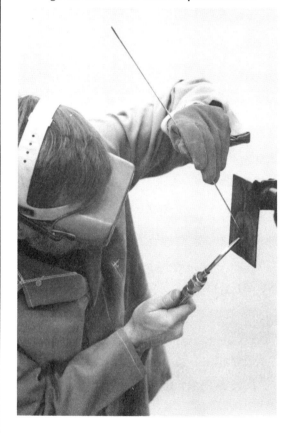

Overhead Welding a Butt Joint **Exercise 4**

1 Tack two ¹⁄₁₆″ or ⅛″ plates together and set the work up on a suitable jig that will permit sufficient freedom for manipulating the torch.

2 Use the same semicircular motion of the torch as previously described. To help keep the puddle shallow, move the filler rod slowly in a circular or swinging motion. The movement of the rod will dis-

tribute the molten puddle and prevent it from forming into large drops and falling off.

3 Watch the flame very closely and, if the puddle has a tendency to run, pull the torch away slightly.

4 Continue practicing overhead welds until satisfactory joints are obtained.

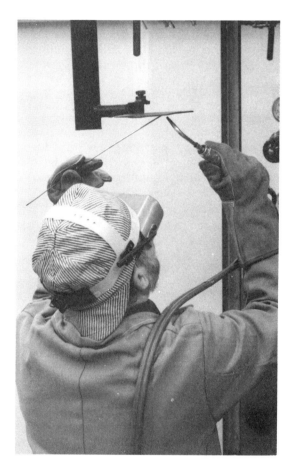

Points to Remember

1 Use a semicircular torch movement for vertical, horizontal, and overhead welding.

2 Do not allow the puddle to become too large as you will lose control of it and it will have a tendency to sag.

3 If the puddle has a tendency to become too fluid, raise the flame slightly.

4 In horizontal welding direct the flame more on the edge of the lower plate.

5 On overhead welds move the filler rod slowly in a circular or swinging motion.

QUESTIONS FOR STUDY AND DISCUSSION

1 What can be done to prevent the puddle from sagging in vertical welding?

2 At what angle should the torch be held for horizontal welding?

3 How should the torch be manipulated for vertical, horizontal, and overhead welding?

4 In horizontal welding, why should the flame be directed more on the edge of the lower plate?

5 What should be done to prevent the puddle from becoming too fluid?

6 Why is overhead welding somewhat more difficult to perform?

7 How can the puddle be prevented from dropping off in overhead welding?

8 How should the filler rod be manipulated in overhead welding?

9 What is done to provide maximum penetration of a horizontal butt joint before welding?

10 What can be done to prevent undercutting of the weld when welding a horizontal T-joint?

11 What can be done to maintain a shallow puddle when welding in the overhead position?

12 Explain why welding with an acetylene torch cannot always be done with an oxyacetylene torch.

HEAVY STEEL PLATE

Oxyacetylene Welding—OAW

Heavy steel plate is rarely welded with oxyacetylene unless other types of welding equipment are not available. Compared with other methods, welding heavy plate with oxyacetylene is a much slower and less cost efficient process. It is included here in the event that necessity demands use of oxyacetylene equipment to weld or repair a particular structure.

Rolled steel stock is commonly labeled as sheet or plate, depending upon the thickness. Generally speaking, if the metal is ⅛″ or less in thickness, it is referred to as *sheet.* If the thickness is over ⅛″ it is designated as *plate.*

Although the technique for welding heavy plate is about the same as in welding light material, the problems associated with heavy plate welding are somewhat more complicated. For this kind of welding more attention has to be given to the manner in which the joints are prepared and to the amount of heat required for sufficient penetration.

Welding a Single-V Butt Joint

To have the greatest maximum strength, a weld must possess complete penetration. Full penetration in stock ⅛″ or less in thickness is reasonably easy to achieve. When the thickness exceeds ⅛″, penetration cannot be obtained if the edges are left square. On such metal the edges must be beveled. Methods of beveling edges include using a cutting torch, a beveling machine, or a grinder.

For plates up to ½″ in thickness, the single-V, as shown in Figure 9-1, is ample. In the single-V, the included angle should be 60°. The bottom of the V can have a ¼6″ or ⅛″ square root face (unbeveled) or the edges feathered to a sharp point. See Exercise 1.

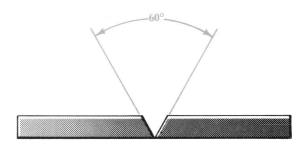

Figure 9-1. In a single-V butt joint, the included angle is 60°.

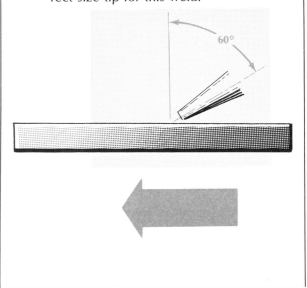

Welding a Single-V Butt Joint **Exercise 1**

1 Obtain two pieces of ¼″ plate and bevel the edges.

2 Separate the plates about ¼6″ and tack weld them together. Use a filler rod that is approximately ¾6″ in diameter.

3 Hold the torch so the flame will be at an angle of 60° from the vertical rather than the 45° angle used on light stock as previously described. Determine the correct size tip for this weld.

Exercise 1, cont.

4 Direct the flame on the V and, as the edges begin to melt, dip the tip of the rod in the puddle. Before adding filler rod make certain that the sides of the V are in a molten state all the way to the bottom of the V. Fill in the bottom of the V for a length of about ½″ with the puddle extending upward to one-half the depth of the V. While the puddle is still in a plastic state, swing the torch in a semicircular motion and fill the V. The completed bead should be between ⅜″ to ½″ in width and project slightly above the surface of the plate. Return the flame to the bottom of the V, advance another ½″, and again raise this section of the bead to the top of the V. Continue to do this until the weld is finished.

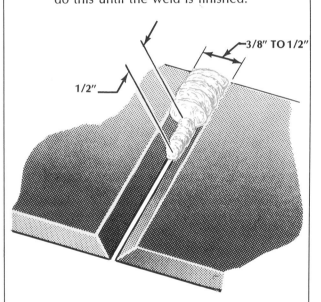

5 To test the weld, cut off several 1″ strips. Grind off the surplus weld metal so that the top of the weld (face) is flush with the top of the plate. The grind marks should be running lengthwise on the piece to prevent premature failure when testing. Place the specimen in the guided bend tester. Put on eye protection and then apply pressure on the specimen. If the weld is good there should be no indications of cracking or fracturing.

Exercise 1, cont.

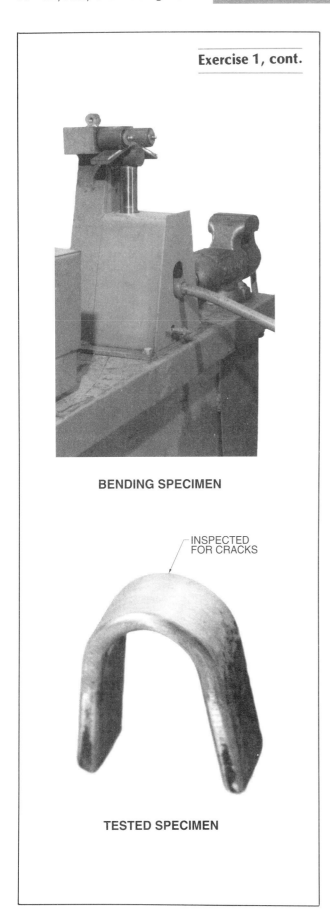

BENDING SPECIMEN

INSPECTED
FOR CRACKS

TESTED SPECIMEN

Backhand Welding

All of the welding that has been described so far is known as forehand welding. As you know, in this method the welding rod precedes the welding torch in the direction of the weld.

In the *backhand* technique, the weld is carried from left to right (right to left for a left-handed person). The welding flame is directed backward at the completed portion of the weld and the rod is between the flame and the completed weld section. See Figure 9-2. Since the flame is constantly directed on the edges of the V ahead of the puddle, no sidewise motion of the torch is necessary. As a result, a narrower V can be utilized than in forehand welding.

In backhand welding the puddle is less fluid. This results in a slightly different appearance of the weld surface. The ripples are heavier and spaced further apart.

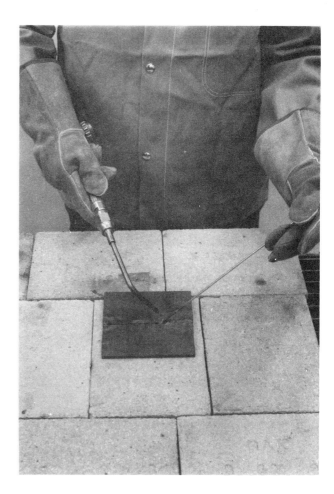

Figure 9-2. In backhand welding the welding torch precedes the welding rod in the direction of the weld.

Welding a Backhand Weld — Exercise 2

1 Place two ¼″ thick plates on the bench with the beveled edges separated about ¹⁄₁₆″ and tack them together.

2 Start the weld at the left (right, if left-handed) and bring the edges of the V to a molten puddle. While doing so, hold the end of the rod in the outer envelope of the flame so it will be ready to melt as soon as the puddle forms.

3 At the start, concentrate the flame a little more on the bottom of the V. As soon as the puddle is fluid enough, dip the rod into it. Once the puddle begins to move, direct the flame more on the filler rod and build up the puddle to the top of the V. As the molten metal begins to fill up the V, move the filler rod slightly from side to side to make sure that the weld metal fuses evenly with the edges of the base metal.

4 Test the joint in the same manner as described in making a single-V butt weld with the forehand method.

Welding a Double-V Butt Joint

When a plate is ½″ or more in thickness, it is much better to run a weld on both sides. For this purpose a double-V is required. See Figure 9-3. Notice that in a double-V a ¹⁄₁₆″ or ⅛″ root face is provided in the center. Either forehand or backhand method can be used to make the weld.

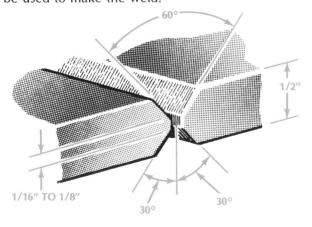

60°
1/2″
1/16″ TO 1/8″
30°   30°

Figure 9-3. Use a double-V butt joint to weld plates 1/2″ or more in thickness.

In making a double-V weld it will be necessary to build up the weld in layers, because it is too difficult to control the molten puddle and secure good penetration in attempting to fill the V in one lap. The usual practice is to first run one layer near the bottom of the V on both sides of the plate. Then successive layers are added to fill the V. Extreme care must be taken when applying the successive layers to fuse each layer with the one already deposited and with the sides of the base metal V.

Welding High-Carbon Steel

Any metal having a carbon content of 0.89 percent carbon or more is considered to be a high-carbon steel. This includes all carbon machine steels, spring steel, and common tool steels.

Whether a steel is weldable or not depends upon the effects the heating and cooling cycle has on the physical properties of the weld area and the base metal itself. If the metal can be taken through the heating and cooling cycle without cracks developing in the weld zone or without seriously affecting the characteristics of the original metal, a steel is assumed to be weldable. See Chapter 3 for a more complete discussion of the effects of heat on steel.

To weld a high-carbon steel, it must be preheated first. This can be done by applying the oxyacetylene flame uniformly over the metal until it reaches a faint red color. The welding should be done with an excess acetylene flame with just enough heat to secure fusion between the weld metal and the base metal. For thin parts where there is likely to be considerable mixing of base metal and weld metal, a low-carbon filler rod should be used. On heavier sections and especially when the parts are to be heat-treated again, a high-carbon rod is advisable.

Points to Remember

1 Plate thicker than ⅛″ should be beveled before welding.

2 Plate ½″ or more in thickness should have a double-V.

3 When beveling plate, the included angle should be 60°

4 Hold the torch at a 60° angle when welding heavy plate.

5 Be sure the bottom surfaces of the V have reached the proper temperature before adding filler rod.

6 The backhand welding technique is particularly adaptable for welding heavy plate.

7 In backhand welding do not swing the torch; move the filler rod instead.

8 On plates ½″ or more in thickness do not try to fill the V in a single pass. Use several passes.

Points to Remember, cont.

9 Remember that a high-carbon steel will lose its hardness if welded. To restore its hardness it must be put through a heat-treating process after welding.

10 Use a slightly excess acetylene flame for welding high-carbon steel.

11 Use a high-carbon filler rod on steels that are to be heat-treated after welding.

QUESTIONS FOR STUDY AND DISCUSSION

1 Why is some metal called sheet and other plate?

2 Why should the edges be beveled when the plate is ⅛″ or more thick?

3 At what angle should the torch be held when welding heavy plate?

4 Why should the edges of heavy plate be spaced for welding?

5 How should test specimens of heavy plate be prepared for checking strength of weld?

6 How does the backhand welding technique differ from the forehand?

7 How should the torch and rod be handled in backhand welding? At what angle?

8 At what angle should heavy plates be beveled in preparation for welding?

9 Why is a double-V used on some plates?

10 In welding plates over ½″ thick, why use more than one pass?

11 Why must extreme care be taken when welding high-carbon steel that has been heat-treated?

12 What kind of flame should be used in welding high-carbon steel?

13 What kind of filler rod is needed to weld high-carbon steel?

14 How is high-carbon steel heated in preparation for welding?

15 How are heavy steel plate welds tested?

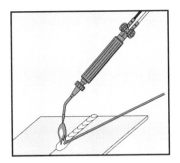

GRAY CAST IRON

Oxyacetylene Welding–OAW

Gray cast iron may be fusion-welded, except that greater caution must be taken to offset expansion and contraction forces than if it were braze welded. See Chapter 27. Since gray cast iron is unusually brittle, it is very susceptible to temperature changes. Consequently, preheating and postweld heat treatment are necessary when welding.

To maintain the gray iron structure throughout the weld area, the weld has to be made with the correct filler rod and the parts cooled slowly. If the casting is cooled rapidly, the weld area is likely to turn into white cast iron, thereby making the weld section not only extremely brittle but so hard that machining might be impossible.

Preparing the Edges

The edges of the casting should be beveled to have a 90° included angle, but the V should extend only to ⅛″ from the bottom of the break. Beveling in this manner makes it easier to build up a sound weld near the bottom and lessens the danger of melting through. Placing carbon back-up blocks underneath the joint also helps to prevent the molten cast iron from running out the seam.

Precautions must be taken to clean the surfaces of the joint before welding. The weld area should be cleaned at least one inch on both sides of the V as shown in Figure 10-1. Improperly cleaned surfaces will result in porous spots and blowholes in the weld, even though sufficient flux is used.

Preheating and Postweld Heat Treatment

One important rule that must be followed for successful fusion-welding of cast iron is to preheat the entire casting to a dull red. Uniform preheating will equalize the expansion and contraction forces and thus minimize the possibility of cracks developing. On a small section, the heating can be carried out by playing the flame over the casting. A large casting may have to be placed in a preheating furnace. The temperature has to be watched very carefully on a

Figure 10-1. Clean the surface surrounding the weld joint before welding.

heavy casting, especially if it has thin members. The thin members will heat more rapidly, so care must be taken not to get them too hot.

After the welding is completed, postweld heat treat by bringing the entire casting up to a uniform temperature. Use the same techniques used for preheating the casting.

Using Filler Rod and Flux

A special cast iron filler rod having the same composition as the base metal is needed to weld cast iron. It is important that the rod contains sufficient silicon to assure soft, machinable weld deposits when used with correct preheat and postweld heat treatment.

A flux is essential in welding cast iron to keep the molten puddle fluid. Otherwise, infusible slag will mix with the iron oxide that forms on the puddle. When this happens, the weld will contain inclusions and blowholes.

Oxyacetylene Welding Gray Cast Iron — Exercise 1

1. Prepare the edges to be welded. Bevel the joint if necessary and remove all foreign matter from the surface.

2. Slowly heat the entire metal to a dull red.

3. Concentrate the flame near the starting point of the weld until the metal begins to melt. Keep the torch in the same position as in welding mild steel with the inner cone of the flame about 1/8" to 1/4" from the seam.

4. When the bottom of the V is thoroughly fused, move the flame from side to side, melting down the sides so the molten metal runs down and combines with the fluid metal in the bottom of the V. Rotate the torch in a circular motion to keep the sides and bottom of the V in a molten condition. If the metal gets too hot and tends to run away, raise the torch slightly.

5. Once you have a molten puddle, bring the filler rod into the outer envelope of the flame and keep it there until the rod is fairly hot. Then dip it into the flux. Now insert the fluxed end of the rod into the molten puddle. The heat of the puddle will melt the rod. Never keep dipping the rod in and out of the puddle. As the rod melts, the molten metal will rise in the groove. When it has been built up slightly above the top surface move the puddle forward about one inch and repeat the operation. Be sure not to move the puddle before the sides of the V have been broken down, as this will force the molten puddle ahead on the cold metal.

6. When gas bubbles or white spots appear in the puddle or at the edges of the seam, add more flux and play the flame around the specks until the impurities float to the top. Skim these impurities off the puddle with the rod. By tapping the rod against the bench the impurities can be removed.

7. After the weld is completed, postweld heat treat the entire piece to a dull red. Allow the casting to cool slowly. A proper cast iron weld should look like the specimen shown in the photograph.

8. To test your weld sample, place it in a vise with the weld flush with the top of the jaws. Put on proper eye protection and then strike the upper end with a heavy hammer until the piece breaks. If the metal has been welded properly, the break should not occur along the fused line but in the base metal.

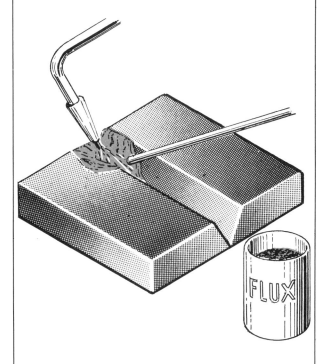

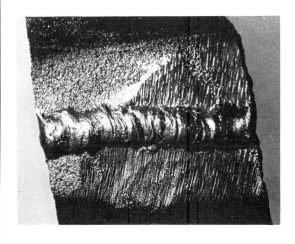

Points to Remember

1 If possible, use carbon back-up blocks when welding cast iron.

2 Clean the surfaces at least one inch around the seam which is to be welded.

3 Preheat the cast iron to a dull red before welding.

4 Use the correct grade of cast iron filler rod.

5 Apply flux to the molten metal as the weld is being made.

6 Keep the torch moving in a circular motion to distribute the heat evenly.

7 Postweld heat treat the entire piece after the weld is completed and then allow it to cool slowly.

QUESTIONS FOR STUDY AND DISCUSSION

1 Why should a rod having the same properties as the base metal be used for this type of welding?

2 Why should cast iron pieces be preheated before welding?

3 When has a casting been preheated enough to weld satisfactorily?

4 Why should welded cast iron pieces be postweld heat treated?

5 What are the various steps of preparation to be followed before welding a joint?

6 Why is a flux necessary in welding cast iron?

7 How is the flux manipulated in order to deposit it in the weld?

8 How should the rod be introduced into the puddle?

9 What precaution should be taken in moving the puddle forward?

10 If the metal has been properly welded, where should the break occur when the completed weld is tested?

11 How are small and large castings preheated differently?

12 What steps are taken to prevent weld metal from flowing out the bottom of the joint during welding?

13 How is the torch moved during the welding process to maintain a molten puddle in the sides and bottom of the joint?

14 When is flux applied when welding gray cast iron?

15 How is a small, gray cast iron weld specimen tested?

16 How are impurities in the molten weld metal removed?

17 What happens to gray cast iron that is cooled too rapidly?

18 What should be done if the weld metal becomes too hot during the welding process?

19 What could occur if the weld area is not thoroughly cleaned?

20 What are indications of impurities in the weld area?

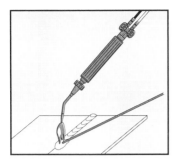

Although the gas shielded-arc processes are the most practical for welding commercially pure aluminum, there are occasions when oxyacetylene welding is used (See Figure 11-1).

Welding Considerations

The following must be kept in mind when welding aluminum with a gas flame:

1. Aluminum has a relatively low melting point compared to other metals that are welded. Pure aluminum melts at 1220°F (660°C).

2. The thermal conductivity of aluminum is high, almost four times that of steel.

3. Due to its light color, there is practically no indication when the melting point is reached; it collapses suddenly into liquid.

4. Molten aluminum oxidizes very rapidly, forming a heavy coating on the surface of the seam, which necessitates the use of a good flux.

5. Aluminum when hot is very flimsy and weak, and care must be taken to support it adequately during the welding operation.

6. Aluminum welds should be made in a single pass if possible.

Choosing the Correct Joint Design

In general, the same principles of joint design for welding steel apply to aluminum. On thin material up to about 1/16" thick, the edges should be formed to a 90° flange at a height equal to the thickness of the material. Flanges will prevent excessive warping and buckling and also serve as filler metal when the flange is melted in the welding operation. Usually no additional filler rod is necessary.

Aluminum from 1/16" to 3/16" in thickness can be butt welded, providing the edges are notched with a saw or cold chisel. See Figure 11-2. Notching minimizes the possibility of burning holes through the joint, permits full penetration, and prevents local distortion.

As a rule the lap joint is not recommended for welding aluminum because of the danger of flux and oxide being trapped between the surfaces of the joint. When this happens the aluminum is likely to corrode.

Figure 11-1. Because of the special characteristics of aluminum, extra care is necessary when using the oxyacetylene process. (*Alcoa*)

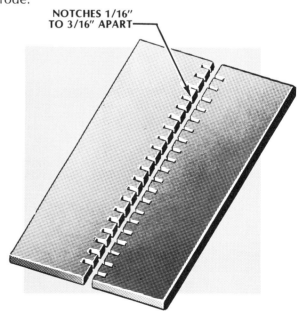

NOTCHES 1/16" TO 3/16" APART

Figure 11-2. Notch the edges of 1/16" to 3/16" aluminum plates before welding with a butt joint.

For welding heavy aluminum plate ³⁄₁₆″ or more in thickness, the edges should be beveled to form a 90° to 120° V. Allow a ¹⁄₁₆″ to ⅛″ root face. The root face should be notched as shown in Figure 11-3.

Aluminum that is greater than ⅜″ in thickness should be prepared as a double-V butt joint with the root face notched. See Figure 11-4.

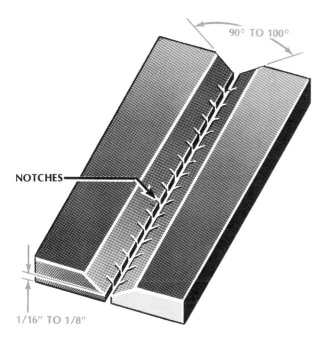

Figure 11-3. When welding 3/16″ to 3/8″ aluminum plates, bevel the edges and notch the root face.

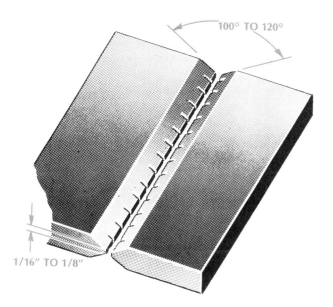

Figure 11-4. Aluminum plates greater than 3/8″ in thickness should be prepared as a double-V butt joint with the root face notched.

Using flux. One of the most important procedures in welding aluminum is to thoroughly clean the edges to be joined. All grease, oil, and dirt must be removed with an appropriate solvent or by rubbing the surface with steel wool or a wire brush. See Figure 11-5.

Since all aluminum oxidizes very rapidly, a layer of flux must be used to insure a sound weld. The flux is sold as a powder, which is usually mixed with water to a consistency of a thin paste (approximately two parts of flux to one part of water).

If the welding does not require the addition of any filler rod, the flux is applied to the joint by means of a brush. When a filler rod is used, the rod also is coated with flux. On heavy sections it is advisable to coat the metal as well as the rod for greater ease in securing better fusion. See Figure 11-6.

When the welding has been completed, it is very important that all traces of the flux be washed away. Otherwise, the remaining flux will subsequently cause corrosion. The flux is removed by washing the piece in hot water or by immersing the weld in a 10 percent cold solution of sulfuric acid, followed by rinsing in hot or cold water.

Selecting the filler rod. As in welding other metals, the proper selection of a filler rod is important for welding aluminum. The composition of the rod should compare to that of the metal to be welded. The three most common rods for welding nonheat-treatable aluminum are 1100, 4043, and 5356. The 4043 and 5356 rods are recommended when greater strength is required.

Figure 11-5. The weld area must be cleaned thoroughly before welding aluminum.

Figure 11-6. When welding heavy aluminum with a filler rod, apply flux both to the rod (top) and the metal (bottom).

Welding rods are obtainable in sizes of ⅟₁₆″, ⅛″, ³⁄₁₆″, and ¼″ diameter. As a rule, a rod whose diameter equals the thickness of the metal should be used.

Preheating aluminum. All aluminum to be welded, including thin sheet, should be preheated as this will decrease the effects of expansion and minimize cracks. Aluminum plate ¼″ or more in thickness should be preheated to a temperature of 300° to 500°F (149° to 260°C). Preheating to these temperatures can usually be done by playing the flame of the oxyacetylene torch over the work. For large or complicated parts, the preheating is done in a furnace.

It is very important that the preheating temperature does not exceed 500°F (260°C). If the temperature goes beyond this point, the alloy may be weakened or parts of the aluminum may collapse under its own weight.

The correct preheating temperature may be determined by the use of a temperature indicating crayon or by any of the following methods:

1. If a mark is made on the metal with a carpenter's blue chalk, it will turn white.

2. If a pine stick is rubbed on the metal, a char mark will be left on it.

3. If the metal is struck with a hammer, no metallic ring will be heard.

Selecting the correct tip size. Since aluminum has such a high thermal conductivity, it is necessary to use a tip slightly larger than the one ordinarily used for steel of the same thickness. Table 11-1 shows the recommended sizes and gas pressures to use for welding aluminum of varying thicknesses.

TABLE 11-1. TORCH WELDING DATA FOR VARIOUS THICKNESSES OF ALUMINUM.

| | OXYACETYLENE | | OXYHYDROGEN | | |
ALUMINUM THICKNESS (inches)	OXYGEN PRESSURE (psi)	ACETYLENE PRESSURE (psi)	ACETYLENE PRESSURE (psi)	TIP ORIFICE dia (inches)	HYDROGEN PRESSURE (psi)
1/16	1+	0.021-0.031	1	0.031 -0.0465	1-3
1/8	1-2	.025- .038	1-2	.038 - .055	2-4
3/16	1-3	.031- .0465	1-3	.0465- .067	3-5
1/4	2-4	.038- .055	2-4	.055 - .076	4-6
3/8	5-7	.067- .086	5-7	.086 - .110	7-9
1/2	6-8	.076- .098	6-8	.098 - .1285	8-10

Reynolds Metals Company

Using the torch. Many operators use hydrogen instead of acetylene for welding aluminum, and this in many cases is more desirable, especially for welding light gauge material. In either case, the torch should be adjusted so it will have a neutral flame. Some authorities recommend a slightly reducing flame, but usually a neutral flame will be found to be very satisfactory in producing a clean, sound weld. Whether using acetylene or hydrogen, the flame should be adjusted to a low gas velocity to permit a soft and not a blowy flame.

It must be remembered that the angle of the torch has much to do with welding speed. Instead of lifting the flame from time to time in order to avoid melting holes in the metal, it will be found advantageous to hold the welding torch at a flatter angle, thus increasing the speed. The speed of welding should also be increased as the edge of the sheet is approached.

CAUTION: The flame should never be permitted to come in contact with the molten metal. Also, hold it no farther away than 1 inch from the material.

Welding Aluminum-Alloy Castings

In general, the welding of aluminum-alloy castings requires techniques similar to those used on aluminum sheet and other wrought sections. However, the susceptibility of many castings to thermal strains and cracks, because of their intricate design and varying section thickness, should be carefully considered. In addition, many castings in highly stressed structures depend upon heat treatment for their strength, and welding tends to destroy the effect of such initial heat treatments. If satisfactory facilities for heat treatment are not available, the welding of such heat-treated castings is not recommended.

When a broken aluminum casting is to be welded, it is first cleaned carefully with a wire brush and gasoline to remove every trace of oil, grease, and dirt. Unless the casting has a very heavy cross-section, it is not necessary to tool the crack or cut out a V, as this can be accomplished by means of the torch and puddling iron. (A puddling iron is a piece of low-carbon stainless steel rod with a flattened end.) It is

---

Welding Aluminum Using the Oxyacetylene Process   Exercise 1

1 Prepare the pieces to be welded.

2 Flux the pieces using the recommended flux.

3 Pass the flame over the starting point until the flux melts.

4 Scrape the surface with the rod at about 3- or 4-second intervals, permitting the rod to come clear of the flame each time; otherwise, the rod will melt before the parent metal. The scraping action will indicate when welding should begin without overheating the aluminum.

5 Using the forehand welding technique, angle the torch at a low angle (less than 30° above horizontal when welding thin material). The torch should be moved forward without any motion.

6 Dip and withdraw the rod in and out of the weld puddle periodically with a forward motion. This method of withdrawal closes the puddle, prevents porosity, and assists the flux in removing the oxide film.

7 Maintain the same procedure throughout the course of the weld.

8 A correct oxyacetylene weld on aluminum will have necessary penetration with correct bead ripple and contour, as shown in the photograph.

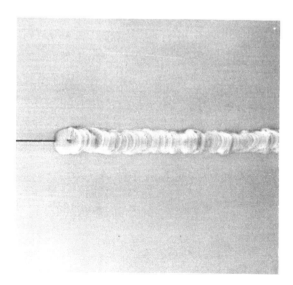

necessary, however, that the stock surrounding the defect be completely melted or cut away before proceeding with the weld. If a piece is broken out, hold it in the correct position by light iron bars and appropriate clamps. The clamps should be so attached that the casting will not be stressed during heating.

If the casting is large or one with intricate sections, it should be preheated slowly and uniformly in a suitable furnace prior to welding. If the casting is small, or if the weld is near the edge and in a thin-walled section, the casting may be preheated in the region of the weld by means of a torch flame. Cast aluminum should be heated slowly to avoid cracking in the section of the casting nearest the flame.

Broken pieces are tack welded into place as soon as the casting has been preheated. The actual welding of the piece should commence at the middle of the break, and should be continued toward the ends. The welding rod must be melted by the torch, as the heat of the molten metal is not sufficient to melt it. When the weld is finished, the excess molten metal is scraped off with a puddling iron, and the casting allowed to cool slowly.

Holes in castings are welded in much the same manner as are cracked and broken castings. But it is necessary to melt away or cut away the sides of the hole in order to remove all pockets and to permit proper manipulation of the torch.

For welding ordinary castings, an aluminum-silicon or aluminum-copper-silicon welding rod is necessary. A flux must also be used. Puddling alone will merely break up the oxide film and leave it incorporated in the weld, while fluxing will cause the oxide particles to rise to the surface, resulting in a clean, sound weld. It is important that the added metal be completely melted and the molten metal thoroughly turned with the end of the welding rod or with a puddling iron. Thus the flux and oxide are worked to the surface of the molten metal, and there is very little danger of the finished weld becoming contaminated with particles of flux or other foreign material.

Points to Remember

1. Always use the recommended flux to weld aluminum.

2. Use a 1100, 4043, or 5356 rod for welding aluminum.

3. Preheat the work before welding.

---

4. Keep preheat below 500°F (260°C).

5. Use a temperature indicating crayon or rub a pine stick or blue chalk on the aluminum to determine the correct preheating temperature.

6. Use a slightly larger tip than for welding steel.

7. Use a neutral or slightly reducing flame for all aluminum welding.

8. Hold the torch at an angle of 30° or less when welding thin material.

9. For thicknesses 3/16″ and above, increase the torch angle to nearer the vertical.

10. Clean the surfaces thoroughly before starting the welding operation.

QUESTIONS FOR STUDY AND DISCUSSION

1 What are some of the characteristics that must be taken into consideration when welding aluminum?

2 How can you determine when aluminum has reached its preheating temperature?

3 Why must flux be used to weld aluminum?

4 How is this flux applied?

5 What type of welding rod is recommended for welding aluminum?

6 At what angle is the torch inclined when welding aluminum? Why?

7 What determines the diameter of the welding rod that should be used?

8 Why are the edges of the joint notched?

9 What is the melting point of aluminum?

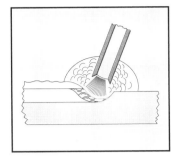

MACHINES AND ACCESSORIES

Shielded Metal-Arc Welding—SMAW

Shielded metal-arc welding, sometimes referred to as arc welding, metallic-arc welding, or just stick welding, is widely used in the construction of many products, ranging from steamships, tanks, locomotives, and automobiles to small household appliances. Shielded metal-arc welding machines are designed to join light and heavy gauge metals of all kinds. The process of shielded metal-arc welding not only simplifies the maintenance and manufacture of goods and machines, but it permits the skilled operator to perform welding operations quickly and easily.

Welding Current

When an electrical current moves through a wire, heat is generated by the resistance of the wire to the flow of electricity. The greater the current flow the greater the resistance and the more intense the heat.

The heat generated for welding comes from an arc which develops when electricity jumps across an air gap between the end of an electrode and the base metal. The air gap produces a high resistance to the flow of current and this resistance generates an intense arc heat which may be anywhere from 6000°F to 10,000°F (approximately 3300°C to 5500°C).

Welding current is provided by an AC or DC machine. The primary current (input) to a welding machine is either 220 or 440 volts. *Since voltage of this magnitude is always dangerous, extreme care must be taken to insure that the motor and frame are well grounded.*

The actual voltage used to provide welding current is low (18 to 36V) whereas high amperage is necessary to produce the heat required for welding. Although low voltage and high amperage for welding are not particularly dangerous if there is adequate grounding and proper insulation, both should be treated with care to avoid any electrical accident.

Electrical Terms

To understand the correct operation of an electric arc welding machine, you must know a few basic electrical terms and electric principles. The following are especially important:

Alternating current (AC). An electrical current having alternating positive and negative values. In the first half cycle the current flows in one direction then reverses direction and for the next half cycle flows the opposite way. See Figure 12-1. The rate of change is referred to as frequency. This frequency is indicated as 25, 40, 50, and 60 cycles per second. In the United States, alternating current is usually established at 60 cycles per second.

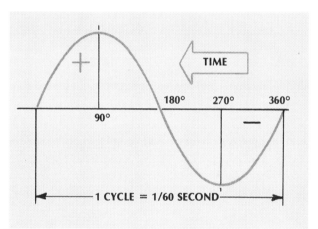

Figure 12-1. Alternating current has alternate positive and negative values. (*Miller Electric Manufacturing Company*)

Direct current (DC). Electrical current which flows in one direction only.

Conductor. A conductor is any material in the form of a wire, cable or bus bar which allows a free passage of, or conducts, an electrical current.

Electrical circuit. Path taken by an electric current in flowing through a conductor from one terminal of the source of supply to the other. It starts from the negative terminal of the power supply where the current is produced, moves along the wire or cable to the load or working source, and then returns to the positive terminal. See Figure 12-2.

Ampere. Amperes (abbreviated amp), or amperage, refers to the amount or rate of current that flows in a circuit. The instrument that measures this rate is called an ammeter. See Figure 12-3.

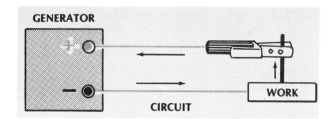

Figure 12-2. Electric current flows from the negative terminal of the power supply, moves along the wire or cable to the work, and then returns to the positive terminal of the power supply.

Volt. The force (emf, or electro-motive force) that causes current to flow in a circuit is known as voltage. This force is similar to the pressure used to make water flow in pipes. In a water system, the pump provides the pressure, whereas in an electrical circuit the power supply produces the force that pushes the current through the wires. Voltage does not flow; only current flows. The force is measured in volts and the instrument used to measure voltage is called a voltmeter. See Figure 12-4.

Resistance. Resistance is the opposition of the material in a conductor to the passage of an electric current causing electrical energy to be transformed into heat.

Static electricity. Static electricity refers to electricity at rest or electricity that is not moving.

Dynamic electricity. Dynamic electricity is electricity in motion in an electrical current.

Constant potential. Potential is synonymous with voltage. It refers to the generation of a stable voltage regardless of the amperage output produced by the welding power supply. This characteristic is particularly important in gas metal arc welding. See Chapter 25.

Voltage drop. Just as the pressure in a water system drops as the distance increases from the water pump, so does the voltage lessen as the distance increases from the generator. This fact is important to remember in using a welding machine because if the cable is too long, there will be too great a voltage drop. When there is too much drop, the welding machine cannot supply enough current for welding.

Open-circuit voltage and arc voltage. Open-circuit voltage is the voltage produced when the machine is running and no welding is being done. This voltage varies from 50 to 100V. After the arc is struck, the voltage drops to what is known as the *arc or working voltage,* which is between 18 and 36V. An adjustment is provided to vary the open-circuit voltage so welding can be done in different positions. See Figures 12-5 and 12-6.

Figure 12-3. An ammeter indicates the amount of current that is flowing. (*Miller Electric Manufacturing Company*)

Figure 12-4. A voltmeter measures voltage (the force of electricity) flowing in a circuit. (*Miller Electric Manufacturing Company*)

Variable voltage. A control which spans a range of voltages is used to set the open-circuit voltage on a welding machine.

Polarity. Polarity indicates the direction of the current in that circuit. Since the current moves in one direction only in *DC welders,* polarity is important because for some welding operations the flow of current must be changed. When the electrode holder cable is connected to the negative pole of the welding machine and the work to the positive pole, the polarity is direct current negative (DC−) or more commonly referred to as *straight polarity.* See Figure 12-7. If the electrode holder cable is connected to the positive pole of the welding machine and the cable leading to the work to the negative pole, the circuit is called direct current positive (DC+) or *reverse polarity.* See Figure 12-8.

the area of the work is greater and more heat is required to melt the metal than the electrode. If heavy deposits are to be made, the work should be hotter than the electrode. For this purpose, straight polarity would be more effective.

On the other hand, in overhead welding it is necessary to quickly freeze the molten filler metal in position against the force of gravity. By using reverse polarity, less heat is generated at the workpiece, thereby giving the filler metal greater holding power.

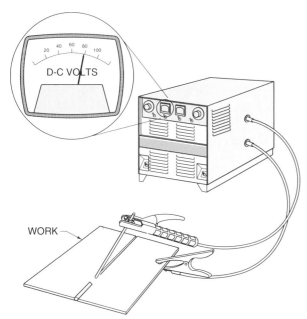

Figure 12-5. Open-circuit voltage is the voltage produced when the machine is running and no welding is being done.

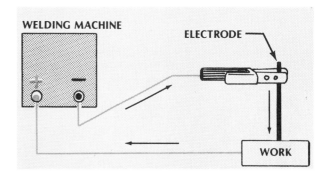

Figure 12-7. In the straight polarity circuit, current flows from the electrode to the work.

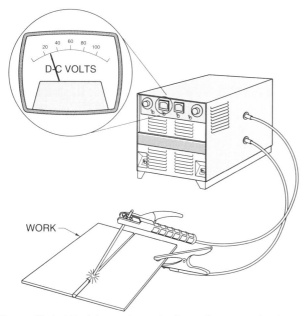

Figure 12-6. Working voltage is the voltage used when welding is in process.

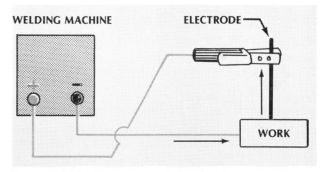

Figure 12-8. In the reverse polarity circuit, current flows from the work to the electrode.

Polarity has an effect on the amount of heat going into the base metal. By changing polarity the greatest heat can be concentrated where it is most needed. With straight polarity more heat is directed to the workpiece. When using reverse polarity more of the heat generated is directed to the electrode.

For some types of welding situations, it is preferable to have more heat at the workpiece because

In other situations, it may be expedient to keep the workpiece as cool as possible, such as in repairing a cast-iron casting. With reverse polarity less heat is produced in the base metal and more heat at the electrode. The result is deposits can be applied rapidly while the base metal is prevented from overheating.

On early DC welders, the change of polarity involved reversing the cables. Modern machines equipped with a *polarity switch* eliminate disconnecting the cables. Moving the switch to straight or reverse changes the polarity. See Figure 12-15.

Because current is constantly reversing in AC welders, polarity is of no consequence.

WELDING MACHINES

To supply the current for welding, three types of units are available: transformers, motor generators, and rectifiers. Power to run these welding machines may come from regular electrical lines or from gasoline or diesel engines. The gasoline or diesel engines are especially useful for field work where electrical power is not available. See Figures 12-9 and 12-10.

cycle for welding machines is based on a ten-minute period of time. Every welding machine is rated at a certain amperage output and voltage output for a given period of time. The National Electrical Manufacturers Association (NEMA) has set a standard based on the ten-minute period. Thus a welder rated at 300 amperes, 32 volts, 60 percent duty cycle will put out the rated amperage at the rated voltage for six minutes out of every ten. The machine must idle and cool the other four of every ten minutes.

Some machines used for automatic welding are rated at 100 percent duty cycle and as such can be run continuously without overheating. The size of

Figure 12-9. A DC welding machine with power supplied by a gasoline engine is useful for work away from a source of electricity. (*Hobart Brothers Company*)

Sizes of machines. Sizes of welding machines are rated according to their approximate amperage capacity at *60 percent duty cycle,* such as 150, 200, 250, 300, 400, 500 or 600. This amperage is the rated current output at the working terminal. *Thus a machine rated at 150 amperes can be adjusted to produce a range of power up to 150 amperes.* The duty

the welding machine to be used is governed largely by the kind of welding that is to be done. The following serves as a general guide to size and service.

150–200 Ampere. For light to medium duty welding. Excellent for all fabrication purposes, and rugged enough for continuous operation on light or medium production work.

250–300 Ampere. For average welding requirements. Used in plants for production, maintenance, repair, tool room work and all general shop welding.

400–600 Ampere. A machine for heavy duty welding with large capacity, and for a wide range of purposes. It is used extensively in heavy structural work, fabricating heavy machine parts, heavy pipe and tank welding, and for cutting scrap and cast iron.

CLASSIFICATION OF WELDING MACHINES

Welding machines are classified into two main groups—*constant current* and *constant potential* (voltage). Constant current machines are designed primarily for manual stick welding, whereas constant potential machines are used mostly for gas metal arc welding. Constant potential machines are described in Chapter 25.

Constant Current Machines

Constant current simply means that a steady supply of current is produced over a wide range of welding voltages regardless of changes in arc length. In manual stick welding, whether an AC or DC machine is used, it is difficult to hold the arc length truly constant. However, with a constant current machine there will be relatively small changes in current with changes in arc length. The result is that welding heat and burn-off rate of the electrode are influenced very little, thereby permitting the welder to maintain good control of the weld puddle.

Constant current welding machines have a sloping volt-amp characteristic. The volt-amp characteristic is actually a curve which shows how the voltage varies in its relationship to amperage between the open-circuit (where there is static electrical potential but no current is flowing) and short-circuit (when the electrode touches the work). See Figure 12-11.

Figure 12-10. An AC welding machine with power supplied by a diesel engine permits efficient repair of agricultural equipment. (*Miller Electric Manufacturing Company*)

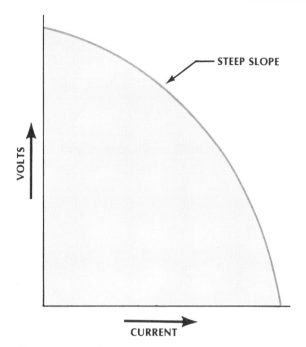

VOLTS

STEEP SLOPE

CURRENT

Figure 12-11. The constant current welding machine has a steep sloping volt-amp curve to control the arc and welding heat.

Under normal welding operations, the open circuit is between 50 and 100 volts where the output voltage is between 18 and 36 volts. By having a high open-circuit voltage, easier arc starting is possible with all types of electrodes. As the welding proceeds the high voltage drops to the arc (welding) voltage, but regardless of the arc length, due to raising or lowering the electrode, the current output will not fluctuate appreciably. The actual arc voltage will vary, depending on the length of the arc. Thus to strike an arc, the electrode must be shorted to the work. At the moment of contact (short circuit) the amperage shoots up while the voltage drops. Then, as the electrode moves away from the work, the voltage rises to maintain the arc while the amperage drops to the required working level.

During welding if the arc length increases, the voltage increases. Conversely, if the arc length decreases the arc voltage decreases. This enables the operator to vary the heat by lengthening or shortening the arc.

Transformers

The transformer type of welding machine produces AC current and is considered to be the least expensive, lightest, and smallest machine. It takes power directly from a power supply line and transforms it to the voltage required for welding.

The welding current output may be adjusted by plugging leads of the electrode holder into sockets on the front of the machine in various locations, or by rotating a handwheel or crank. See Figures 12-12, 12-13, and 12-14.

Some AC transformers also have an arc booster switch which supplies a burst of current for easy arc starting as soon as the electrode comes in contact with the work. After the arc is struck, the current automatically returns to the amount set for the job.

One outstanding advantage of the AC welder is the freedom from *magnetic arc blow* which often occurs when welding with DC machines. Arc blow is a condition that causes the arc to wander while welding in corners on heavy metal or if using large electrodes. Since the current in a DC welder flows in one direction, metal being welded becomes magnetized. When this happens, the arc is often deflected, resulting in excessive spatter. Moreover, arc blow breaks the continuity of the deposited metal, making it necessary to refill the crater. The process of refilling the crater not only slows down the welding but very often leaves weak spots in the weld.

Other features of the AC welder include its low operating and maintenance cost, its high overall

AMPERAGE CONTROL

Figure 12-12. This welding machine has a single dial for controlling amperage. (*The Lincoln Electric Company*)

Figure 12-13. On some machines the amperage range is determined by the receptacle used. (*Hobart Brothers Company*)

Figure 12-14. On some welding machines amperage is selected from either high or low range with fine adjustments made by rotating the dial. (*Miller Electric Manufacturing Company*)

electrical efficiency, and its noiseless operation.

DC Welding Machines

Motor generators are designed to produce DC current in either straight or reverse polarity. The polarity selected for welding depends on the kind of electrode used and the material to be welded. A switch on the machine can be turned for straight or reverse polarity. See Figure 12-15.

Present day motor generators for manual stick welding are usually of the constant current, dual control type.

With a dual control machine welding current is adjusted by two controls. One control provides an approximate or coarse amperage setting. The second control is usually a rheostat that can be turned to provide a fine adjustment of the welding amperage to increase or decrease the heat. See Figure 12-16.

On dual control machines the slope of the output current can be varied to produce a soft or harsh arc. By flattening the volt-amp curve (increasing the amperage) a digging arc can be obtained for a deeper penetration. With a steeper curve (reduced amperage in relation to voltage) a soft or quiet arc results which is useful for welding light gauge materials. In other words, a machine with dual control allows

greater flexibility for welding materials of different thicknesses.

Rectifiers

Rectifiers are essentially transformers containing an electrical device which changes alternating cur-

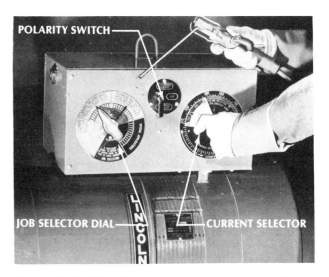

Figure 12-15. This DC motor generator has a switch that can be turned for straight or reverse polarity. (*The Lincoln Electric Company*)

Figure 12-16. The constant-current DC welding machine can change arc characteristics by varying the volt-amp curve. (*The Lincoln Electric Company*)

rent into direct current. Some types are designed to provide a choice of low voltage for gas metal arc welding and submerged-arc welding, and others are designed for a high open circuit with drooping voltage characteristics for gas tungsten arc welding and shielded metal-arc (stick) electrode welding.

Rectifier welding machines are also available to produce both DC and AC current. By turning a switch the output terminals can be changed to the transformer or to the rectifier and produce either AC or DC straight or reverse polarity current.

The transformer rectifier is usually considered more efficient electrically than the generator and provides quiet operation. Current control is achieved by a switching arrangement where one switch can be set for the desired current range and a second dial for securing the fine adjustment desired for welding. See Figure 12-17.

PERSONAL EQUIPMENT

Helmet and hand held helmet. An electric arc not only produces a brilliant light but it also gives off invisible ultraviolet and infrared rays which are extremely dangerous to the eyes and skin.

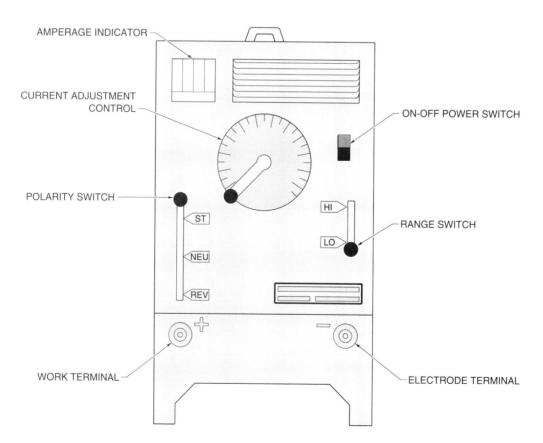

Figure 12-17. The rectifier type welding machine can be adapted to a variety of welding applications.

WARNING: You should never look at an arc with the naked eye!

The *helmet* fits over the head. See Figure 12-18. Adjustable headgear provides a comfortable fit. The helmet may be swung up when you are not welding. Research and technology have combined to produce welding helmets far superior to the conventional welding helmets of a few years ago.

Auto-darkening shades instantly darken when arc light strikes the filter. These auto-darkening shades darken in less than 100 millionth of a second. Helmets may have fixed or adjustable shades. Either helmet allows both hands to be used on the welding project. Helmets also weigh less and are cooler than in the past. The *hand held mask* is frequently used by an observer.

Lenses come in different shades, and the type used depends on the kind of welding done. In general, the recommended practice is as follows:

always replace the cover glass when enough spatter has accumulated on it to interfere with your vision. To protect your eyes, always insert a clear plastic lens inside the welding helmet.

Goggles. Because the welding helmet does not provide total protection, goggles or safety glasses should be worn at all times when welding. See Figure 12-19. In shielded metal-arc welding a thin crust forms on the deposited bead. This substance, known as slag, must be removed. While removing slag, tiny particles are often deflected upward. Because of stress built up in the weld, slag will on occasion pop off the weld by itself. Without proper eye protection, these particles may cause a serious eye injury. Therefore, *always wear goggles when welding.*

Gloves. Another important item you will need for arc welding is a pair of gloves. *You must wear gloves to protect your hands from the ultraviolet rays and spattering hot metal.*

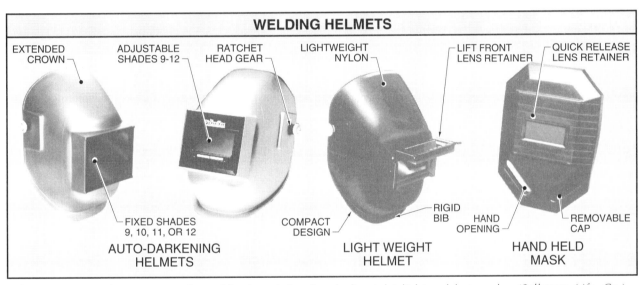

WELDING HELMETS

EXTENDED CROWN — ADJUSTABLE SHADES 9-12 — RATCHET HEAD GEAR — LIGHTWEIGHT NYLON — LIFT FRONT LENS RETAINER — QUICK RELEASE LENS RETAINER

FIXED SHADES 9, 10, 11, OR 12 — COMPACT DESIGN — RIGID BIB — HAND OPENING — REMOVABLE CAP

AUTO-DARKENING HELMETS — **LIGHT WEIGHT HELMET** — **HAND HELD MASK**

Figure 12-18. Helmets protect the welder from infared and ultraviolet light and hot sparks. *(Sellstrom Mfg. Co.)*

Shade 10 for arc cutting and welding beyond 75 amperes and up to 200 amperes.

Shade 12 for arc welding and cutting beyond 200 amperes and up to 400 amperes.

Shade 14 for arc welding and cutting over 400 amperes.

During the welding process, small particles of metal fly upward from the work and may lodge on the lens. Colored lenses are protected by use of *clear glass* or *plastic cover plates.* Although the methods of inserting the cover plates vary on different makes of helmets, the change can be made easily and quickly.

These clear glasses are inexpensive and can be purchased from any welding supply dealer. *Since you must have clear vision at all times during welding,*

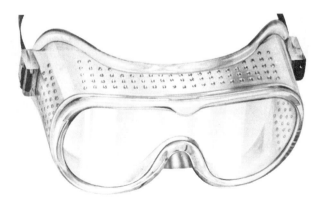

Figure 12-19. Eye protection must be worn at all times when welding. *(Purity Cylinder Gases Inc.)*

Several different kinds of gloves are available. As a rule, the leather gauntlet gloves provide ample protection. See Figure 12-20. Regardless of the type used, they should be flexible enough to permit proper hand movement, yet not so thin as to allow the heat to penetrate easily.

Apron. An apron is recommended when welding as spattering metal might cause personal injury. Since the spattering particles are hot, a leather apron offers the best protection. See Figure 12-21.

In situations where there may not be an excessive amount of metal spatter, suitable coveralls (fire retardant) may be worn to protect clothing.

In some instances it is necessary to wear a leather jacket or leather sleeves for protection from hot metal spatter. Coveralls or clothing should be sufficiently heavy to prevent infrared and ultraviolet rays from penetrating to the skin. Cuffs on overalls should be turned down or eliminated, and pockets removed so they will not serve as lodging places for falling globules of molten metal. Sleeves and collars should be kept buttoned. Leather boots are preferable to Oxford-type shoes.

SHOP EQUIPMENT

Electrode holder. To do a good welding job, a properly designed electrode holder is essential. The electrode holder is a handle-like tool attached to the cable that holds the electrode during welding. See Figure 12-22.

A well-designed holder can be identified by these features:

1. It is reasonably light, to reduce excessive fatigue while welding.
2. It does not heat too rapidly.
3. It is well balanced.
4. It receives and ejects the electrodes easily.
5. All exposed surfaces, including the jaws, are protected by insulation.

Check to see that the jaws of the electrode holder are properly insulated. Laying a holder with uninsulated jaws on the bench plate while the machine is running will cause a flash.

Cleaning tools. To produce a strong welded joint, the surface of the metal must be free of all foreign matter such as rust, oil, and paint. A steel brush is used for cleaning purposes.

After a bead is deposited on the metal, the slag which covers the weld is removed with a chipping hammer. See Figure 12-23. The chipping operation is followed by additional wire brushing. Complete removal of slag is especially important when several

Figure 12-20. Always wear gloves when welding to protect hands from ultraviolet rays and hot metal. (*The Lincoln Electric Company*)

Figure 12-21. The properly dressed welder is adequately protected from hot metal spatter and harmful rays.

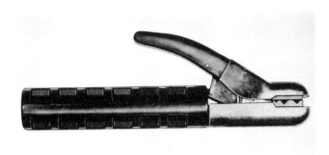

Figure 12-22. The electrode holder grips the electrode at the correct angle for welding. (*The Lincoln Electric Company*)

Figure 12-23. Chipping hammers and wire brushes are used to clean a weld. (*Hobart Brothers Company*)

passes must be made over a joint. Otherwise, gas holes will form in the bead, resulting in porosity which weakens the weld.

Welding screen. Whenever welding is done in areas where other people may be working, the welding operation should be enclosed with screens so the ultraviolet rays will not injure those nearby. These screens can easily be constructed from heavy fire-resistant canvas painted with black or gray ultraviolet protective paint.

When welding is to be done in a permanent location, a booth is desirable. See Figure 12-24.

Cables. The cables carry the current to and from the work. One cable runs from the welding machine to the holder, and the other cable is attached to the work or bench. The cable connected to the work is called the *ground cable.* Thus when the welding machine is turned on and the electrode in the holder comes in contact with the work, a circuit is formed, allowing the electricity to flow.

It is important to use the correct diameter cable specified for the welding machine. If the cable is too small for the current, it overheats and power is lost. A larger cable is necessary to carry a required voltage any distance from the machine. Otherwise, there will be an excessive voltage drop. With larger diameter cables, you must take precaution not to exceed the recommended lengths because the voltage drop will lower the efficiency of your welding. Check with the welding machine manufacturer for the proper cable size and for specific lengths and usage.

All cable connections should be tight because any loose connection will cause the cable or clamp to overheat. A loose connection may even produce arcing at the connection. Cables should be kept clean and handled as to avoid damage to the insulation.

Ground connections. Proper ground connections can be made in several ways. The ground cable can be fastened to the work or bench by a *C* clamp, a special ground clamp, or by bolting or welding the lug on the end of the cable to the bench. See Figures 12-25 and 12-26.

Providing Ventilation

Electrodes used in shielded metal-arc welding give off a great deal of smoke and fumes. These fumes are not harmful if the welding area is properly ventilated. There should be either a suction fan or other adequate source of air circulation. *Unless there is sufficient movement of air in the room, shielded metal-arc welding should not be done.*

Permanent welding booths should be equipped with a sheet metal hood mounted directly above the welding table, and an exhaust system to draw out the smoke and fumes. See Figure 12-24.

The general recommendation for adequate ventilation is a minimum of 2000 cubic feet of air flow per minute per welder. If individual movable exhaust hoods can be placed near the work the rate of air flow in the direction of the hood should be approximately 100 linear feet per minute in the welding zone.

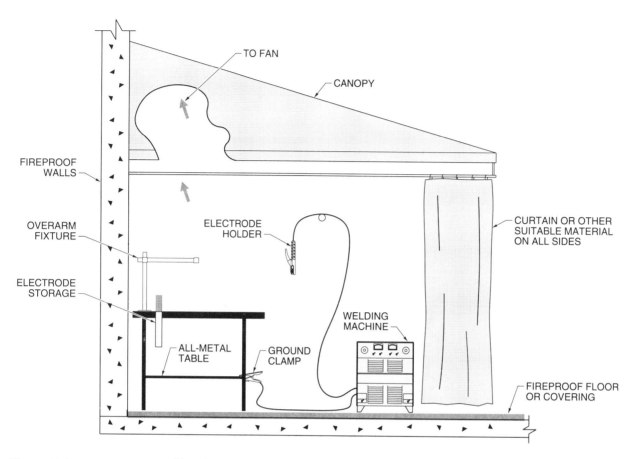

Figure 12-24. A permanent welding booth provides a safe work area.

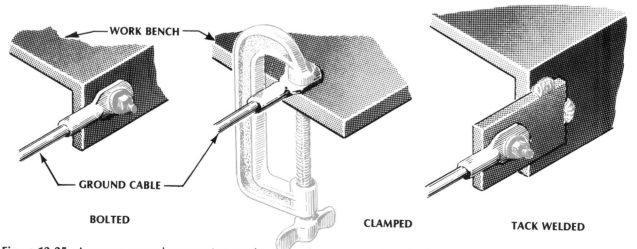

Figure 12-25. A proper ground connection can be secured using several methods.

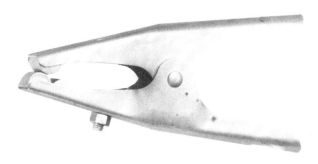

Figure 12-26. A removable ground clamp is commonly used when shielded metal-arc welding.

13 Always turn off the machine when leaving the work.

14 Never stand in water or on a wet floor or use wet gloves when welding. Water is an electrical conductor and any wet surface will carry current. Always dry out the work pieces or bench if there is any evidence of moisture.

Points to Remember

1 Never look at a welding arc without a welding helmet.

2 Always replace the clear cover glasses when they become spattered.

3 Examine the colored lenses in the helmet. Replace cracked ones at once. Place a clear lens over the colored lens inside the helmet.

4 Wear goggles when welding.

5 Always wear gloves and an apron when welding.

6 Use an electrode holder that is completely insulated.

7 Weld only in areas where there is adequate ventilation.

8 If you weld outside of a permanent booth, be sure to have screens so the arc will not harm persons nearby.

9 Prevent welding cables from coming in contact with hot metal, water, oil, or grease. Avoid dragging the cables over or around sharp corners.

10 Make sure that you have a good ground connection.

11 Keep cables in an orderly manner to prevent them from becoming a stumbling hazard. Fasten the cables overhead whenever possible.

12 Do not weld near flammable materials.

QUESTIONS FOR STUDY AND DISCUSSION

1 What is a circuit?

2 What is voltage? What instrument is used to measure voltage?

3 What term is used to indicate rate of current flow in a circuit?

4 What is voltage drop? What effect does it have on welding current?

5 What is the difference between AC and DC current?

6 What is the difference between static and dynamic electricity?

7 What is meant by open-circuit voltage and arc voltage?

8 What is polarity?

9 What determines whether the machine is to be set for straight or reversed polarity?

10 How are sizes of welding machines rated?

11 How does a DC motor generator differ from a transformer type unit?

12 What is meant by a constant current, dual control machine?

13 What is meant when a welding machine has a sloping volt-amp curve?

14 What is one main advantage of an AC welder?

15 Why is a rectifier type welding unit often preferred?

16 Why should you never look at an electric arc without eye protection?

17 What determines the correct shade of lens to use for welding?

18 Why should shaded lenses be covered with clear plastic lenses?

19 When does a short circuit occur in the shielded metal-arc welding process?

20 Why should welder's gloves be worn when arc welding?

21 What are some of the requirements of a good electrode holder?

22 Why is it important to weld only where there is adequate ventilation?

23 What is likely to happen if cable connections are loose?

24 How is a hand-held helmet used by the welder?

25 What is duty cycle when specifying welding machine ratings?

26 What temperatures are produced in the shielded metal-arc welding process?

27 What effect does welding polarity have on where heat is directed?

28 How does arc length affect voltage?

29 Why are goggles important when welding?

30 What is the difference between open-circuit voltage and working voltage?

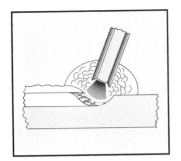

Shielded Metal-Arc Welding—SMAW

There are many different kinds and sizes of electrodes and unless the correct one is selected you will have difficulty in doing a good welding job.

In general, all electrodes are classified into five main groups: *mild steel, high-carbon steel, special alloy steel, cast iron, and nonferrous.* The greatest range of arc welding is done with electrodes in the *mild steel* group. Special alloy steel electrodes are made for welding various kinds of steel alloys. Cast-iron electrodes are used for welding cast iron, and nonferrous electrodes for welding such metals as aluminum, copper, and brass. In this chapter we will discuss mild steel electrodes. Other types of electrodes will be covered in subsequent chapters dealing with the welding of special metals.

THE ELECTRODE

An electrode is a coated metal wire having approximately the same composition as the metal to be welded. When the current is produced by the generator or transformer and flows through the circuit to the electrode, an arc is formed between the end of the electrode and the work. The arc melts the electrode and the base metal. The melted metal of the electrode flows into the molten crater and forms a bond between the two pieces of metal being joined.

Electrodes are not only manufactured to weld different metals, but they are also designed for DC or AC welding machines. A few electrodes work equally well on either DC or AC. Electrode usage also depends on the welding position. Some electrodes are best suited for flat position welding and horizontal fillet welding, while other types are used for welding in any position.

The two kinds of mild steel electrodes are known as *bare* and *shielded.* Originally, bare electrodes were uncoated metal rods; today they have a very light coating. Their use for welding is very limited because such electrodes are difficult to weld with and they produce brittle welds with low strength. Practically all welding is done with shielded electrodes.

Shielded electrodes have heavy coatings of various substances such as cellulose sodium, cellulose

potassium, titania sodium, titania potassium, iron oxide, iron powder as well as several other ingredients. Each of the substances in the coating is intended to serve a particular function in the welding process.

1. Act as a cleansing and deoxidizing agent in the molten crater.

2. Release carbon dioxide to protect the molten metal from atmospheric oxides and nitrides. See Figure 13-1. Since oxygen and nitrogen weaken a weld if allowed to come in contact with the molten metal, the exclusion of these contaminants is important.

3. Form a slag over the deposited metal which further protects the weld until the metal cools sufficiently so it is no longer affected by atmospheric contamination. The slag also slows the cooling rate of the deposited metal thereby permitting a more ductile weld to form.

4. Provide easier arc starting, stabilize the arc better, and reduce splatter.

5. Permit better penetration and improve the X-ray quality of the weld.

The coating of some electrodes contains powdered iron which converts to steel and becomes a part of the weld deposit. The powdered iron also

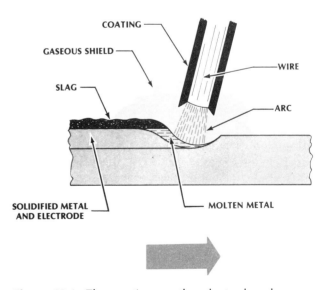

Figure 13-1. The coating on the electrode releases a gaseous shield to protect the molten metal from atmospheric contamination. (*The Lincoln Electric Company*)

helps to increase the speed of welding and improve the weld appearance.

A group of electrodes known as low-hydrogen electrodes have coatings high in limestone and other ingredients with low-hydrogen content, such as calcium fluoride, calcium carbonate, magnesium-aluminum silicate, and ferrous alloys. These electrodes are used to weld high-sulfur and high-carbon steels that have a great affinity for hydrogen which often causes porosity and underbead cracking in a weld.

Identifying Electrodes

You will find that electrodes are referred to by a manufacturer's trade name. To insure some degree of uniformity in manufacturing electrodes, the American Welding Society has set up certain requirements for electrodes. Thus different manufacturers' electrodes which are within the classification established by the AWS may be expected to have the same welding characteristics. All electrodes manufactured in the United States are imprinted with an AWS symbol. See Figure 13-2.

In this classification, each type of electrode has been assigned specific symbols, such as E-6010, E-7010, E-8010, etc. The *prefix E* identifies the electrode for electric arc welding as illustrated in Figure 13-3. The *first two digits* in the symbol designate the minimum allowable tensile strength of the deposited weld metal in thousands of pounds per square inch (psi). For example, the 60 series electrodes have a minimum pull strength of 60,000 psi; the 70 series, a strength of 70,000 psi.

The *third digit* of the symbol indicates possible welding positions. Two numbers are used for this purpose: 1 and 2. Number 1 is for an electrode which can be used for welding in any position. Number 2 represents an electrode restricted for welding in flat position and horizontal fillets only.

The *fourth digit* of the symbol simply shows some special characteristic of the electrode, such as type of coating, weld quality, type of arc, and amount of penetration. The fourth digit may be 0, 1, 2, 3, 4, 5, 6, 7, or 8. Because the welding position is contingent upon the manufacturer's characteristics of the electrode coating, the third and fourth digits are often identified together. See Table 13-1. The significance of the various fourth digits in the AWS electrode classification system are:

0. Direct current with reverse polarity only. Produces high quality deposits with deep penetration and flat or concave beads, cellulose sodium coating.

1. Alternating current or direct current with reverse polarity. Produces high quality deposits with

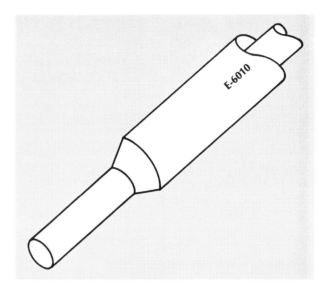

Figure 13-2. The American Welding Society (AWS) numerical electrode classification identifies the characteristics and usage of the electrode.

deep penetration and flat to slightly concave beads, cellulose potassium coating.

2. Direct current with straight polarity only or AC current. Medium quality deposits, medium arc, medium penetration and convex heads, titania sodium coating.

3. Alternating current or direct current with either polarity. Medium to high quality deposits, soft arc, shallow penetration and slightly convex beads, titania potassium coating.

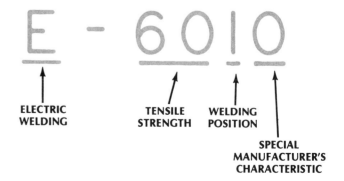

Figure 13-3. The AWS electrode identification uses a letter and numbers, each having a specific meaning.

TABLE 13-1. ELECTRODE CHARACTERISTICS.

AWS CLASSIFICATION	COATING	WELD CURRENT	POSITION	WELD CHARACTERISTICS
EXX 10	Cellulose	DCR	ALL	Deep penetration, flat or concave beads
EXX 20	sodium		FLAT, HOR. FILLET	
EXX 11	Cellulose potassium	AC, DCR	ALL	Deep penetration, flat or concave beads
EXX 12	Titania sodium	AC, DCS	ALL	Medium penetration, convex beads
EXX 13	Titania potassium	AC, DCR DCS	ALL	Shallow penetration, convex beads
EXX 14	Titania		ALL	Medium penetration, fast deposit
EXX 24	iron powder	AC, DCR, DCS	FLAT, HOR. FILLET	
EXX 15	Low-hydrogen sodium	DCR	ALL	Moderate penetration, convex beads
EXX 16	Low-hydrogen potassium	AC, DCR	ALL	Moderate penetration, convex beads
EXX 27	Iron powder, iron oxide	AC, DCR, DCS	FLAT, HOR. FILLET	Medium penetration, flat beads
EXX 18	Iron powder,		ALL	Shallow to medium penetration, convex beads
EXX 28	low-hydrogen	AC, DCR	FLAT, HOR. FILLET	

4. AC or DC either polarity, fast deposition rate, deep groove butt, fillet and lap welds, medium penetration, easy slag removal, iron powder, titania coating.

5. Direct current with reverse polarity, high quality deposits, soft arc, moderate penetration, flat to slightly convex bead, low-hydrogen content in weld deposits, low-hydrogen sodium coating.

6. AC or DC reverse polarity with qualities similar to number 5. Low-hydrogen, potassium coating.

7. AC or DC straight polarity, iron powder iron oxide, fast fill, fast deposition rate, medium penetration, low spatter, flat beads.

8. AC, DC reverse polarity, iron powder, low hydrogen, fill freeze, shallow to medium penetration, high deposition, easy slag removal, convex beads.

Thus for mild steel, the complete classification number E-6010 would signify an electrode that (a) has a minimum tensile strength of 60,000 for the as-welded deposited weld metal, (b) is usable in all welding positions, and (c) can be used with DC reverse polarity only. In the same way, E-7024 designates an electrode with 70,000 psi minimum tensile strength usable for welding in flat position and horizontal fillets only, and operates on DC, either po-

larity, or AC, and its coating contains iron powder.

It is significant to note that the fourth digit cannot be considered individually; it must be associated with the third digit since in this way it identifies both the polarity and position of the electrode.

In the past, some electrodes were identified using a color code established by the National Electrical Manufacturers Association (NEMA). The identification code is no longer used except for some surfacing electrodes.

Selecting the Correct Electrode

The ideal electrode is one that will provide good arc stability, smooth weld bead, fast deposition, minimum spatter, maximum weld strength, and easy slag removal. To achieve these characteristics seven factors should be considered in selecting an electrode. These are:

Properties of the base metal. A top quality weld should be as strong as the parent metal. This means that the electrode to be used must produce a weld metal with approximately the same mechanical properties as the parent metal.

Electrodes are available for welding different classifications of metal. Some electrodes are designed to weld carbon steels; others are best suited for low-alloy steels and some are intended specifically for high-strength alloy steels. Therefore, in undertaking any welding operation, the first consideration is to check the chemical analysis of the metal and then select an electrode that is recommended for that metal.

Electrode diameter. As a rule, an electrode is never used that has a diameter larger than the thickness of the metal to be welded. Some operators prefer larger electrodes because they permit faster travel along the joint and thus speed up the welding operation; but this requires considerable skill. For example, it takes approximately half the time to deposit a quantity of weld metal from ¼" coated mild steel electrodes than from ¾6" electrodes of the same type. The larger sizes not only make possible the use of higher current but require fewer stops to change the

electrode. Therefore, from the standpoint of economy it is always a good practice to use the largest size electrode that is practical for the work at hand.

When making vertical or overhead welds, ¾6" is the largest diameter electrode that should be used regardless of plate thickness. Larger electrodes make it too difficult to control the deposited metal. Ordinarily, a *fast-freeze* type of electrode is best for vertical and overhead welding. See Table 13-2.

The diameter of the electrode is also influenced by the factors of the joint design. Thus, in a thick metal section with a narrow V, a small diameter electrode is used to run the first weld bead or root pass. This is done to insure thorough penetration at the root of the weld. Successive passes are then made with larger diameter electrodes if necessary.

Joint design and fit-up. Joints with insufficient beveled edges require deep penetrating, fast-freeze electrodes. Some electrodes have this particular dig-

TABLE 13-2. ELECTRODE CHARACTERISTICS.

TYPE	AWS CLASS	CURRENT TYPE	WELDING POSITION	WELD RESULTS	ELECTRODE GROUP
	E-6010 E-6011	DCR (DC-EP) AC	ALL	Deep penetration, flat beads	Fast-freeze
Mild Steel	E-6012 E-6013	DCS (DC-EN) AC (DC-EN) (DC-EP)	ALL	Shallow penetration, good bead contour, minimum spatter, for poor fit-up	Fill-freeze
	E-6020	DCR (DC-EP) DCS (DC-EN) AC	F,H	High deposition, deep groove single pass welds	Fast-fill
	E-6027	DCR (DC-EP) DCS (DC-EN) AC	F,H	High deposition, deep penetration	Fast-fill
Iron Powder	E-7014	DCR (DC-EP) DCS (DC-EN) AC	All	Low penetration, high speed	Fill-freeze
	E-7024	DCR (DC-EP) DCS (DC-EN) AC	F,H	High deposition, single and multiple passes	Fast-fill
Low Hydrogen	E-7016	DCR (DC-EP) AC	ALL	Welding of high-sulfur and high-carbon steels that tend to develop porosity and crack under weld bead	Fill-freeze
	E-7018	DCR (DC-EP) AC	ALL		
	E-7028	DCR (DC-EP) AC	F,H		Fast-fill

DCR—Direct Current Reverse Polarity   F—Flat, H—Horizontal
DCS—Direct Current Straight Polarity   DC-EP—Direct Current Electrode Positive
AC—Alternating Current   DC-EN—Direct Current Electrode Negative

ging characteristic and may require more skillful electrode manipulation by the operator. On the other hand, joints with open gaps need a mild penetrating fill-freeze electrode that rapidly bridges gaps.

Welding position. The position of the weld joint is an important factor in the type of electrode to be used. Some electrodes produce better results when

current straight polarity, and DCRP means direct current reverse polarity. To minimize electrode polarity confusion, some manufacturers are now designating straight polarity electrodes as DC-EN (direct current — electrode negative) and reverse polarity electrodes as DC-EP (direct current — electrode positive). See Table 13-2.

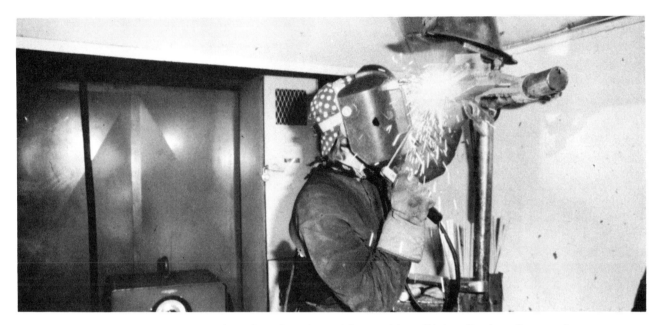

Figure 13-4. Match the electrode to the job and to the welding position. (*Hobart Brothers Company*)

the welding is done in a flat position. Other electrodes are designed for vertical, horizontal, and overhead welding. See Figure 13-4.

Welding current and polarity. Electrodes are made for use with either AC current or DC current reverse polarity or DC current straight polarity, although some electrodes function as well on both AC and DC current. When referring to polarity, DCSP means direct

Production efficiency. Deposition rate is extremely significant in any production work. The faster a weld can be made the lower the cost. Not all electrodes have a high-speed high-current rating and still produce smooth, even bead ripples. Unless electrodes are noted for a fast deposition rate they may prove very difficult to handle when used at high speed travel.

Service conditions. The service requirements of the part being welded may demand special weld deposits. For example, high-corrosion resistance, or ductility, or high strength may be important factors. In such cases electrodes must be selected that will produce these specific characteristics.

Conserving and Storing Electrodes

Most electrodes are costly; therefore, every bit of the electrode should be consumed. Do not discard stub ends until they are down to only 1½" to 2" long. See Figure 13-5.

Always store electrodes in a dry place at a normal

TABLE 13-3.　CURRENT SETTINGS FOR E-6010 ELECTRODES.

ELECTRODES DIA (INCHES)	AMPERES*
3/32	60-90
1/8	80-120
5/32	110-160
3/16	150-200
7/32	175-250
1/4	225-300
5/16	250-450

*These ranges may vary slightly for electrodes made by different manufacturers.

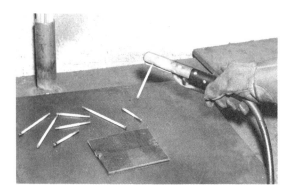

Figure 13-5. Use electrodes until the stub ends are down to less than 2″ long.

room temperature and 50 percent maximum relative humidity. When exposed to moisture, the coating has a tendency to disintegrate. Low-hydrogen electrodes are especially vulnerable to moisture. These electrodes, after being removed from their moisture-proof container, should be stored at 250° to 400° F. Stationary and portable drying ovens are often used for storing electrodes at specified holding temperatures.

In storing electrodes, be sure they are not bumped, bent, or stepped on, since this will remove the coating and make the electrode useless.

Deposition Classification of Electrodes

Electrodes for welding mild steel are sometimes classified as fast-freeze, fill-freeze, and fast-fill[1] (See Table 13-2.) The *fast-freeze* electrodes are those which produce a snappy, deep penetrating arc and fast-freezing deposits. They are commonly called reverse polarity electrodes even though some can be used on AC. These electrodes have little slag and produce flat beads. They are widely used for all types of all-position welding for both fabrication and repair work.

Fill-freeze electrodes have a moderately forceful arc and deposit rate between those of the fast-freeze and fast-fill electrodes. They are commonly called straight polarity electrodes even though they may be used on AC. These electrodes have complete slag coverage and weld beads with distinct, even ripples. They are a general-purpose electrode for production shops and are particularly useful for joints with poor fit-up and repair work. They can be used in all positions, though the fast-freeze electrodes are preferred for vertical and overhead welding.

The *fast-fill* group includes the heavy coated, iron powder electrodes with soft arc and fast deposit rate. These electrodes have a heavy slag and produce exceptionally smooth weld beads. They are generally

used for production welding where all work can be positioned for downhand (flat) welding.

Types of Mild Steel Electrodes

E-6010. This is an all-position, fast-freeze electrode. It is suitable only on DC machines with reversed polarity, and is designed primarily for welding mild and low-alloy steels. It should be used only where there is an absolutely good fit-up. The E-6010 electrode has wide applications in ship construction, buildings, bridges, tanks, and piping. Table 13-3 shows the amperage settings for different sizes of this electrode.

E-6011. The E-6011 electrode is similar to the E-6010 except that it can also be used on AC machines. Although the electrode can be used on DC machines with reversed polarity, it does not work quite as well as the E-6010. Its amperage setting is slightly lower than for E-6010. See Table 13-4.

TABLE 13-4. CURRENT SETTINGS FOR E-6011 ELECTRODES.

ELECTRODES DIA (INCHES)	AMPERES *
3/32	50-90
1/8	80-130
5/32	120-180
3/16	140-220
7/32	170-250
1/4	225-325

*These ranges may vary slightly for electrodes made by different manufacturers.

E-6012. This is a fill-freeze electrode and may be used on either DC or AC welders. When employed on DC welders the current must be set for straight polarity. The electrode provides medium penetration, a quiet arc, slight spatter, and dense slag. Although it is considered an all-position electrode, it is used in greater quantities for flat and horizontal position welds. This electrode is especially useful to bridge gaps under conditions of poor *fit-up work;* that is, joints where the edges do not fit closely together. Higher currents can be used with the E-6012 electrodes than with any other type of all-position electrodes. See Table 13-5.

E-6013. Electrodes of this type are very similar to E-6012 with a few slight exceptions. Slag removal is better and the arc can be maintained easier, especially with small diameter electrodes. This permits better operation with lower open-circuit voltage. The

TABLE 13-5. CURRENT SETTINGS FOR E-6012 ELECTRODES.

ELECTRODES DIA (INCHES)	AMPERES*
3/32	40-90
1/8	80-120
5/32	120-190
3/16	140-240
7/32	180-315
1/4	225-350

*These ranges may vary slightly for electrodes made by different manufacturers.

bead deposited is noticeably flatter and smoother but with shallower penetration than the E-6012 class. Although the electrode is used particularly for welding sheet metal, it has many other applications. It works well in all positions and it functions best on AC welders. When used with DC machines the polarity may be straight or reverse. Current settings for E-6013 electrodes are shown in Table 13-6.

TABLE 13-6. CURRENT SETTINGS FOR E-6013 ELECTRODES.

ELECTRODES DIA (INCHES)	AMPERES*
1/16	20-40
5/64	25-50
3/32	30-80
1/8	80-120
5/32	120-190
3/16	140-240
7/32	225-300
1/4	250-350

*These ranges may vary slightly for electrodes made by different manufacturers.

Iron Powder Electrodes

Iron powder electrodes are those which contain a high content of iron powder. They are designed for welding mild steels where high speed and fast deposition rate are required. The three principal types are E-6027, E-7014, and E-7024. All of them produce low spatter with easy slag removal. Typical application includes railroad cars, earth-moving equipment, positioned welds in pressure vessels, piping, and ships. The E-7014 and E-7024 are often used where higher strength joints are necessary.

E-6027. Produces high quality welds for high-speed deposition of ¼″ and ⁵⁄₁₆″ horizontal fillets and for butt and fillet welds in the flat position and for

cover beads on butt welds where complete coverage and good bead appearance are required. Current may be AC or DC with either polarity. A drag welding technique is recommended to keep the cover over both legs of fillet welds. See Table 13-7 for approximate current settings.

TABLE 13-7. CURRENT SETTINGS FOR E-6027 ELECTRODES.

ELECTRODES DIA (INCHES)	AMPERES*
3/16	225-300
7/32	275-375
1/4	350-450

*These ranges may vary slightly for electrodes made by different manufacturers.

E-7014. This is a fast-fill and fast-freeze electrode where high speed is necessary. May be used in all positions with AC or DCR and DCS current. The E-7014 electrode deposits much more metal than an E-6012 or E-6013 type. It is particularly effective in vertical downhill welding. See Table 13-8 for approximate current settings.

TABLE 13-8. CURRENT SETTINGS FOR E-7014 ELECTRODES.

ELECTRODES DIA (INCHES)	AMPERES*
3/32	80-110
1/8	110-150
5/32	140-190
3/16	180-260
7/32	250-325
1/4	300-400
5/16	400-500

*These ranges may vary slightly for electrodes made by different manufacturers.

E-7024. The E-7024 is a fast-fill electrode which provides exceptional economy for single or multiple-pass welds and is excellent on build-up applications because of the high deposition rate and easy slag removal. It is recommended only for flat and horizontal positions with AC or DC straight or reverse polarity. See Table 13-9 for approximate current settings.

Low-Hydrogen Electrodes

Low-hydrogen electrodes are designed for welding high-sulfur and high-carbon steels. When such

TABLE 13-9. CURRENT SETTINGS FOR E-7024
ELECTRODES.

ELECTRODES DIA (INCHES)	AMPERES *
3/32	90-120
1/8	120-150
5/32	180-230
3/16	250-300
7/32	300-350
1/4	350-400
5/16	400-500

*These ranges may vary slightly for electrodes made by different manufacturers.

steels are welded they tend to develop porosity and cracks under the weld bead because of the hydrogen absorption from arc atmospheres. Low-hydrogen electrodes were developed to prevent the introduction of hydrogen in the weld.

The basic low-hydrogen electrodes are E-7016, E-7018, and E-7028.

E-7016. This is an all-position electrode suitable for AC or DC reverse polarity current. It is especially recommended for welding hardenable steels where no preheat is used, and where stress-relieving normally would be required but cannot be effected. See Table 13-10 for approximate current settings.

TABLE 13-10. CURRENT SETTINGS FOR E-7016
ELECTRODES.

ELECTRODES DIA (INCHES)	AMPERES *
3/32	75-105
1/8	100-150
5/32	140-190
3/16	190-250
7/32	250-300
1/4	300-375

*These ranges may vary slightly for electrodes made by different manufacturers.

E-7018. The E-7018 is a low-hydrogen type electrode but also contains iron powder. It is a high-speed fast-deposition rate electrode designed to pass the most severe X-ray requirements when applied in all welding positions, using either AC or DC reverse polarity current. Its puddle fluidity permits gases to escape when the lowest currents are used for out-of-position welding. See Table 13-11 for approximate current settings for low-hydrogen electrodes.

E-7028. The E-7028 is a low-hydrogen electrode with a heavy iron powder type covering and is con-

TABLE 13-11. CURRENT SETTINGS FOR E-7018
ELECTRODES.

ELECTRODES DIA (INCHES)	AMPERES *
3/32	70-120
1/8	100-150
5/32	120-200
3/16	200-275
7/32	275-350
1/4	300-400

*These ranges may vary slightly for electrodes made by different manufacturers.

sidered the counterpart of E-7018 but for flat and horizontal positions only. See Table 13-12 for approximate current settings.

TABLE 13-12. CURRENT SETTINGS FOR E-7028
ELECTRODES.

ELECTRODES DIA (INCHES)	AMPERES *
5/32	175-250
3/16	250-325
7/32	300-400
1/4	375-475

*These ranges may vary slightly for electrodes made by different manufacturers.

Variables Governing Electrode Selection

Although there are a variety of electrode classification charts which list the basic characteristics or differences in electrodes, many of the variables encountered in production will often require the use of tests to determine the suitability of an electrode for a specific application. By first analyzing the variables in terms of their importance in a welding situation, considerable time and effort can be saved.

The variables normally associated with most types of welding are listed in Table 13-13. These are shown with a relative rating ranging from 1 to 10, with 10 as the highest value and 1 the lowest. These variables and their corresponding ratings are based on experience and intended primarily as an aid in the electrode selection process. For example, notice in Table 13-13 that if high-sulfur steel is to be welded, either a E-7016 or E-7018 electrode should be used. If poor fit-up is the problem, electrode E-6012 is considered best for this condition. On the other hand, if deposition rate is the primary factor then either E-6027 or E-7024 is the most suited for this purpose.

TABLE 13-13.   MILD STEEL ELECTRODE SELECTION CHART.*

	ELECTRODE CLASS										
	E6010	E6011	E6012	E6013	E7014	E7016	E7018	E6020	E7024	E6027	E7028
Groove butt welds, flat (< 1/4″)	5	5	3	8	9	7	9	10	9	10	10
Groove butt welds, all positions (< 1/4″)	10	9	5	8	6	7	6	(b)	(b)	(b)	(b)
Fillet welds, flat or horizontal	2	3	8	7	9	5	9	10	10	9	9
Fillet welds, all positions	10	9	6	7	7	8	6	(b)	(b)	(b)	(b)
Current(c)	DCR	AC DCR	DCS AC	AC DC	DC AC	DCR AC	DCR AC	DC AC	DC AC	AC DC	DCR AC
Thin material (1/4″)	5	7	8	9	8	2	2	(b)	7	(b)	(b)
Heavy plate or highly restrained joint	8	8	8	8	8	10	9	8	7	8	9
High-sulfur or off-analysis steel	(b)	(b)	5	3	3	9	9	(b)	5	(b)	9
Deposition rate	4	4	5	5	6	4	6	6	10	10	8
Depth of penetration	10	9	6	5	6	7	7	8	4	8	7
Appearance, undercutting	6	6	8	9	9	7	10	9	10	10	10
Soundness	6	6	3	5	7	10	9	9	8	9	9
Ductility	6	7	4	5	6	10	10	10	5	10	10
Low-temperature impact strength	8	8	4	5	8	10	10	8	9	9	10
Low spatter loss	1	2	6	7	9	6	8	9	10	10	9
Poor fit-up	6	7	10	8	9	4	4	(b)	8	(b)	4
Welder appeal	7	6	8	9	10	6	8	9	10	10	9
Slag removal	9	8	6	8	8	4	7	9	9	9	8

(a) Rating is on a comparative basis of same size electrodes with 10 as the highest value. Ratings may change with size.
(b) Not recommended.
(c) DCR—direct current reverse, electrode positive; DCS—direct current straight, electrode negative; AC—alternating current; DC—direct current, either polarity.
*AWS

Special Electrodes

Standard electrodes are used for most general types of welding. With these electrodes fusion is achieved by changing the solid state of metal into a molten mass, which combines with the metal deposit of the electrode to form a permanent bond. The success of the bond depends on the generation of sufficient heat to produce a completely molten puddle. However, metals often undergo unfavorable changes when subjected to high heat, and precautions must be taken to avoid stresses, distortion, warpage, and metallurgical structural changes.

Points to Remember

1   Use the correct type electrode for the welding to be done.

2   Remember, some electrodes can be used only on DC machines and others on AC machines.

3   If welding is to be done on a DC machine, check whether straight or reverse polarity is needed for the particular electrode to be used.

4   Select an electrode with a diameter that is about one-half the thickness of the plate to be welded.

5   Always use up the electrode until the stub is down to 1½″ to 2″ long.

6   Store electrodes in a dry place where the coating cannot be damaged.

QUESTIONS FOR STUDY AND DISCUSSION

1 What is the difference between bare and shielded electrodes?

2 Why are bare electrodes rarely used?

3 What are the functions of the heavy coating on shielded electrodes?

4 What has been done to insure uniformity of electrode specifications?

5 What symbols have been adopted to identify different types of electrodes?

6 Explain the identifying symbols of the electrode classification E-6010.

7 What is an all-position electrode?

8 How is it possible to determine what current and polarity an electrode is designed for?

9 What factors should be taken into consideration when selecting an electrode for the job to be done?

10 Why are smaller diameter electrodes used for overhead welding?

11 What precautions must be taken in storing electrodes?

12 What is the specific feature of electrodes with coatings containing powdered iron?

13 Why are low-hydrogen electrodes used?

14 What are some of the specific characteristics of electrodes designated as fast-freeze?

15 Some electrodes are classified as fill-freeze. What does this mean?

16 Fast-fill electrodes are intended for what types of welding?

17 What is the function of slag in the welding process?

18 How does joint design affect the diameter of the electrode used?

19 What organization is responsible for establishing a standard numerical electrode classification?

20 Which electrodes provide deep penetration?

STRIKING ARCS

Shielded Metal-Arc Welding – SMAW

Learning to arc weld involves mastery of a specific series of operations. Skill in performing these operations requires practice. Once this skill has been acquired, the operations can be applied on any welding job. The first basic operation is learning to strike an arc and run a straight bead.

BASIC PRINCIPLES OF SUSTAINING A WELDING ARC

Before proceeding with the first welding operation, it is well to review several basic principles for maintaining an arc since the success of any welding operation depends upon a stable arc. To sustain a stable arc, four basic elements will be necessary:

1. *Machine setting* involves the adjustment of the welding machine to the correct current setting required. Current represents the actual flow of electricity and is regulated by a control on the power supply. The current used depends on the size and type of electrode used, the position of the weld, and the base metal. The machine setting also depends on the design of the welding machine. See Chapter 12.

2. A proper *arc length* between the electrode and the work is essential to generate the heat needed for welding. An arc that is too long produces an unstable welding arc, reduces penetration, increases spatter, causes flat and wide beads, and prevents the gas shield from protecting the molten puddle from atmospheric contamination. Too short an arc will not create enough heat to melt the base metal, the electrode will have a tendency to stick, penetration will be poor, and uneven beads with irregular ripples will result.

3. *Electrode angle* is the angle at which the electrode is held during the welding process. Using the correct angle will insure proper penetration and bead formation. As different positions and weld joints become necessary, electrode angle becomes an increasingly important factor in obtaining a satisfactory weld.

4. *Speed of travel* is the rate at which the electrode is moved across the weld area. Factors such as size and type of electrode, current, position, base metal, and amperage all affect the speed of travel necessary for completing a sound weld.

Checking and Adjusting the Equipment

To start your first welding operation, proceed as follows:

1. Inspect the cable connections to make certain that they are all tight.

2. Make sure the bench top and metal to be welded are dry and free from dirt, rust, and grease.

3. Select proper polarity.

4. Adjust the control unit for the amperage needed for the selected electrode. *Remember,* the recommended current setting specified for the electrode is only approximate. Final adjustment of current value is made after beginning the welding operation. For example, the amperage range for the electrode may be 90-100. It is best for the beginner to set the control midway between the two limits, which in this case is 95 amperes. If after the welding is started the arc is too hot, turn the control to reduce the amperage. Increase the current setting if the arc is not hot enough for penetration. No specific rules can be given for the final setting because many factors are involved, such as the skill of the operator, welding position, type of metal, and the nature of the welding job. Ability to make the final adjustment comes as experience is gained.

Gripping the Electrode

Place the bare end of the electrode in the holder as shown in Figure 14-1. By gripping the electrode near the end, most of the coated portion can be used. *Always keep the jaws of the holder clean* to insure good electrical contact with the electrode. *Be careful not to touch the welding bench with an uninsulated holder, as this will cause a flash. When not in use, hang the holder in the place provided for it.*

Grip the holder lightly in your hand. If you hold it too tightly, your hand and arm will tire quickly.

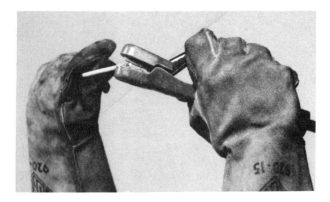

Figure 14-1. Place the bare end of the electrode in the electrode holder.

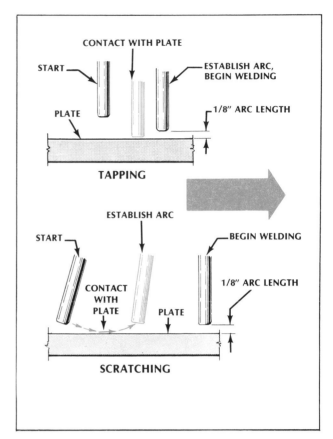

Striking the Arc | Exercise 1

1 Obtain a piece of ¼″ or thicker steel plate and lay it flat on the bench top. Insert a ⅛″ or ⁵⁄₃₂″ E-6012, E-6013, or E-7024 electrode in the holder and set the machine for the correct current.

2 There are two methods which can be used to start, or strike, the arc—the *tapping* and the *scratching motion*. The tapping method is the one preferred by experienced welders, whereas the scratching motion is found to be easier for the beginner.

In the tapping motion, the electrode is brought straight down and withdrawn instantly. With the scratching method, the electrode is moved at an angle in contact with the plate in a scratching motion much as in striking a match. Regardless of the motion used, upon contact with the plate, promptly raise the electrode a distance equal to the diameter of the electrode. Otherwise, the electrode will stick to the metal. If it is allowed to stick to the metal with the current flowing, the electrode will become red hot. *Should the electrode weld fast to the plate, break it loose by quickly twisting or bending the holder. If it should fail to dislodge, disengage the electrode by releasing it from the holder.*

3 Practice starting the arc until this operation can be performed quickly and easily.

Running Short Beads | Exercise 2

1 With a soapstone, which is a marking chalk used to draw lines on metal, draw a series of lines on a steel plate, each line approximately 2″ in length and ⅜″ apart, as illustrated.

2 Run a continuous bead over each line, moving the electrode from left to right. Hold the electrode in a vertical position and angle the electrode holder slightly toward the end of the weld. This is referred to as *travel angle*.

3 Move the electrode just rapidly enough so deposited metal has time to penetrate into the base plate. If the current is set properly and the arc is maintained at the correct length, there will be a continuous crackling or frying noise. Learn to recognize this sound. An arc that is too long will have somewhat of a humming sound. Too short of an arc makes a pop-

Exercise 2, cont.

ping sound. Notice the action of the molten puddle and how the trailing edge of the puddle solidifies as the electrode travels forward.

The appearance of the puddle is often an indication of how well a weld is being made. If the molten metal is clear and bright it means that no molten slag is mixing with the puddle. Slag is brittle and when it flows in the molten metal the weld is weakened. Normally, if the edges of the weld bead have a dull irregular appearance it means that slag is being trapped into the puddle.

4 Complete the exercise by running straight, consistent, parallel beads.

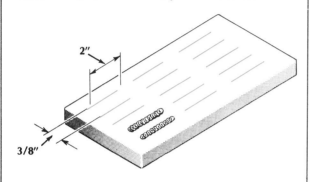

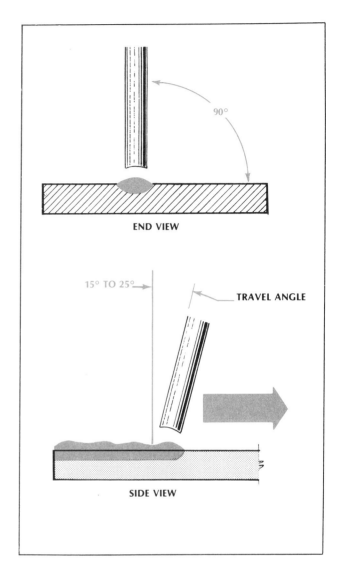

Adjusting the Amperage

After you have become accustomed to striking an arc and running short weld beads, vary the welding current to see how it affects the welding heat. First turn the machine down about five amperes and check if there is any difference when you run a bead. Then turn it down another five amperes and again try to run a bead. As you reduce the amperage it soon becomes apparent that there is insufficient heat to melt the base metal. Furthermore, you will find that as the electrode burns off, it does not fuse with the base metal but lies on the surface as spatter which easily scrapes off.

Now reverse the process by gradually raising the amperage. Turn the machine up five amperes in several steps and each time run short beads. It will soon become obvious that as the amperage is increased the arc gets hotter and the electrode melts faster.

From this experiment you can appreciate the importance of having the correct welding heat to make a sound weld. However, as you gain experience in welding, proper adjustment of welding current becomes relatively easy.

Points to Remember

1 Inspect the equipment before starting to weld.

2 See that the polarity switch is set in the recommended position.

3 See that no combustible materials are near where the welding is to be done, as flying sparks from the spatter of the arc may easily ignite the materials.

4 Do not lay the holder on the bench while the current is flowing.

5 Release the electrode if it sticks to the plate.

6 Always shut off the machine when leaving the welding bench.

7 Adjust the amperage as necessary after starting with the recommended amperage range.

QUESTIONS FOR STUDY AND DISCUSSION

1 How does an arc length affect a weld?

2 What causes the arc voltage to fluctuate?

3 Why must the current be adjusted for a particular welding operation?

4 What equipment checks are made before proceeding to weld?

5 Why should the electrode be clamped at its extreme end?

6 Why should the holder never be placed on the work bench while the current is on?

7 What two methods may be used in striking an arc?

8 In striking an arc, why should the electrodes be withdrawn instantly?

9 What should be done if the electrode welds fast to the plate?

10 The arc should be maintained at approximately what length?

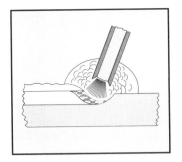

To produce good welds you must not only know how to manipulate the electrode, but you need to know certain weld characteristics. Especially important is a knowledge of what constitutes a good weld and what causes a poor weld. Some of the more important elements affecting good welds are discussed in this chapter.

FIVE ESSENTIALS OF ARC WELDING

To secure a weld that has proper penetration, you must keep in mind the following five factors: (1) Correct electrode, (2) Correct arc length, (3) Correct current and amperage, (4) Correct speed of travel, and (5) Correct electrode angle.

Correct electrode. The choice of an electrode involves such items as position of the weld, properties of the base metal, diameter of the electrode, type of joint, and current value. Since many different kinds of electrodes are manufactured, you must know the results that can be expected from different electrodes. If the characteristics of the electrodes are known, then you have greater assurance that a correct weld will be made. Without the right kind of electrode it is almost impossible to get the results desired, regardless of the welding technique used.

Correct arc length. The significance of *arc length* was briefly mentioned in Chapter 14. Since it is one of the essentials for good welding, further amplification of correct arc length is included here.

If the arc is too long, the metal melts off the electrode in large globules which wobble from side to side as the arc wavers. This produces a wide, spattered, and irregular bead without sufficient fusion between the base metal and the deposited metal. An arc that is too short fails to generate enough heat to melt the base metal properly. This produces high, uneven beads with irregular ripples. Running with the arc length too short also increases the possibility of the electrode sticking to the work.

The length of the arc depends on the size of electrode used and the kind of welding done. For small diameter electrodes, a shorter arc is necessary than for larger electrodes. As a rule, the length of the arc should be approximately equal to the diameter of the electrode. For example, an electrode ⅛″ in diameter should have an arc length of about ⅛″. You will find, too, that a shorter arc is required more for horizontal, vertical, and overhead welding than in most flat position welds because it gives better control of the puddle.

The proper arc length is also important because it prevents impurities from entering a weld. A long arc allows the atmosphere to flow into the arc stream, thereby permitting impurities of nitrides and oxides to form. Moreover, when the arc is too long, heat from the arc stream is dissipated too rapidly, causing considerable metal spatter. See Figure 15-1. A correct bead has the proper height and width, and the ripples are uniformly spaced. If the arc length is too long, the bead will be coarse and excess spatter will appear around the weld. If the arc length is too short, the bead will have a narrow width and excessive height.

Correct current and amperage. For the desired weld characteristics, check to see that the correct current is being used for the particular electrode. Chapter 13 contains the currents and polarities necessary for the most commonly used electrodes. If the amperage is too high, the electrode melts too fast and the molten pool is large and irregular. When the welding amperage is too low, there is not enough heat to melt the base metal and the molten pool will be too small. The result is not only poor fusion but the beads will pile up and be irregular in shape. See Figure 15-2.

Correct speed of travel. Where the speed is too fast, the molten pool does not last long enough and impurities are locked in the weld. The bead is narrow and the ripples pointed. If the rate of travel is too slow, the metal piles up excessively and the bead is high and wide with straight ripples as illustrated in Figure 15-2.

Correct electrode angle. The angle at which the electrode is held will greatly affect weld bead shape and is particularly important in fillet and deep groove welding. Electrode angle involves two positions— travel angle and work angle. *Travel angle* is the angle in the line of welding and may vary from 5° to 30° from the vertical, depending on welder preference

CORRECT BEAD FORMATION

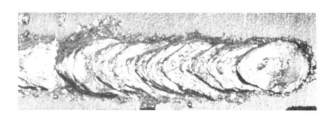

ARC LENGTH TOO LONG

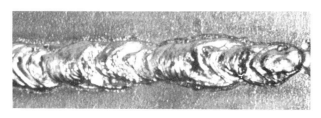

ARC LENGTH TOO SHORT

Figure 15-1. Correct arc length is necessary for proper bead formation.

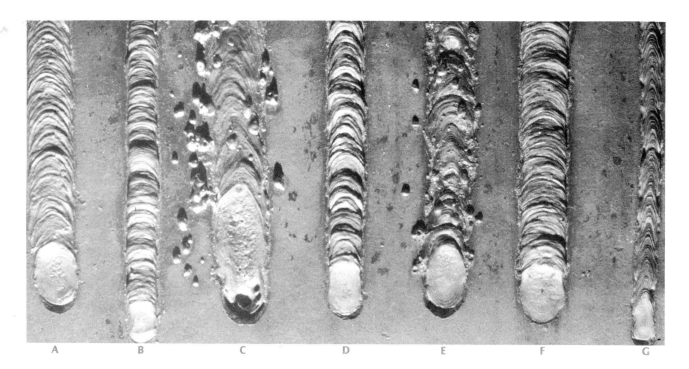

Figure 15-2. Examples of properly and improperly formed beads. (*The Lincoln Electric Company*)

A. Amperage, voltage, and speed normal
B. Amperage too low
C. Amperage too high
D. Voltage too low
E. Voltage too high
F. Speed too slow
G. Speed too fast

and welder conditions. *Work angle* is the angle from horizontal, measured at right angles to the line of welding, which normally splits the angle of the weld joint. See Figure 15-3.

Ordinarily, a slight angle of the electrode in either direction from the work angle will not affect weld appearance or quality. However, whenever undercuts occur in the vertical member of a fillet weld the angle of the arc should be lowered and the electrode directed more toward the vertical member. Work angle is especially important in multiple-pass fillet welding. See Figure 15-3.

Crater formation

As the arc comes in contact with the base metal, a pool, or pocket, is formed. As previously stated, this pool is known as a *crater*. The size and depth of a crater indicate the amount of penetration. In general, the depth of penetration should be from one-third to one-half the total thickness of the bead, depending upon the size of the electrode. See Figure 15-4.

To secure a sound weld, the metal deposited from the electrode must fuse completely with the base metal. Fusion will result only when the base metal has been heated to a liquid state and the molten metal from the electrode readily flows into it. If the arc is too short, there will be insufficient *spread* of heat to form the correct size crater of molten metal. When the arc is too long, the heat is not centralized or intense enough to form the desired crater.

Controlling the crater. An improperly filled crater may cause a weld to fail when a load is applied on a welded structure. Therefore, be sure to fill a crater properly.

Occasionally, you will find that the crater is getting too hot and the fluid metal has a tendency to run. When this happens, lift the electrode slightly and quickly, and shift it to the side or ahead of the crater. Such a movement reduces the heat, allows the crater to solidify momentarily, and stops the deposit of metal from the electrode. Then return the electrode to the crater and shorten the arc.

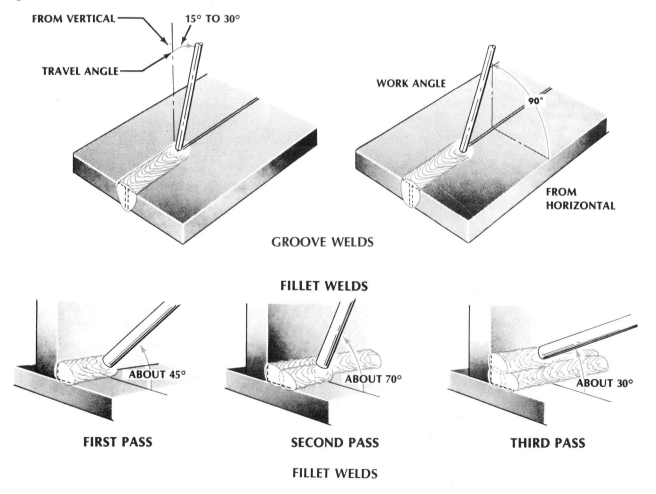

FROM VERTICAL — 15° TO 30°

TRAVEL ANGLE

WORK ANGLE — 90°

FROM HORIZONTAL

GROOVE WELDS

FILLET WELDS

FIRST PASS — ABOUT 45°

SECOND PASS — ABOUT 70°

THIRD PASS — ABOUT 30°

FILLET WELDS

Figure 15-3. Correct electrode angle is important to making proper welds. (*LTV Steel*)

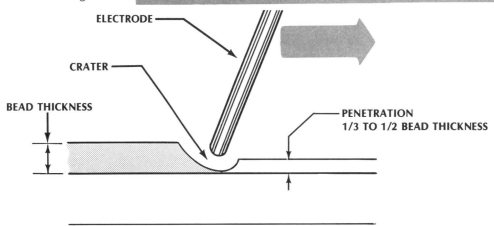

Figure 15-4. The depth of the crater determines the amount of penetration.

Another method used by welders to control the temperature of the molten puddle is a *whipping* motion of the electrode. This motion technique is used with E-6010 and E-6011 electrodes and is especially helpful when welding pieces that do not have a tight fit and large openings have to be filled. It is also used in overhead and vertical welding to better control the weld puddle.

In a whipping motion the electrode is struck and held momentarily. Then it is moved forward about ¼″ or ⅜″ and is hesitated. Hesitating the electrode temporarily reduces the heat of the crater by maintaining the arc away from the crater. Then just as the puddle begins to freeze the electrode is moved back into the center of the puddle and the sequences repeated. The movement of the electrode is done by pivoting the wrist and not moving the arm while making the pass.

Remelting the crater.   There is always a tendency when starting an electrode for a large globule of metal to fall on the surface of the plate with little or no penetration. This is especially true when beginning a new electrode at the crater left from a previously deposited weld. To fill the crater and secure proper fusion, strike the arc approximately ½″ in front of the crater as shown at *A* in Figure 15-5. Then bring it back through the crater to point *B* just beyond the crater, and weld back through the crater.

Undercutting and Overlapping

Undercutting is a condition that results when the welding amperage is too high. The excessive amperage leaves a groove in the base metal along both sides of the bead which greatly reduces the strength of a weld as illustrated in Figure 15-6. Undercutting may also occur when there is insufficient deposition of metal on a vertical plate. This can be corrected by slightly changing the electrode angle.

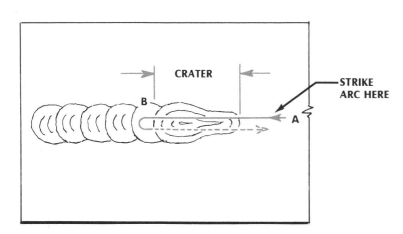

Figure 15-5. To continue a bead, strike the arc and move the electrode as shown to properly fill the crater left from the previously deposited weld.

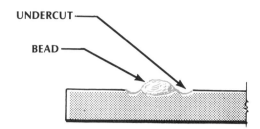

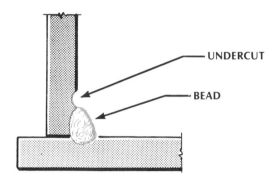

Figure 15-6. Undercutting is caused by excessive amperage and insufficient metal deposition.

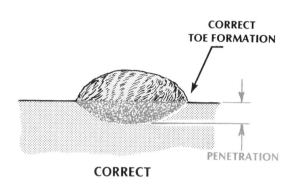

CORRECT

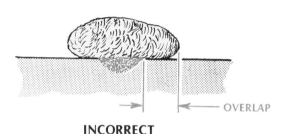

INCORRECT

Figure 15-7. Welding amperage that is too low results in poor penetration, causing overlapping.

Overlapping occurs when the amperage is set too low. In this instance, the molten metal falls from the electrode without actually fusing into the base metal as shown in Figure 15-7.

Cleaning a Weld

It has already been mentioned that when a weld is made, a layer of slag covers the deposited bead. If additional layers of weld metal are deposited, this slag must be removed; otherwise it will flow in with the deposited metal and cause a weak weld. To remove the slag, strike the weld with a chipping hammer. Hammer the bead so the chipping is directed away from the body, and away from the eyes and face as pictured in Figure 15-8.

WARNING: Always wear safety glasses when chipping. Do not pound the bead too hard; otherwise the structure of the weld may be damaged. After the slag is loosened, drag the point end of the hammer along the weld where it joins the plate. This will remove the remaining particles of slag. Follow the chipping with a good, hard brushing, using a stiff wire brush as illustrated in Figure 15-9.

Figure 15-8. Strike the weld with a chipping hammer to remove slag.

Figure 15-9. After chipping, brush the weld with a wire brush. (*Hobart Brothers Company*)

Weaving the Electrode

Weaving is a technique used to increase the width and volume of the bead. Enlarging the size of the bead is often necessary on deep groove or fillet welds where a number of passes must be made. Figure 15-10 illustrates several weaving patterns. The pattern used depends to some extent on the position of the weld. In subsequent chapters additional instructions will be given as to the most suitable weaving pattern for specific weld joints.

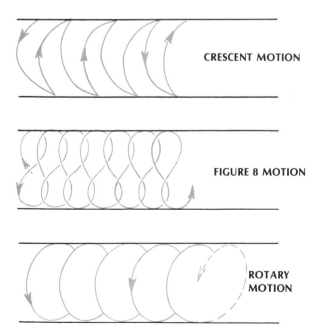

CRESCENT MOTION

FIGURE 8 MOTION

ROTARY MOTION

Figure 15-10. The crescent, figure 8, and rotary motions are three typical weaving patterns used to increase the width and volume of the bead.

Running Continuous Beads | Exercise 1

1 After you have mastered the operation of depositing short beads, secure a steel plate ¼″ × 4″ × 6″.

2 With a soapstone, draw a series of lines approximately ¾″ apart as shown.

3 Use a ⅛″ diameter E-6010, E-6011, E-6012, or E-6013 electrode and run continuous beads on these lines, starting from the left edge and working to the right.

4 After the plate is filled, remove the slag and examine the beads.

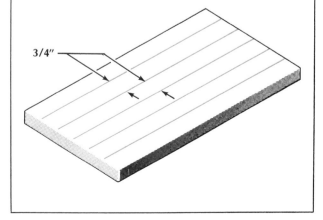

3/4″

Moving the Electrode in Several Directions — Exercise 2

1 Draw the lines on the plate as shown.

2 Then deposit a continuous bead, moving the electrode from left to right, bottom to top, right to left, and top to bottom.

3 Carefully control the bead formation by using the correct arc length, travel angle, and speed of travel.

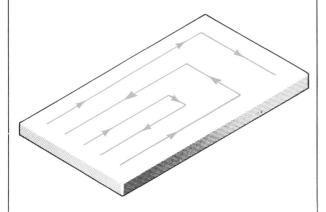

Restarting the Arc — Exercise 3

1 Draw a series of straight lines on the plate and divide these lines into 2-inch sections as shown.

2 Run a bead over the first 2-inch section, but break the arc.

3 Continue the deposit for another two inches; then repeat the practice of breaking the arc and refilling the crater.

4 Follow this procedure until skill is mastered in depositing uniform and continuous beads with properly filled craters.

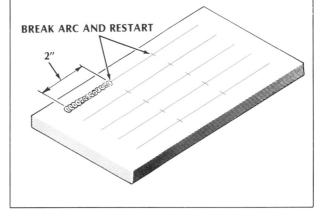

BREAK ARC AND RESTART

2"

Weaving the Electrode — Exercise 4

1 Lay out a series of straight lines on a ¼″ × 4″ × 6″ plate.

2 Run continuous beads over these lines and then clean each bead.

3 Now proceed to practice weaving by depositing a weld back and forth between the first pair of continuous beads as illustrated.

4 Use one type of weaving motion to fill the first space and then try several other weaves on the remaining sections. Make certain that the short beads are fused into the long, straight beads.

5 Continue the weaving practice on several plates until a satisfactory plate is completed.

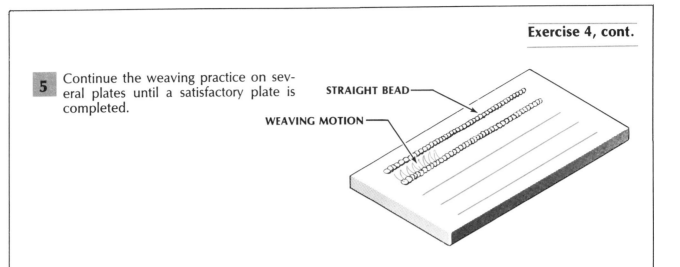

STRAIGHT BEAD

WEAVING MOTION

Depositing a Metal Pad   **Exercise 5**

Padding or surfacing is a process for building worn surfaces of shafts, wheels, and other machine parts. The operation consists of depositing several layers of beads, one on top of the other as pictured.

1 Obtain a plate ¼″ (or thicker) × 3″ × 5″.

2 As shown, fill the first section completely with beads.

3 Clean the weld thoroughly, and then deposit a second layer of weaving beads about ½″ wide at right angles to the first layer. After the weld is cleaned, deposit a third layer at right angles to the second layer as shown.

4 Clean this surface and deposit a fourth and fifth layer.

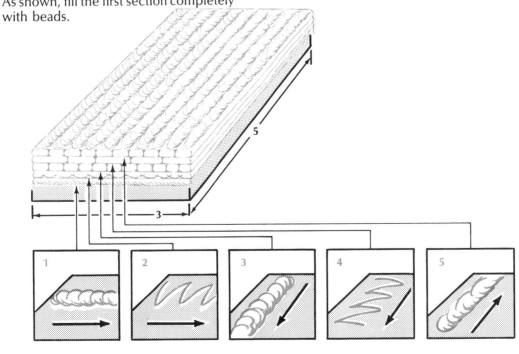

Exercise 5, cont.

WORN SHAFT BUILT UP BY PADDING

Points to Remember

1 Use the correct type of electrode.

2 Use an arc that is about as long as the diameter of the electrode.

3 Use the correct welding current and amperage.

4 Move the electrode just fast enough to produce evenly spaced ripples.

5 Keep the penetration to a depth equal to one-third to one-half the total thickness of the weld bead.

6 Be sure the molten metal from the electrode fuses completely with the base metal.

7 Always restart the electrode ½″ in front of the previously made crater.

8 Avoid undercutting and overlapping by using the correct amperage and electrode angle.

9 When cleaning the slag from a weld, chip away from your body to prevent the slag from flying up into your face.

QUESTIONS FOR STUDY AND DISCUSSION

1 What causes a bead to overlap the base metal? Why does this usually indicate a poor weld?

2 What causes undercutting? How can undercutting be avoided?

3 What is meant by a crater?

4 What should be the depth of penetration?

5 How is a crater affected when the arc is too short? What happens when the arc is too long?

6 What are the five essentials for securing a sound weld?

7 What factors must be taken into consideration when selecting an electrode?

8 When an arc is too long, what happens to the metal as it melts from the electrode?

9 How is it possible to identify a weld that has been made with an arc that is too long?

10 What is likely to happen to the electrode when the arc is too short?

11 What are some of the characteristics of a weld made with an arc that is too short?

12 What are some of the factors that must be considered when deciding the length of an arc?

13 In what way does the amount of amperage affect a weld?

14 What determines the speed at which an electrode should be moved?

15 How should an electrode be restarted to fill a crater left from a previously deposited weld?

16 What should be done when the crater gets too hot and the metal has a tendency to run over the surface?

17 How should slag from a weld be removed from a workpiece?

18 What is meant by weaving?

19 When is a weaving motion used?

20 What is the purpose of padding?

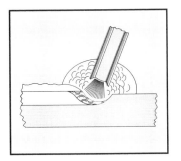

Shielded Metal-Arc Welding – SMAW

Although welding can be done in any position, the operation is simplified if the joint is flat. When placed in this position, the welding speed is increased, the molten metal has less tendency to run, better penetration can be secured, and the work is less fatiguing. At first glance some structures may appear to require horizontal, vertical, or overhead welding, but upon more careful examination you may be able to change them to the easier and more efficient flat position. See Figure 16-1.

WELD PASSES

In carrying out some welding operations, very often the pieces have to be tack-welded. Tack welds are simply short sections of weld beads ¼" to ½" long used to maintain the proper position and root opening between the metal being welded. See Chapter 3. These tack welds are spaced along the joint and must be consumed into the joint.

Once the joint is tacked, the necessary weld passes are made. Sometimes more than one pass is necessary to fill the joint. If this is the case, the first pass, known as the *root pass* or *stringer bead* is laid in the bottom of the V. It is made with a small diameter electrode by moving it straight down into the groove without any weaving motion. See Figure 16-2. Its principal function is to fill the root opening and join the two metal pieces. Since it serves as the base for the other passes it is very important that it produces complete penetration. Complete penetration is more assured if the root pass penetrates the bottom surface of the groove but not more than ⅟₁₆", and consumes all tack welds previously made.

The next layer is called the *fill* or *filler pass*. One or more filler passes may be needed to fill the groove, depending on the thickness of the metal. In depositing the filler passes a slight weaving motion is generally advisable to insure proper fusion to the previously laid beads and sides of the groove joint.

The final pass, called the *cap* or *cover pass* (sometimes referred to as the *wash bead*), is intended to

Figure 16-1. Welding in the flat position is easier and produces better results at less expense. *(Miller Electric Manufacturing Company)*

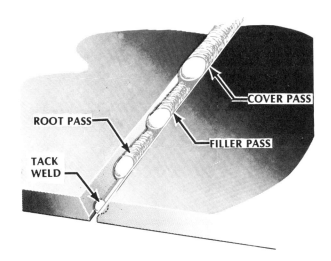

Figure 16-2. Sometimes more than one pass is necessary to fill the joint. The root pass is first, followed by filler passes and the cover pass.

provide additional reinforcement to the weld and give it a nice appearance. The cover pass should not extend more than ⅟₁₆" above the plate surface. A weaving motion is used to obtain the necessary weld width when covering the filler passes.

Joints Commonly Welded in the Flat Position

The butt joint is often used when structural pieces have flat surfaces, such as in tanks, boilers, and a variety of machine parts. See Figure 16-3. The joint may be opened, closed, or the edges beveled, as in Figure 16-4. A *closed butt joint* has the edges of the two plates in direct contact with each other. See Exercise 1. This joint is suitable for welding steel plates that *do not exceed ⅛″ to ³⁄₁₆″ in thickness.* Heavier metal can be welded but only if the machine has sufficient amperage capacity and if heavier electrodes are used. *Remember that on thick metal it is difficult to secure ample penetration to produce a strong weld with a single pass bead.*

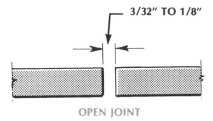

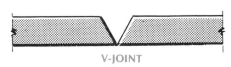

Figure 16-4. A butt joint can be closed, open, or it can have beveled edges.

In the *open butt joint* the edges are placed slightly apart, usually ³⁄₃₂″ to ⅛″, to allow for expansion and penetration. As a rule, a back-up strip or block of scrap steel or copper is placed under this joint. See Figure 16-5. A back-up strip prevents the bottom edges from burning through.

When the thickness of the metal exceeds ⅛″ the edges of a butt joint should be beveled. The beveling can be done by cutting the edges with a flame torch

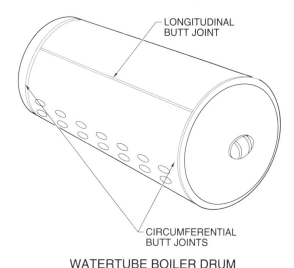

WATERTUBE BOILER DRUM

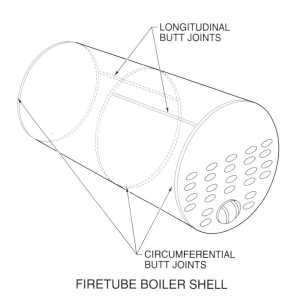

FIRETUBE BOILER SHELL

Figure 16-3. Butt joints are often used on boilers.

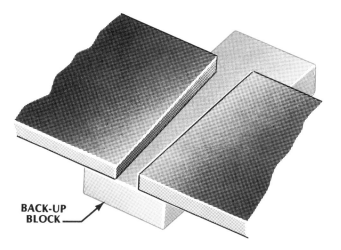

Figure 16-5. A back-up strip or block prevents the bottom edges in an open butt joint from burning through.

or by grinding them on a grinder. The included angle of the V should not exceed 60°, to limit the amount of contraction that usually results when the metal cools. The edges may be prepared in several ways as shown in Figure 16-6. Notice that on heavy metal ³⁄₈″ or more in thickness the edges are beveled on both sides. Beveling in this manner will insure better penetration, requires less weld metal, and contraction forces are better equalized.

When the joint consists of two pieces of different thicknesses, adjust the work angle of the electrode so the greatest portion of the heat is concentrated on the thickest plate. See Figure 16-7.

In welding thin stock with a single pass, as in a closed or open butt joint, simply allow the electrode to travel along the joint without any weaving motion. Move the electrode slow enough to give the arc sufficient time to melt the metal. Take care that the travel is not too slow because the arc will burn through the metal.

When a multiple pass is to be made in a grooved joint, be sure to hold the electrode down in the groove so it almost touches both sides of the joint while depositing the root pass. Move the electrode fast enough to keep the slag flowing back on the finished weld. If the electrode is not moved rapidly enough, the slag may become trapped in the bottom of the weld, thereby preventing proper fusion.

After completing the root pass proceed with the necessary filler passes. Complete the weld with a cover pass. See Figure 16-8. Remember, always remove the slag completely after each pass. If any slag particles are allowed to remain they will weaken the weld.

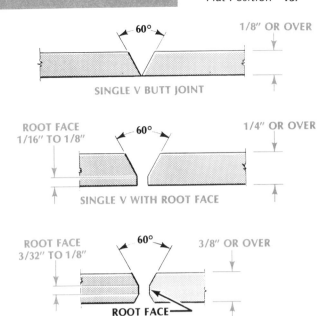

Figure 16-6. Depending on the thickness of the metal, the edges in a V butt joint are prepared in different ways.

The lap joint is one of the most frequently used joints in flat position welding. It is a relatively simple joint, since no beveling or machining is necessary. One standard requirement is to have clean, evenly aligned surfaces. See Figure 16-9. The joint consists of lapping one edge over another and joining. The amount the pieces should overlap depends upon the thickness of the plates and the strength required of the welded piece. Usually the thicker the plates the greater the amount of overlap.

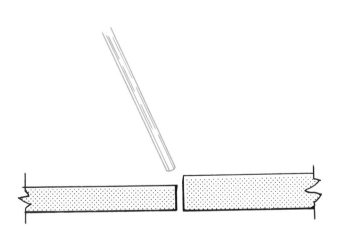

Figure 16-7. When welding plates of different thicknesses, direct more heat to the thicker plate.

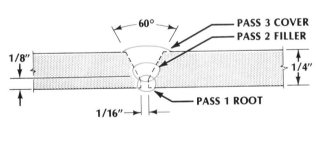

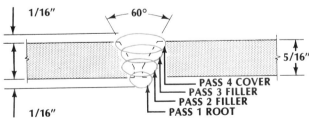

Figure 16-8. After completing the root pass in a multiple-pass butt joint, proceed with the necessary filler passes and finish with a cover pass.

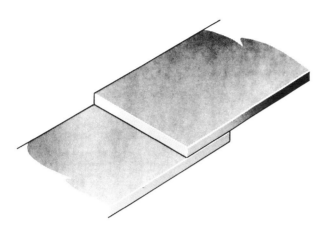

Figure 16-9. The lap joint requires no edge preparation before welding.

When the structure is subjected to heavy bending stresses, it is advisable to weld the edges of both sides of the joint. See Figure 16-10.

The lap joint is adaptable for a variety of new construction work as well as for making numerous types of repairs. For example, such a joint can be employed when joining a series of metal plates together or in reinforcing another structural member. Since a lap weld stiffens the structure where the plates are lapped, this joint is used a great deal in tank and ship building. See Exercise 2.

When an exceptionally strong lap joint is required, especially on heavy plates ⅜″ and over in thickness, a multiple-pass fillet weld is recommended. This joint has two or more layers of beads along the seam, with each bead lapping over the other. See Exercise 3.

The T-fillet joint is frequently used in fabricating straight and rolled shapes. See Figures 16-11 and 16-12. The strength of this joint depends considerably on having the edges of the joint fit close together. See Exercise 4. The T-joint should not be used if it is subjected to heavy stresses from the opposite direction of the welded joint. This weakness can be partially overcome by using a double fillet—that is, welding both sides of the joint. See Figures 16-13 and 16-14.

When welding heavy plate or when extra strength is required a wider fillet is necessary. You can make a wider fillet by running several layers of beads. See Exercise 5.

Figure 16-11. The strength of the single-pass T-fillet joint depends greatly on having the edges of the joint fit close together. (*The Lincoln Electric Company*)

Figure 16-12. The T-fillet joint is used frequently in fabrication of steel structures. (*Hobart Brothers Company*)

Figure 16-10. Weld both sides of the lap joint when greater strength is required.

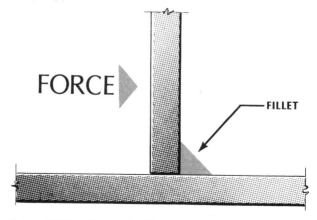

FORCE

FILLET

Figure 16-13. The single fillet T-joint should not be used where heavy stresses will come from the opposite direction of the welded joint.

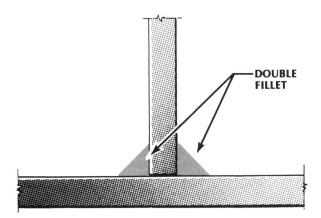

Figure 16-14. A double fillet T-joint resists weld failure when force is applied to either side.

The outside corner weld is often used in constructing rectangular shaped objects such as tanks, metal furniture, and other machine sections where the outside corner must have a smooth radius. See Figure 16-15 and Exercise 6.

A butt joint is generally used to weld round shafts or rods. See Exercise 7. To consider some of the problems that may be encountered when shielded metal-arc welding in the flat position see Figure 16-16.

Figure 16-15. An outside corner weld is often used in constructing rectangular shaped objects.

ARC WELDING PROBLEMS

CHARACTERISTICS	CAUSE	REMEDY
1. Unstable arc, arc goes out. Spatters over work.	Arc too long.	Shorten arc.
2. Poor or no penetration. Arc goes out often.	Not enough current for size of electrode. Wrong electrode.	Increase amperage. Use proper electrode.
3. Loud crackling from arc. Flux melts too rapidly. Wide bead, spatter in large drops.	Too much current for electrode. May be moisture in electrode cover.	Decrease amperage.
4. Difficulty in striking arc. Poor penetration.	Wrong polarity. Too little amperage.	Change polarity. Or, increase amperage.
5. Weak weld. Arc hard to start. Arc keeps breaking.	Dirty work.	Clean work. Remove slag from previous weld.
6. Arcing at ground clamp.	Poor ground.	Correct poor ground.

Figure 16-16.

Welding a Butt Joint
Exercise 1

1 Obtain two ³⁄₁₆″ or ¼″ plates.

2 Butt the two pieces together and tack weld.

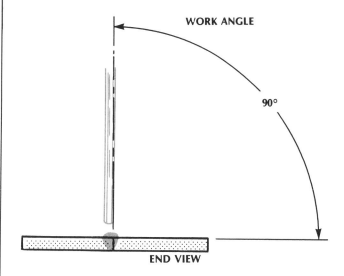

WORK ANGLE

90°

END VIEW

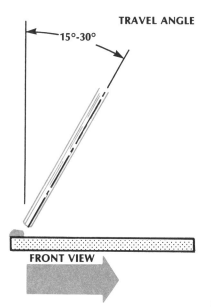

TRAVEL ANGLE

15°-30°

FRONT VIEW

3 Using a work angle of 90° and a travel angle of 15°-30° deposit a bead across the piece.

4 Let cool and then repeat the procedure on the reverse side.

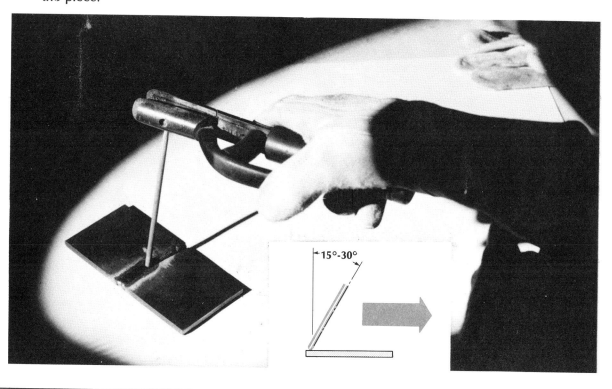

15°-30°

Welding a Single-Pass Lap Joint

1 Obtain two pieces of ³⁄₁₆″ or ¼″ steel plate. Remember, a *single pass* simply means depositing one layer of beads. When a weld is built up of more than one layer it is known as a *multiple pass* weld.

2 Use ⅛″ electrodes and adjust the machine for the correct amperage.

3 Tack the plates at each end. With the plates properly tacked, run a ¼″ fillet along the joint. Hold the electrode at a 45° angle. Position the tacked pieces against a firebrick to achieve the correct welding position.

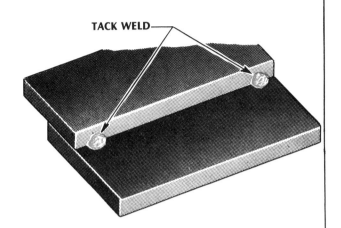

TACK WELD

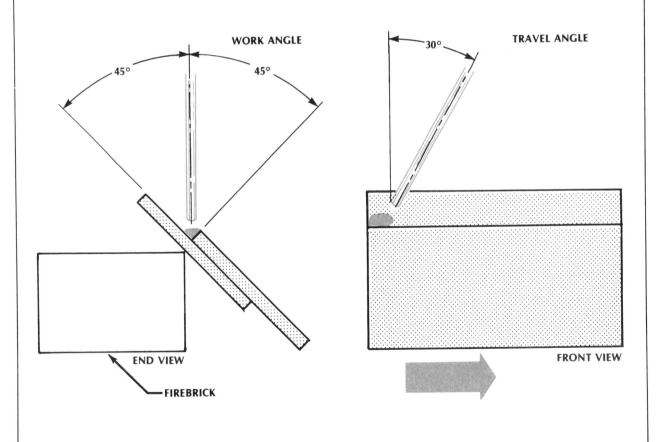

WORK ANGLE

45°   45°

END VIEW

FIREBRICK

TRAVEL ANGLE

30°

FRONT VIEW

Exercise 2, cont.

4 Weave the electrode slightly, maintaining the arc for a little longer period on the lower plate.

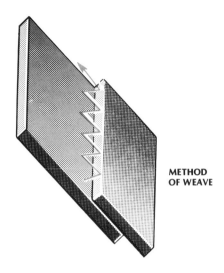

METHOD
OF WEAVE

5 Be sure to get complete fusion at the *root,* or joining point, of the joint and avoid overlaps on the top surface. A weld made with a concave fillet usually is too weak because it lacks sufficient reinforcing material, while a weld with a convex bead has too much waste metal, which is of no value to the joint.

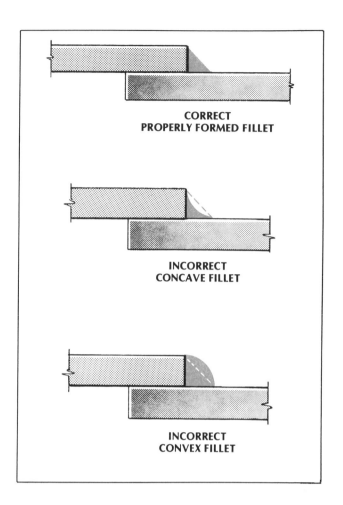

CORRECT
PROPERLY FORMED FILLET

INCORRECT
CONCAVE FILLET

INCORRECT
CONVEX FILLET

Welding a Multiple-Pass Fillet Lap Weld   Exercise 3

1 Deposit the first bead by moving the electrode straight down the joint without any weaving motion.

2 Clean the weld carefully and lay the second pass over this *stringer bead.*

3 During the second pass, weave the electrode, pausing for an instant at the top of the weave to deposit extra metal on the vertical edge of the upper plate.

Exercise 3, cont.

4 Maintain a consistent bead width across the piece.

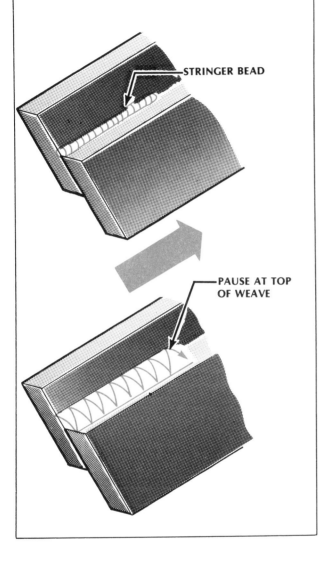

STRINGER BEAD

PAUSE AT TOP OF WEAVE

Welding a Single-Pass T-Fillet Joint   Exercise 4

1 Obtain two 3/16″ or 1/4″ plates.

2 Set the vertical plate on the middle of the horizontal plate and tack weld each end.

3 Position the tacked pieces against a firebrick to achieve the correct welding position.

4 Deposit a 1/4″ fillet bead along the edge.

5 Hold the electrode at a work angle of 45° and a travel angle of 30° and advance it in a straight line without any weaving motion.

6 Travel rapidly enough to stay ahead of the molten pool. Concentrate the arc more on the lower plate to prevent undercutting the upper plate. Watch the crater closely so it will form a bead with the correct contour.

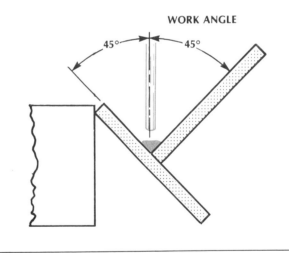

WORK ANGLE

45°   45°

Exercise 4, cont.

TRAVEL ANGLE

30°

Welding a Multiple-Pass T-Fillet Joint   Exercise 5

1 Deposit the first pass as described in Exercise 3. Remove the slag carefully and completely.

2 Angle the electrode with a 70° work angle and a 30° travel angle. Deposit a bead which partially covers the first pass. Remove the slag carefully and completely.

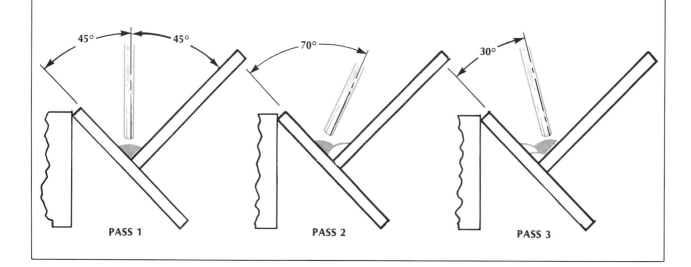

45°   45°   70°   30°

PASS 1   PASS 2   PASS 3

3 Angle the electrode with a 30° work angle and a 30° travel angle. Deposit a bead which covers the first pass and part of the second pass.

4 If more or less weld metal is required, make multiple passes using different bead configurations.

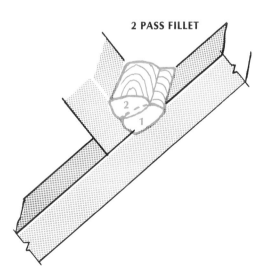

2 PASS FILLET

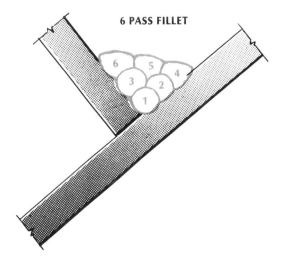

6 PASS FILLET

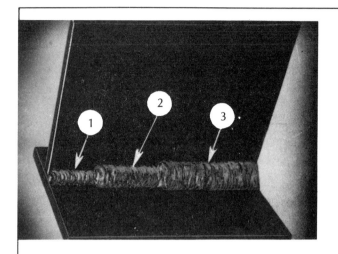

Exercise 5, cont.

5 Additional weld metal can be deposited on a multiple-pass T-fillet joint by weaving the electrode.

Welding an Outside Corner Weld   Exercise 6

1 To make an outside corner weld, tack the two plates and run a bead along the edge with the electrode held at a work angle of 45° and a travel angle of 30°.

2 On light stock, one bead is usually enough.

3 Heavy stock will probably require a series of passes to fill in the corner.

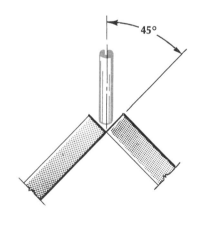

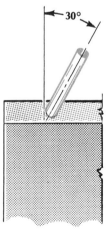

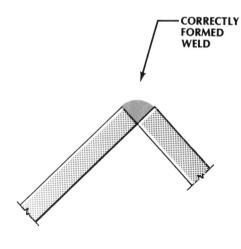

CORRECTLY FORMED WELD

Welding Round Stock — Exercise 7

1 Bevel opposite sides of the stock, leaving a root face in the center.

2 Be sure to grind the edges so they have the same groove angle.

3 Place the pieces in a vise or section of angle iron to hold them in position.

4 To prevent the shaft from warping, deposit a small bead on one side and then lay a similar bead on the opposite side.

5 Use a slight weaving motion on the final pass.

Points to Remember

1 When making a root pass use smaller diameter electrodes and do not use any weaving motion while making the pass.

2 Use a slight but not heavy weaving motion in making a filler pass.

3 Place the edges of an open butt joint slightly apart to allow for joint expansion and penetration.

4 When the thickness of the metal exceeds ⅛″ on a butt joint, bevel the edges.

5 When welding a lap or T-joint, weld the joint on both sides if the structure is subjected to heavy stresses.

6 Tack the two pieces before starting to weld. This will keep them in position.

7 Hold the electrode at a 45° work angle when welding a lap joint, keeping the arc more on the upper plate.

8 For an exceptionally strong lap or T-joint on heavy metal, use multiple passes.

9 When a butt joint is used to weld a round shaft, bevel opposite sides to the same angle.

QUESTIONS FOR STUDY AND DISCUSSION

1 What is the function of a root pass?

2 How many filler passes are used on a groove joint?

3 What is a cover pass and why is it used?

4 What is an advantage of welding in flat position rather than in other positions?

5 What is meant by single-pass fillet weld?

6 When making a lap weld, what determines how much the pieces should overlap?

7 How is it possible to avoid undercutting when welding a lap joint?

8 What is the purpose of using a double fillet on a lap joint?

9 When should a multiple pass be used on a lap joint?

10 What are some of the factors that must be considered when welding a T-joint?

11 When welding a T-joint, why should the arc favor the bottom plate?

12 How many passes should be made on an outside corner weld?

13 When is a butt joint used in welding?

14 What is the difference between an open and closed butt joint?

15 When should the edges of butt joints be beveled?

16 How should the edges of round stock be prepared for welding?

17 What work angle is used to weld a lap joint in the flat position?

18 How are tack welds used to weld a T-joint?

19 What determines the edge preparation required for welding a butt joint?

20 What work and travel angles are used when welding a butt joint in the flat position?

21 What procedures are followed when welding plates of different thicknesses?

22 What welding technique is recommended when making a root pass?

23 What work angles are used to weld a multiple-pass T-fillet weld?

24 What are some common applications of outside corner welds?

25 What are the characteristics of too much amperage used?

Shielded Metal-Arc Welding – SMAW

On many jobs it is practically impossible to weld pieces in the flat position. Occasionally the welding operation must be done while the work is in a horizontal position. A weld is in a horizontal position when the joint is on a vertical plate and the line of weld runs on a line with the horizon as in Figure 17-1. A fill-freeze or fast-freeze type of electrode should be used.

To perform welds of this kind, use a slightly shorter arc length at a slight reduction in amperage setting than used for flat position welding. The shorter arc length will minimize the tendency of the molten puddle to sag and cause overlaps. An overlap occurs when the puddle runs down to the lower side of the bead and solidifies on the surface without actually penetrating the metal. See Figure 17-2. A sagging puddle usually leaves an undercut on the top side of the seam as well as an improperly shaped bead, all of which weaken a weld.

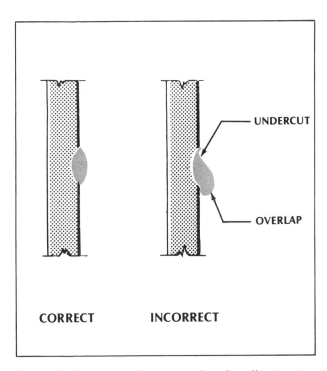

Figure 17-2. Using a shorter arc length will minimize undercut and overlap caused by a sagging puddle.

Figure 17-1. Welding a horizontal butt joint.

HOW TO HOLD THE ELECTRODE

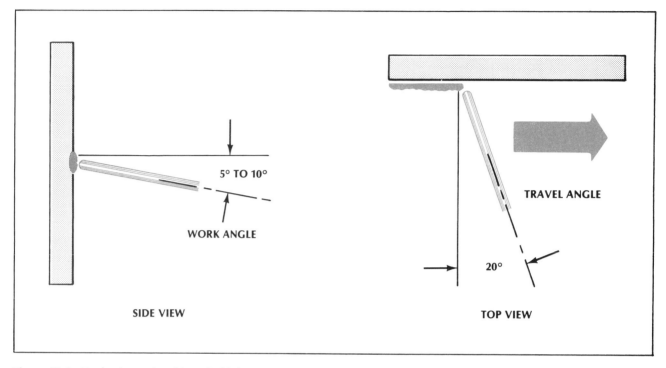

SIDE VIEW

TOP VIEW

Figure 17-3. For horizontal welding, hold the electrode with a work angle of 5° to 10° and a travel angle of 20°.

For horizontal welding, hold the electrode so that it points upward 5° to 10° and has a 20° travel angle as illustrated in Figure 17-3. In laying the bead, use a narrow weaving motion as shown in Figure 17-4. By weaving the electrode, the heat will be distributed more evenly, thereby reducing still further any tendency for the puddle to sag. Keep the arc length as short as possible. If the force of the arc has a tendency to undercut the plate at the top of the bead, drop the electrode holder a little to increase the upward angle.

As the electrode is moved in and out of the crater, pause slightly each time it is returned. This keeps the crater small and the bead is less likely to sag.

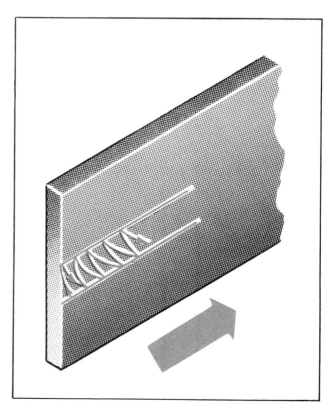

Figure 17-4. Using a slight weaving motion will distribute the heat more evenly.

Depositing Straight Beads in a Horizontal Position | Exercise 1

1 Obtain a ¼″ plate and draw a series of parallel lines ½″ apart.

2 To keep the piece in place, tack it to a flat piece. Secure this piece on the bench in a vertical position.

3 Adjust the machine to the correct current and, with a slight weaving motion, deposit beads between the guide lines. Lay one bead by starting from the left side of the plate and working to the right.

4 Skip a ½″ space and reverse the direction of travel on the next bead.

5 Continue this operation until uniform beads can be made without overlapping and undercutting.

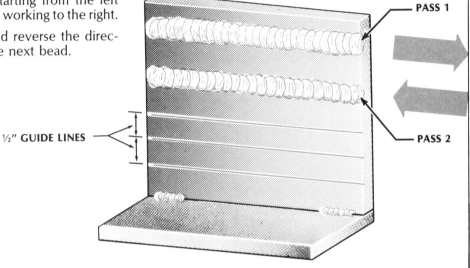

PASS 1

½″ **GUIDE LINES**

PASS 2

Making a Single-Pass Lap Joint in a Horizontal Position | Exercise 2

1 Tack two ¼″ plates to form a lap joint. Secure the piece in a vertical position.

2 Angle the electrode to a 45° angle and run a single bead along the edge, using a slight weaving motion.

Watch the formation of the bead closely for any undercutting. Continue this operation on additional lap joints until a satisfactory weld is made.

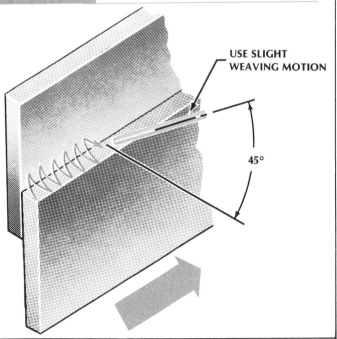

USE SLIGHT WEAVING MOTION

45°

Welding a Multiple-Pass Fillet T-joint in a Horizontal Position

1 Tack two ¼″ plates to form a T-joint and secure the base plate in a vertical position.

2 Angle the electrode to a 45° angle and run a bead along the root of the joint without any weaving motion. Clean the slag off the first bead. Angle the electrode to a 70° angle and deposit a second bead. Use a slight weaving motion, penetrating the first bead and the plate.

3 Clean the slag off the second bead. Angle the electrode to a 30° angle and deposit a third bead, using a slight weaving motion. Notice that the third bead penetrates into the first and second layers as well as into the upright plate. This penetration is important; otherwise a weak weld will result, and the layers may separate from each other.

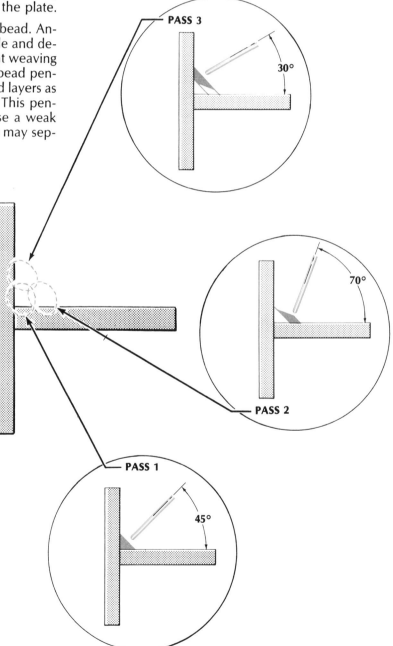

PASS 3

30°

PASS 2

70°

PASS 1

45°

Welding a Multiple-Pass Butt Joint in a Horizontal Position

1 Obtain two pieces of ¼″ steel plate and bevel the edge of one plate.

2 Tack the two plates together to form a single bevel butt joint, allowing a ¹⁄₁₆″ root opening. Secure the plates in a vertical position with the beveled plate on top. The plate that is not beveled should be on the bottom. Its flat edge serves as a shelf, thus helping to prevent the molten metal from running out of the joint.

3 Deposit the first bead deep in the root of the joint. Remove the slag and lay the second bead. Then follow with a third bead. Each bead should penetrate the base metal.

4 On some welding jobs, the practice is to bevel both edges of the joint to form a 60° included angle. This is a single V-butt joint. Since such a joint does not provide a retaining shelf for the bead as the single bevel butt joint, a little more skill is required to produce a satisfactory weld. In practicing this type of weld, the position of the electrode for each pass is changed.

The number of passes on the joint will depend on the thickness of the metal as well as the diameter of the electrode. Sufficient penetration into each adjacent pass is necessary for complete fusion in the weld. It is common practice on a wide joint to finish the weld with a *cover pass,* commonly known as a *wash bead* or *cap pass* to produce a smooth finish. A cover pass is made by using a wide weaving motion that covers the entire area of the deposited beads.

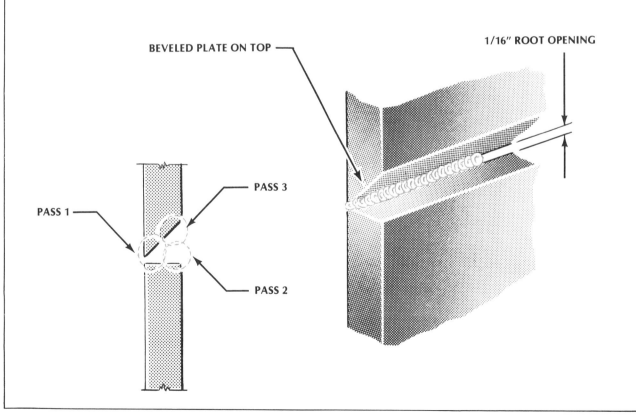

BEVELED PLATE ON TOP

1/16″ ROOT OPENING

PASS 3

PASS 1

PASS 2

Exercise 4, cont.

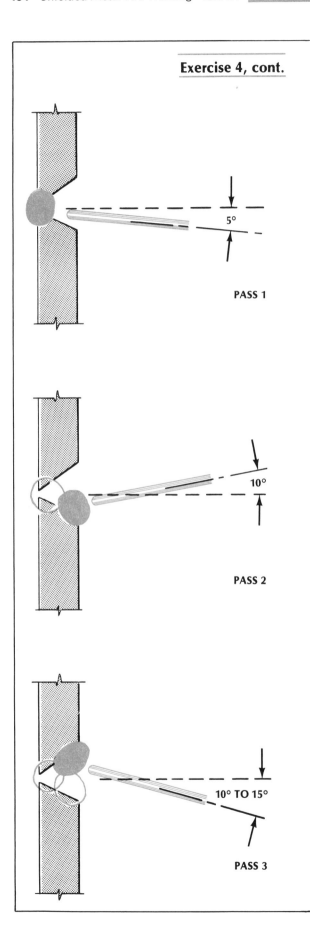

PASS 1

PASS 2

PASS 3

Points to Remember

1 Use a lower welding current and shorter arc length for horizontal welding.

2 For horizontal welding, tilt the electrode upward 5° to 10° and slant it slightly away from the weld.

3 Use a slight weaving motion.

4 Do not allow the molten pool to sag and cause overlaps and undercuts.

5 On a multiple-pass butt joint, use a cover pass on the final pass.

QUESTIONS FOR STUDY AND DISCUSSION

1 Why is it essential to use a lower current and a shorter arc length when welding in a horizontal position?

2 What can be done to avoid overlaps on horizontal welds?

3 In what position should the electrode be held for welding horizontal beads?

4 Why should a weaving motion be used when making horizontal welds?

5 What determines the number of passes that should be made on a weld?

6 What included angle is used when beveling the edges for a single V-butt joint?

7 Where and why is a cover pass used?

8 What work angles are required when welding a three pass T-fillet?

9 Why is a slight weaving motion used when welding the second and third pass of a multiple-pass T-fillet?

10 What must be done between passes of a multiple-pass T-fillet to assure proper penetration?

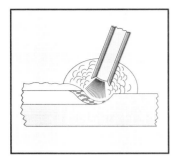

VERTICAL POSITION

Shielded Metal-Arc Welding – SMAW

18

In the fabrication of many structures such as steel buildings, bridges, tanks, pipelines, ships, and machinery, the operator must frequently make vertical welds. A vertical weld is one with a joint or line of weld running up and down as shown in Figure 18-1.

One of the problems of vertical welding is that gravity tends to pull down the molten metal from the electrode and plates being welded. To prevent this from happening, fast-freeze or fill-freeze types of electrodes should be used. Puddle control can also be achieved by proper electrode manipulation and selecting electrodes specifically designed for vertical position welding.

POSITION AND MOVEMENT OF THE ELECTRODE

Vertical welding is done by depositing beads either in an upward or downward direction, (referred to as *vertical up* and *vertical down*). *Vertical down welding* is very practical for welding light gauge metal because penetration is shallow, thereby forming an

adequate weld without burning through the metal. Moreover, vertical down welding can be performed much more rapidly, which is important in production work. Although it is generally recommended for welding lighter materials, vertical down welding is sometimes used for other metal thicknesses.

On heavy plates of ¼″ or more in thickness, *vertical up welding* is often more practical, since deeper penetration can be obtained. Welding upward also makes it possible to create a shelf for successive layers of beads.

For vertical down welding, maintain a 15° to 30° travel angle of the electrode. See Figure 18-2. Start at the top of the joint and move downward with little or no weaving motion. If a slight weave is necessary, manipulate the electrode so the crescent is at the top.

For vertical up welding, start with the electrode at a right angle to the plates. Then lower the electrode holder, keeping the electrode tip in place, until the electrode forms a travel angle of 10° to 15°, pointing away from the weld. See Figure 18-2.

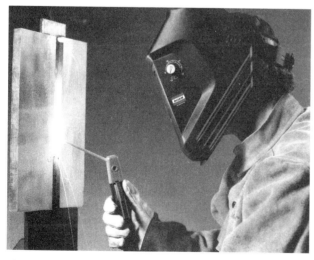

Figure 18-1. Vertical position welding requires special procedures to prevent molten metal from running down the joint. *(The Lincoln Electric Company)*

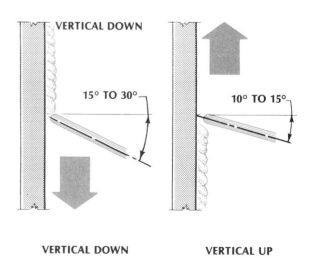

Figure 18-2. Maintain a travel angle of 15° to 30° for vertical down and 10° to 15° for vertical up welding.

Running Straight Beads Vertical Down   Exercise 1

1 Set up a practice piece in a vertical position with a series of straight lines drawn on it. Use E-6012 or E-6013 electrodes.

2 Start at the top of the plate with the electrode pointed upward about 60° from the vertical plate.

3 Keep the arc short and draw the electrode downward to form the bead.

4 Travel just fast enough to keep the molten metal and slag from running ahead of the crater. Do not use any weaving motion to start with.

5 Once this technique is mastered try weaving the electrode but very slightly with the crest at the top of the crater.

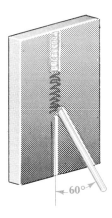

VERTICAL DOWN WELDING WITH NO WEAVE MOTION

VERTICAL DOWN WELDING WITH SLIGHT WEAVE MOTION

Running Straight Beads Vertical Up   Exercise 2

1 Obtain a ¼″ plate and draw a series of straight lines. Then secure the piece so the lines are in a vertical position. Use an E-6010 or E-6011 electrode for the necessary fast-freeze characteristics.

2 Strike the arc on the bottom of the plate. As the metal is deposited, move the tip of the electrode upward in a rocking motion. This is called a whipping motion. In whipping the electrode, do not break the arc but simply pivot it with a wrist movement so the arc is moved up ahead of the weld long enough for the puddle in the crater to solidify.

3 Return it to the crater and repeat the operation, working up along the line to the top of the plate.

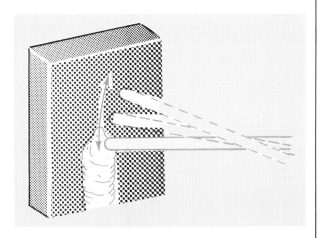

4 *Remember, do not break the arc while moving the electrode upward.* Withdraw it just long enough to permit the deposited metal to solidify and form a shelf so additional metal can be deposited.

5 Continue to lay beads from bottom to top until each line is smooth and uniform in width.

Vertical Beads with a Weaving Motion

On many joints in vertical up welding it is necessary to form beads of various widths. The width of the bead can be controlled by using a weaving motion. Each pattern will produce a bead approximately twice the diameter of the electrode. Notice that each weave is shaped so the electrode can dig into the metal at the bottom of the stroke, and the upward motion momentarily removes the heat until the metal can solidify. When a smooth weld is required on the final pass of a wide joint, a *wash bead* should be used.

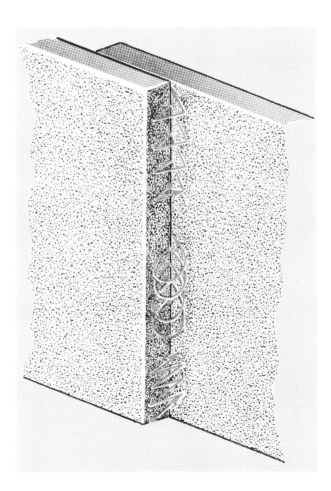

Figure 18-3. Weaving patterns can be used to control the weld bead.

Welding a Vertical Up Lap Joint
Exercise 3

1. Obtain two ¼″ plates and tack them together to form a lap joint.

2. Secure the piece in a vertical position by tacking it to a flat scrap piece or by mounting it in a positioner.

3. Deposit a small stringer bead in the root without any electrode motion.

4. Deposit a cover pass using a weaving motion from the bottom to the top.

5. Make certain that the second pass completely penetrates the first pass.

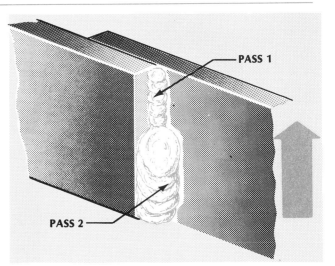

PASS 1

PASS 2

Welding a Vertical Up Butt Joint
Exercise 4

1. Obtain two ¼″ plates and bevel the edges to form a 60° included angle.

2. Tack the plates together with a ¹⁄₁₆″ root opening and fasten them in an upright position to provide a vertical butt joint.

3. Deposit a root pass in the root opening and follow with additional filler passes as necessary. Be sure to clean off the slag after each pass.

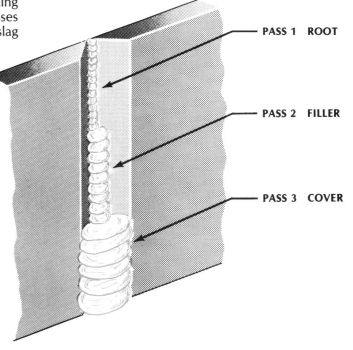

PASS 1   ROOT

PASS 2   FILLER

PASS 3   COVER

Welding a Vertical Up T-Joint | Exercise 5

1 Obtain two ¼″ plates and tack them to form a T-joint. Secure the joint in a vertical position.

2 Deposit a narrow root pass.

3 Remove the slag. Using a weaving motion, deposit a second pass.

4 Repeat step 1 on other side of joint.

5 Remove the slag and deposit two more passes.

6 Check for complete penetration of each pass.

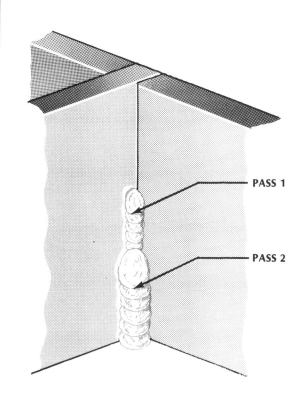

PASS 1

PASS 2

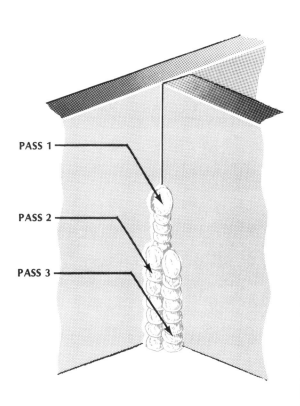

PASS 1

PASS 2

PASS 3

Welding Technique for the E-7018 Electrode

Although the vertical techniques as previously described will generally prevail for all types of electrodes, sometimes a slight modification in procedure is advisable when using E-7018 electrodes.

On vertical down welding, drag the electrode lightly with a very short arc. Avoid a long arc since the weld depends on the molten slag for shielding. Stringer beads or small weaves are always preferred to wide weave passes. Use lower amperage with DC than with AC. Point the electrode directly into the joint and tip it forward only a few degrees in the direction of travel.

With vertical up welding a triangular weave motion will often produce better results. Do *not* use a whipping motion or take the electrode out of the molten pool. Point the electrode directly into the joint and slightly upward to permit the arc force to assist in controlling the puddle. Use amperage in the lower level of the recommended range.

Points to Remember

1 For welding light gauge metal in a vertical position, the vertical down technique is more practical than the vertical up.

2 For plates ¼″ or more in thickness, better results will be obtained by using the vertical up method of welding.

3 Whipping the electrode will often provide better control of the molten puddle in vertical up welding.

4 On grooved joints, deposit the root pass deep into the root opening.

QUESTIONS FOR STUDY AND DISCUSSION

1 In vertical welding, what can be done to prevent the molten puddle from sagging?

2 Why is welding vertical down on a vertical joint more applicable on light gauge metal?

3 In what position should the electrode be held in vertical down welding?

4 What motions should be used in vertical down welding?

5 How should the electrode be held in making a vertical up weld?

6 What is the advantage of using a whipping motion on a vertical weld?

7 How can the width of a bead be increased on a vertical up weld?

8 How does welding with E-7018 electrodes differ from welding with E-6010 electrodes in the vertical up position?

9 What direction of travel provides the most penetration when welding in the vertical position?

10 What types of electrodes are commonly used in vertical welding?

11 What kind of weaving motion is used when welding vertical up welds using an E-7018 electrode?

12 Which is faster, vertical up welding or vertical down welding?

13 What determines if a weld is in the vertical position?

14 What types of electrodes can be used with a whipping motion?

15 What is the advantage of using a weaving motion when welding in the vertical position?

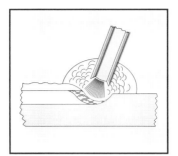

OVERHEAD POSITION

19

Shielded Metal-Arc Welding – SMAW

Figure 19-1. In overhead welding gravity makes it more difficult to secure uniform beads and correct penetration. (*Bernard Welding Equipment Company*)

Welding in an overhead position is probably the most difficult operation to master. It is difficult because you must assume an awkward stance, and at the same time work against gravity, which exerts a downward force. See Figure 19-1. In an overhead position the puddle has a tendency to drop, making it harder to secure uniform beads and correct penetration. Nevertheless with a little practice it is possible to secure welds as good as those made in other positions.

Position For Overhead Welding

In learning to weld in the overhead position, a positioner is needed so the work can be adjusted to any height or position. See Figure 19-2.

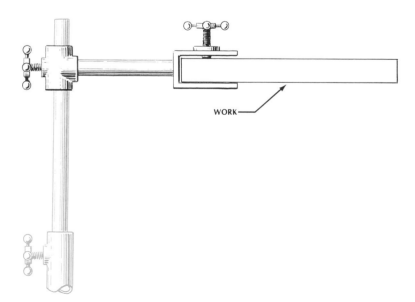

WORK

Figure 19-2. A positioner is necessary to secure the work at the correct height.

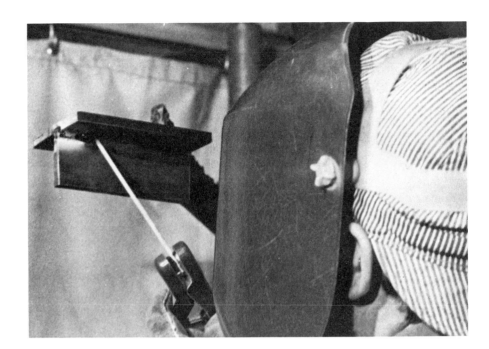

Figure 19-3. A welding cap is essential when welding in the overhead position.

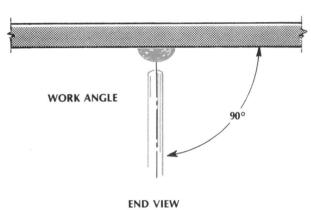

WORK ANGLE

90°

END VIEW

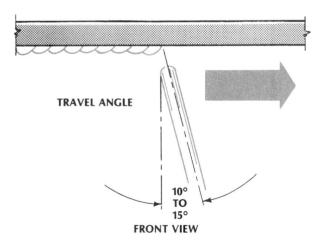

TRAVEL ANGLE

10° TO 15°

FRONT VIEW

Figure 19-4. In overhead welding the electrode should have a work angle of 90° and a travel angle of 10° to 15°.

WARNING: There is the possibility of some falling molten metal. Be sure sleeves are rolled down and a protective garment with a tight-fitting collar is zipped or buttoned up to the neck. Wear a cap and heavy duty shoes. See Figure 19-3.

To start welding, hold the electrode at a right angle to the joint. The electrode should have a work angle of 90° and a travel angle of 10° to 15°. See Figure 19-4.

Grip the electrode holder so the knuckles are up and the palm down. This prevents particles of molten metal from being caught in the hollow palm of the glove and allows the spatter to roll off the glove. Although the electrode can be held in one hand, sometimes it is easier if it is held with both hands. See Figure 19-5. To avoid falling sparks and hot metal drippings, stand to the side rather than directly underneath the arc. The weight of the cable can be minimized by draping it over a shoulder if welding in a standing position, or over a knee if in a sitting position. See Figures 19-6 and 19-7.

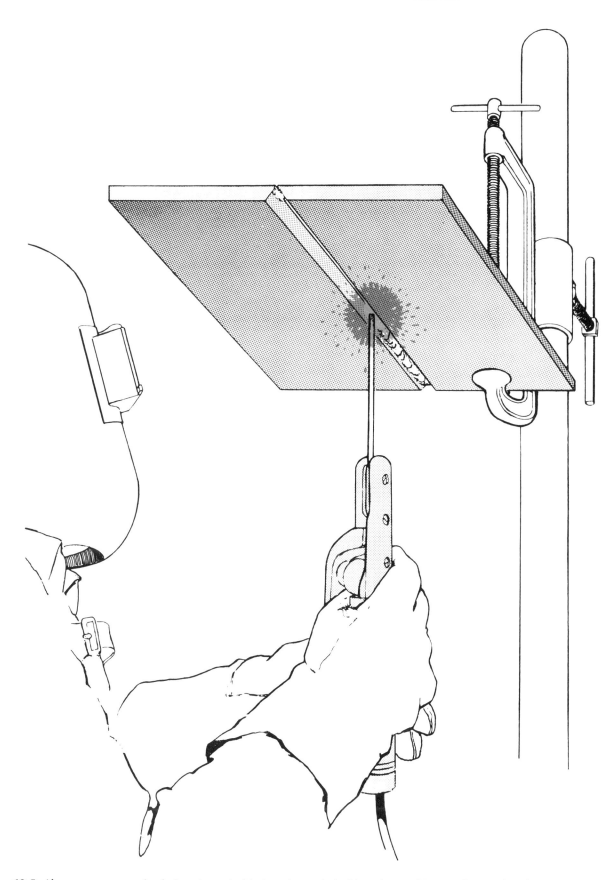

Figure 19-5. If necessary, use both hands to hold the electrode holder when welding in the overhead position.

Figure 19-6. Drape the cable over a shoulder when overhead welding in a standing position. (*Hobart Brothers Company*)

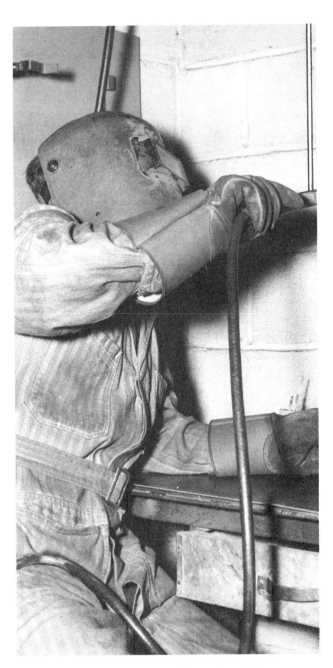

Figure 19-7. If you are sitting, drape the cable over your knee.

Running Beads in Overhead Position	Exercise 1

1. Secure in the positioner a ¼″ plate with guidelines drawn approximately ½″ apart.

2. Reduce the amperage as recommended. Strike the arc and form a puddle as in flat position welding. Move the electrode forward keeping the arc as short as possible.

Exercise 1, cont.

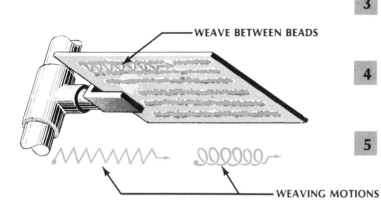

WEAVE BETWEEN BEADS

WEAVING MOTIONS

3 Run a series of straight beads without any weaving motion. To prevent the puddle from dropping, reduce the amperage slightly.

4 Continue to deposit straight beads until proper control of the puddle is mastered. Practice running the beads in one direction and then reverse the direction.

5 Using a weaving motion fill in the space between beads.

Welding a Lap Joint in Overhead Position   Exercise 2

1 Tack two ¼″ plates to form a lap joint and secure in the positioner.

2 Hold the electrode with a 45° work angle and a 15° travel angle.

3 Deposit the root pass deep in the root of the joint.

4 Remove the slag and deposit the second pass penetrating into the root pass and the bottom plate. Clean off the slag and deposit the third pass penetrating into the root pass and the top plate.

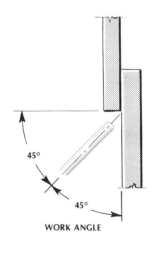

45°

45°

WORK ANGLE

END VIEW

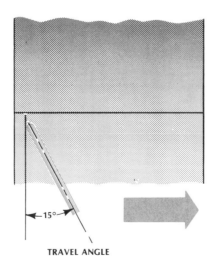

←15°→

TRAVEL ANGLE

FRONT VIEW

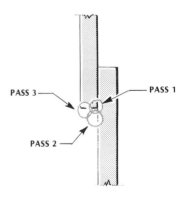

PASS 3

PASS 1

PASS 2

Welding a T-Joint in Overhead Position   Exercise 3

1 Tack two ¼″ plates to form a T-joint and secure in the positioner.

2 Deposit the root pass in the root of the joint. Clean off the slag and deposit the second and third pass.

3 Adjust the work angle of the electrode to insure complete penetration.

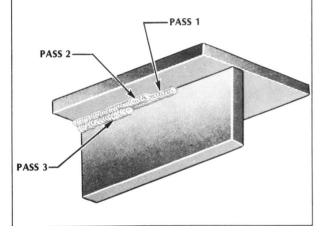

PASS 1
PASS 2
PASS 3

Welding a Single-V Butt Joint in Overhead Position   Exercise 4

1 Bevel the edges of two ¼″ plates and tack them together with a ¹⁄₁₆″ root opening. Secure in the positioner.

2 Deposit the root pass obtaining complete penetration.

3 Clean off the slag.

4 Deposit additional passes to cover the groove faces of the joint.

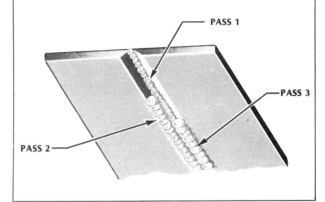

PASS 1
PASS 3
PASS 2

Points to Remember

1 In overhead welding, weld with a travel angle of 10° to 15°.

2 Grip the electrode holder so that knuckles are face up and palm down.

3 Stand to the side of the arc.

4 Drape the cable over a shoulder if welding in a standing position, or over a knee if in a sitting position.

5 Keep the arc length as short as possible.

6 Use a fast-freezing electrode with appropriate motion.

7 Be sure protective clothing is worn.

QUESTIONS FOR STUDY AND DISCUSSION

1 Why is overhead welding more difficult?

2 What is the recommended travel angle for overhead welding?

3 Why should the electrode holder be grasped so the palm of your hand is facing down?

4 What position should you assume for overhead welding?

5 What should be done to prevent the puddle from falling?

6 Why should sleeves be turned down?

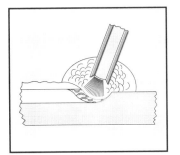

Most structures made of cast iron can be welded successfully. Because of the peculiar characteristics of cast iron, you will find that welding this metal requires a great deal more care than welding mild steel. However, if certain precautions are observed you should be able to arc weld almost any cast-iron piece. See Figure 20-1.

Figure 20-1. This cast-iron frame is being welded using the shielded metal-arc welding process. (*Hobart Brothers Company*)

TYPES OF CAST IRON

Cast iron is an iron-base material with a high percentage of carbon. The five types of cast iron are gray, white, malleable, alloy, and nodular.

Gray cast iron results when the silicon content is high and the iron is permitted to cool slowly. The combination of high silicon and slow cooling forces the carbon to separate in the form of graphite flakes, sometimes called free carbon. It is this separation of the carbon from the iron that makes gray cast iron so brittle.

Gray cast iron is used a great deal for machine castings. It can readily be identified by the dark gray, porous structure when the piece is fractured. If brought in contact with a revolving emery wheel, the metal gives off short streamers that follow a straight line and are brick red in color, with numerous fine, repeating yellow sparklers. See Figure 20-2. Gray cast iron can be arc welded with comparative ease.

White cast iron possesses what is known as combined carbon. Combined carbon means that the carbon element has actually united with the iron instead of existing in a free state as in gray cast iron. This condition is brought about through the process of rapidly cooling the metal, leaving it very hard. In fact it is so hard that it is exceedingly difficult to machine, and special cutting tools or grinders must be used to cut the metal. White cast iron is often used for castings with outer surfaces that must resist a great deal of wear. It is also used to make malleable iron castings.

The fracture of a piece of white cast iron will disclose a fine, silvery white, silky, crystalline formation. The spark test will show short streamers that are red in color. There are fewer sparklers than in gray cast iron and these are small and repeating. See Figure 20-2.

Although white cast iron can be welded, welding is not recommended for this metal.

Malleable cast iron is actually white cast iron which has been subjected to a long annealing process. The annealing treatment draws out the brittleness from the casting, leaving the metal soft but possessing considerable toughness and strength.

The fracture of a piece of malleable cast iron will indicate a white rim and a dark center. The spark test will show a moderate amount of short, straw-yellow streamers with numerous sparklers that are small and continue repeating. See Figure 20-2.

Malleable cast iron can be welded; however, you must be sure that the metal is not heated above its critical temperature (approximately 1382°F or 750°C). If it is heated beyond this point, the metal reverts back to the original characteristics of white cast iron.

Alloy cast irons are those which contain certain alloying elements such as copper, aluminum, nickel, titanium, vanadium, chromium, molybdenum, and magnesium. By adding one or more of these elements to the iron it is possible to improve its tensile strength, machinability, fatigue resistance, and corrosion resistance. The alloy combinations cause the graphite to separate in a fine and evenly distributed structure, resulting in a cast iron possessing much higher mechanical properties.

Most alloy cast irons can be arc welded but greater precautions must be taken in the preheating and postweld heat treatment stages to prevent the destruction of the alloying elements.

Nodular iron sometimes called ductile iron, has the ductility of a malleable iron, the corrosion resistance of alloy cast iron and a tensile strength greater than gray cast iron. These special qualities are obtained by the addition of magnesium to the iron at the time of melting and then using special annealing techniques. The addition of magnesium and control of the cooling rate causes the graphite to change from a stringer structure to rounded masses in the form of spheroids or nodules. It is the formation of these nodular forms that gives the cast iron better mechanical properties.

Nodular iron can be arc welded, providing adequate preheat and postweld heat treatments are used; otherwise some of the original properties are lost.

Preparing Cast Iron for Welding

Follow this procedure to prepare cast iron for welding:

1. Grind a narrow strip along each edge of the

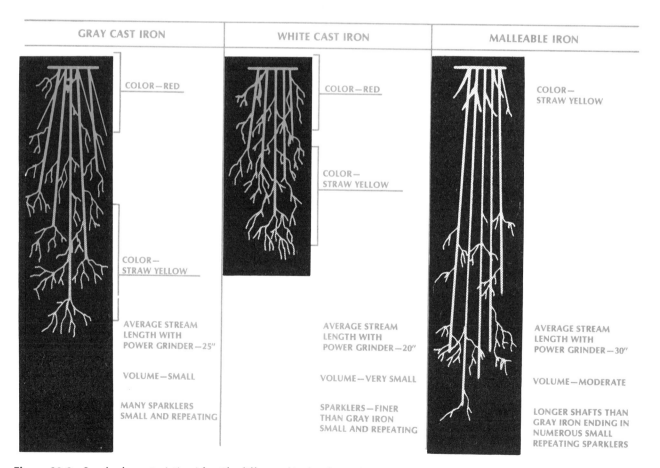

Figure 20-2. Spark characteristics identify different kinds of cast iron.

joint to remove the surface layer known as *casting skin*. See Figure 20-3. Elimination of the surface layer is important because it is full of impurities. These impurities were embedded in the skin when the metal was cast. Unless they are removed, they will interfere with the fusion action of the weld.

2. V the edges as illustrated in Figure 20-3. When the metal does not exceed ³⁄₁₆″ in thickness no V is necessary, but still remove the casting skin. On ³⁄₁₆″ to ³⁄₈″ metal, only a single V is required. The included angle of the V should be approximately 60°. Heavy cast-iron pieces ³⁄₈″ or more in thickness should have a double V with a ¹⁄₁₆″ to ³⁄₃₂″ root face. The included angle should be 60°.

3. If only a crack in a casting is to be welded, V the crack approximately ¹⁄₈″ to ³⁄₁₆″ deep with a diamond point chisel or by grinding. See Figure 20-4. On sections that are less than ³⁄₁₆″ in thickness, V only one-half the thickness.

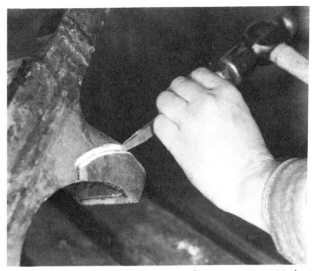

Figure 20-4. Before welding a crack in a casting, V the crack with a diamond point chisel.

4. Be sure the casting is entirely free from rust, scale, dirt, oil, and grease, as these substances, if trapped in the weld, will weaken it. Use a wire brush, and if grease or oil is present, wipe it off with a cleaning solvent.

5. Fine, hairline cracks in a casting can be made more visible by rubbing a piece of white chalk over the surface. The chalk leaves a visible line where the crack is located.

6. Because of expansion from heat during the welding operation, cracks have a tendency to extend. To prevent this from happening, drill a ¹⁄₈″ hole a short distance beyond each end of the crack. See Figure 20-5.

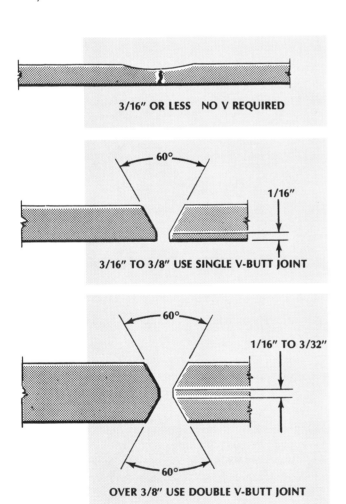

Figure 20-3. Welding cast iron begins with proper edge preparation.

3/16″ OR LESS   NO V REQUIRED

60°

1/16″

3/16″ TO 3/8″ USE SINGLE V-BUTT JOINT

60°

1/16″ TO 3/32″

60°

OVER 3/8″ USE DOUBLE V-BUTT JOINT

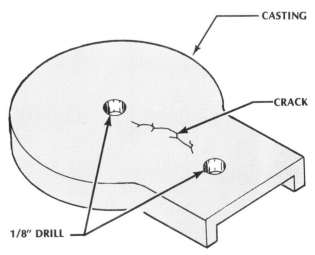

CASTING

CRACK

1/8″ DRILL

Figure 20-5. To prevent cracks from extending during the welding operation, drill a 1/8″ hole beyond each end of the crack.

Keeping Cast Iron Cool

An important point in welding cast iron is to keep the piece as cool as possible to prevent cracks from forming. In the case of malleable cast iron, excessive heat will transform it into white cast iron. If possible, preheat the entire section with an oxyacetylene torch. *CAUTION: Never heat it beyond a dull red color or a temperature exceeding 1200°F (approx. 650°C).* Normally a preheat between 500° and 1200°F (260 to 650°C) is recommended. Correct preheat temperature for a welding can be determined by (1) using a commercial temperature stick called a Tempilstik, or (2) placing a pointed wood stick on the heated surface; if the stick starts to burn the temperature is right for welding.

The preheat should be as uniform as possible over the entire casting and kept at this temperature until the weld is completed. Then the piece must be cooled slowly to room temperature. Very slow cooling is essential with malleable and nodular iron.

When it is impossible to preheat the piece, it can be kept cool by running short beads 2″ to 3″ long. After a bead is deposited, allow it to cool until you can touch it with your hand before starting the next bead. *While cooling, peen the bead by striking it lightly with a hammer.* See Figure 20-6. Peening helps to make the weld tight and relieves stresses. However, peening can only be done on the machinable weld deposits, not the whole casting.

Electrodes for Welding Cast Iron

There are two main groups of electrodes for welding cast iron; machinable and nonmachinable.

Machinable type electrodes are those whose deposits are soft and ductile enough so they can easily

be machined after welding. See Figure 20-7. They are used to repair all kinds of broken castings, correcting for machine errors, filling up defects, or to weld cast iron to steel.

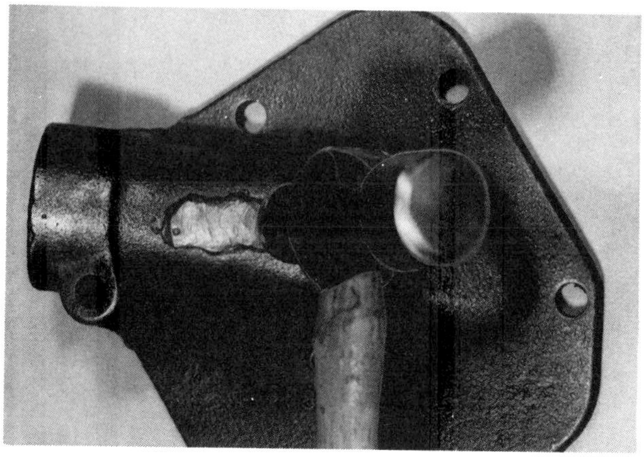

There are two common types of machinable electrodes. Characteristics of these AWS classification electrodes are as follows: (Classification based on chemical composition of metal involved. The suffix CI designates an electrode for cast iron.)

ENi-CI Nickel electrode — DCR or AC, general purpose welding especially for thin and medium sections, and castings with low phosphorus content, and where little or no preheat is used.

ENi-FeCI Nickel iron electrode — DCR or AC for welding heavy sections, high-phosphorus castings, high-nickel alloy castings where high-strength welds are required, and welding nodular iron.

Figure 20-7. This weld was made in a casting using a machinable type electrode. (*The Lincoln Electric Company*)

The *nonmachinable* electrodes have a mild steel core with a heavy coating which melts at low temperatures, allowing the use of low welding current. They leave a very hard deposit and are used only when the welded section is not to be machined afterwards. These electrodes produce a tight and waterproof weld, making them ideal for repairing motor blocks, water jackets, transmission cases, compressor blocks, pulley wheels, pump parts,

Figure 20-6. Peen the bead by striking it lightly with a hammer while it is cooling.

mower wheels, and other similar structures. See Figure 20-8.

Cast-iron electrodes are commonly known by the manufacturer's trade name such as:

Ferroweld—nonmachinable—Lincoln
Softweld —machinable —Lincoln
Strongcast—nonmachinable—Hobart
Nicklecast—machinable —Hobart
Airco 77 —nonmachinable—Airco
Airco 375 —machinable —Airco

Figure 20-8. A nonmachinable electrode was used to weld this crack in a cast-iron cylinder. (*The Lincoln Electric Company*)

WELDING PROCEDURE

1. Set the machine for the correct amperage. Follow the recommendation of the electrode manufacturer. As a rule, the amperage setting for welding cast iron is lower than for welding mild steel.

2. Since it is important to keep the heat to a minimum, always use small diameter electrodes. Operators seldom weld with electrodes greater than ⅛″ in diameter.

3. Tip the electrode 5 to 10 degrees in the direction of travel, and deposit straight, progressive stringer beads. Use a slightly longer arc than in welding mild steel. If more than one layer is to be deposited, use a slight weaving motion after the first bead is made. At no time should the electrode be manipulated so that the width of the deposit is greater than three times the electrode diameter.

4. Use a back-step welding technique when necessary so the crater of each bead lies on top of the previous bead.

5. When welding cracks in castings, start about ⅜″ before the end of the crack and weld back to the hole, filling the hole. See Figure 20-9. Then move slightly beyond the hole. Next move to the other end of the crack and do the same thing. Continue to alternate the weld on each end, limiting the length of each weld 1″ to 1½″ on thin material and 2″ to 3″ on heavier pieces. Always allow each section of weld to cool before starting the next, and peen each short bead.

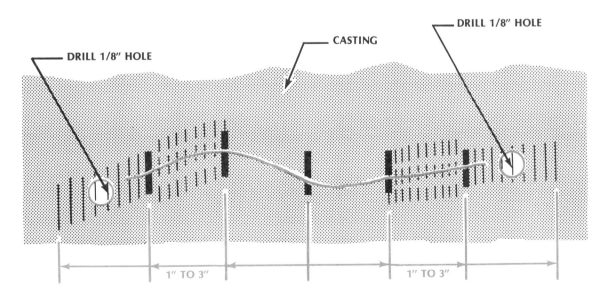

Figure 20-9. Before welding a crack in a casting, drill holes beyond each end of the crack to prevent the crack from extending. Then alternate the weld at each end of the crack to allow each section to cool.

Welding Sections with Pieces Knocked Out

To weld a section with one or more pieces broken out, fit the parts together. V the breaks and tack them before welding.

If the broken part cannot be made to fit in the casting, shape another piece, using mild steel. Place the new section in position and weld.

Studding Broken Castings

When a casting is 1½″ or more in thickness and is subjected to heavy stresses, use steel studs to strengthen the joint. Studding is not advisable on castings lighter than 1½″ because it tends to weaken rather than strengthen the joint. To apply studs proceed as follows:

1. V the crack.

2. Drill and tap ¼″ or ⅜″ holes in the casting at right angles to the sides of the V. Space the holes so the center-to-center distance is equal to three to six times the diameter of the stud. See Figure 20-10.

3. Screw the studs into the tapped holes. The threaded end of the studs should be about ⅜″ to ⅝″ in length and should project approximately ¼″ to ⅜″ above the casting.

4. Deposit beads around the base of the studs, welding them thoroughly to the casting. Remove the

slag and deposit additional layers of beads to fill the V.

Brazing Cast Iron

The three common electrodes for brazing cast iron are ECuSn-A, ECuSn-C (both of which are of the copper-tin classification) and ECuAl-A2 (copper-aluminum classification). The main difference between ECuSn-A and ECuSn-C is in the amount of tin they contain. The ECuSn-C has a higher percentage (8 percent) than the ECuSn-A (5 percent), thereby producing welds of greater hardness, tensile and yield strength. Both are used with DC reverse polarity and normally require that the area to be brazed be preheated to 400°F (205°C).

The ECuAl-A2 electrode has a relatively low melting point and high deposition rate at lower amperage which permits more rapid welding. This minimizes distortion and the formation of white cast iron in the fusion zone. The resulting tensile and yield strength of the deposits is nearly double that of the copper-tin deposits.

The ultimate success of any brazing operation with copper alloy electrodes depends considerably on the following:

1. Use of wide grooves.

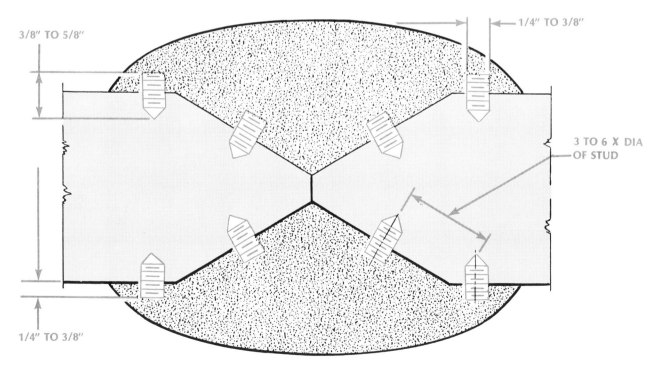

Figure 20-10. Studs can be used to reinforce the weld in a broken casting.

2. Joints free of moisture, grease, oil, dirt.

3. A preheat between 300 to 400°F (about 150 to 205°C).

4. Use of the lowest possible amperage for good bonding.

5. Welding done at a fast rate to minimize dilution from the base metal.

6. Avoidance of puddling.

7. Parts cooled slowly.

Welding Filler Metal

The composition of the welding filler metal is varied to suit the requirements for the weld. The factors that influence the selection of a welding filler metal include the type of cast iron, mechanical properties desired in the joint, need for welding filler metal to deform plastically and relieve welding stresses, machinability of the joint, color matching between base and welding filler metal, allowable dilution (chemical composition change caused by melting and mixing of the welding filler metal and the base metal), and cost.

Cast irons can be welded with cast iron, steel, nickel alloy, and copper alloy electrodes. See Figure 20-11. Cast irons can be braze welded with copper alloy electrodes. Cast iron electrodes are most often used for oxyacetylene welding of gray and ductile iron. Extensive heat input is required before, during, and after welding to prevent cracking and maintain machinability.

WELDING FILLER METALS FOR CAST IRON		
Description	Form	Process
Cast Iron		
Gray iron	Welding rod	OAW
Gray iron	Bare electrode	BMAW
Alloy gray iron	Welding rod	OAW
Ductile iron	Welding rod	OAW
Steel		
Carbon steel	Covered electrode	SMAW
Carbon steel	Bare electrode	GMAW
Nickel Alloys		
93% Ni	Bare electrode	GMAW
95% Ni	Covered electrode	SMAW
53 Ni-45 Fe	Covered electrode	SMAW
53 Ni-45 Fe	Flux cored electrode	FCAW
55 Ni-40 Cu-4 Fe	Covered electrode	SMAW
65 Ni-30 Cu-4 Fe	Covered electrode	SMAW
Copper Alloys		
Low-fuming brass	Welding rod	OAW
Nickel brass	Welding rod	OAW
Copper-tin	Covered electrode	SMAW
Copper-tin	Bare electrode	GMAW
Copper-aluminum	Covered electrode	SMAW

Figure 20-11. Matching and nonmatching welding filler metals may be used to weld cast irons.

Points To Remember

1 Always remove the casting skin before welding cast-iron pieces.

2 V the edges with a single or double V, depending upon the thickness of the metal.

3 Clean an old casting very carefully before welding.

4 Run short beads of 1″ to 1½″ on thin metal and 2″ to 3″ on heavy metal.

5 Keep the casting as cool as possible.

6 Peen the bead to relieve stresses.

7 Use the correct type of electrode.

8 Use small diameter electrodes and maintain a lower amperage setting than for mild steel.

9 Reinforce heavy castings with studs for added strength.

10 Preheat the piece whenever possible, and cool slowly after the weld is completed.

QUESTIONS FOR STUDY AND DISCUSSION

1 What is the difference between gray, white, and malleable cast iron?

2 Why should the outer skin be removed before welding cast iron?

3 How should the joints be prepared for welding cast iron?

4 How should old castings be cleaned before welding?

5 What can be done to make fine cracks in castings more visible?

6 How can cracks in castings be prevented from spreading?

7 Why is it important to keep cast iron cool when welding? How should this be done?

8 If a casting is to be preheated, how can you determine when correct heat is reached?

9 Why should a bead be peened after welding?

10 What type of electrode is used for welding the various types of cast iron?

11 Why is a lower amperage setting required for cast iron welding?

12 Why is it a good practice never to use electrodes greater than $1/8''$ in diameter for welding cast iron?

13 What is the correct procedure for welding cracks in castings?

14 Why is it advisable to use studs in repairing some broken castings?

15 Why should studs be used only on heavy castings $1\frac{1}{2}''$ or more in thickness?

16 How far apart should studs be fastened before welding?

17 How deep should studs be driven in the castings? How much should they project above the surface?

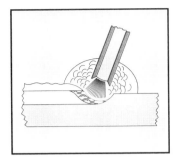

CARBON STEELS

Shielded Metal-Arc Welding – SMAW

Carbon steels can readily be arc welded but some require considerably more control of the welding process. For certain types, special electrodes must be used or some preheating and postweld heat treatment applied to secure sound welds with the required mechanical properties.

Welding Carbon Steels

An important point to remember in welding carbon steels is the effects of heat on the welded area. When steel is heated to a high temperature its structure undergoes a change. Depending on the amount of carbon present, the steel changes from a mixture of *ferrite* (pure iron) and *cementite* (iron and iron-carbide, sometimes called pearlite) to a solid solution known as *austenite,* where the carbon goes into a solution form and becomes evenly distributed. If an austenitic structure is cooled quickly, a *martensitic* condition develops which causes the carbon to precipitate and leave an extremely hard and brittle material. When this happens in a weld area, underbead cracking can be expected as well as cracks alongside the weld in the parent metal. Accordingly a welder must be aware, when welding certain types of carbon steels, of the .effects of welding heat on the structure of the metal and take suitable action to prevent the formation of weld defects.

Carbon steels are those steels whose principal element is carbon. Steels in this group are referred to as low carbon, medium carbon, and high carbon.

Plain carbon steels are made in three grades: *killed, semikilled,* and *rimmed.* A killed steel is one that is deoxidized by adding silicon or aluminum (in the furnace ladle or mold) to cause it to solidify quietly without evolving gases. Killed steel is homogeneous, has a smooth surface and contains no blowholes. A semikilled steel is only partially deoxidized, while rimmed steel receives no deoxidizing treatment.

Welding low-carbon steels. Low-carbon steels are the easiest to weld. No particular control needs to be taken since the welding heat has no appreciable effect on the parent metal. Any mild steel coated electrodes in the E60XX or E70XX series will produce good welds. The choice of electrodes in these series is influenced by specific requirements such as depth of penetration, type of current, position of the weld, joint design, and deposition rate. See Chapter 13.

Welding medium-carbon steels. A medium-carbon steel is one whose carbon content ranges from 0.30 percent to 0.45 percent. Most medium-carbon steels are relatively easy to weld, especially with the availability of the E-70XX type electrodes. The E-7018 and E-7024 electrodes are frequently used because of their higher tensile strength and less tendency to produce underbead cracking particularly when no preheat can be applied. The E-6012 or E-6020 electrodes can also be used if precautions are taken and the cooling rate is sufficiently retarded to prevent excessive hardening of the weld.

Welding high-carbon steels. High-carbon steels are those whose carbon content is 0.45 percent or higher and are readily hardenable. These steels are considered more difficult to weld than other carbon steels. However, with proper care, high-carbon steels can be arc welded successfully. Higher tensile strength electrodes in the E-80XX, E-90XX, or E-100XX class are preferred because they minimize underbead cracking. For welding some high-carbon steels stainless steel electrodes, such as E-310-15 are often recommended. Generally, a preheating and postweld heat treatment are advisable for these steels.

Heat Control Processes

Preheating. Preheating involves heating the base metal to a relatively low temperature before welding is begun. Its main purpose is to lower the cooling rate of the weld, thereby reducing the thermal (heat) conductivity of the metal. The lower thermal conductivity allows a slower withdrawal of heat from the weld zone. This lessens the tendency for martensite to form and consequently, there is less likelihood for hard zones to develop in the surrounding weld area than if a weld joint is made without preheat. Preheating also burns grease, oil, and scale out of the joint and permits faster welding speeds. Preheating can be accomplished by moving an oxyacetylene flame over the surface or placing the part in a heating furnace. Specifically, the advantages of a preheat treatment are:

1. Prevents cold cracks.

2. Reduces hardness in heat-affected zones.
3. Reduces residual stresses.
4. Reduces distortion.

Correct temperature is an important factor in preheating. Preheat temperatures for mild steel should be between 200° and 700°F (94° and 371°C), depending on the carbon content. The greater the carbon content, the higher the preheating temperatures. Preheating temperatures can be measured in various ways. Here are several:

1. Use of surface thermometers or thermocouples.

2. Marking the surface with a carpenter's blue chalk. A mark made with this chalk will turn to a whitish gray when the temperature reaches approximately 625°F (330°C).

3. Rubbing a 50-50 solder on the surface. The solder starts to melt at 360°F (182°C).

4. Rubbing a pine stick on the heated surface. The pine stick chars at about 635°F (335°C).

5. Using tempilstiks which are crayons or liquids. When applied, they melt and change color at a specific temperature.

Process control. For welding most steels, the *control heat process* is often more economical than the use of preheating. The principle of this technique is to induce a large volume of heat into the base metal by welding with high current at low speed or by making multiple-pass welds. The high current and slow welding rate builds up considerable heat in the metal. This naturally slows up the rate of cooling and the subsequent prevention of hard crystalline zones near the weld area.

In multiple-pass welding the deposit of the first layer preheats the base metal. The heat of the next pass tempers the base metal adjacent to the first weld bead. Each successive pass then leaves just enough heat so there is no rapid cooling and thus no appreciable hardening.

Postweld heat treatment. Postweld heat treating is intended primarily as a stress-relief treatment. For welding some of the higher range carbon steels, postweld heat treating is as important as preheating. Although preheating does control the cooling rate, the possibility of stresses becoming locked in the welded area is always a factor. Unless these stresses are removed cracks may develop when the piece cools completely, or the part may become distorted, especially after a machining operation.

Postweld heat treatment temperatures for stress relief should be in the range of 900° to 1250°F (472 to 677°C). Soaking period normally runs about one hour per inch of metal thickness.

Formation of Weld Cracks

Cracks in the weld metal may run in a longitudinal or transverse direction and are often invisible to the naked eye. The defects will usually show up under some type of ultrasonic, magnetic, or radiographic inspection.

Basically weld cracks occur when there is a shrinkage of the weld bead. Since high-carbon steels do not stretch very much because of their hardness, whenever a weld bead is deposited on the hard, rigid surface the generated contraction forces must be absorbed by the more ductile metal in the deposited weld bead.

Weld cracks can be avoided by keeping the weld penetration as low as possible thus minimizing any excessive stressing of the parent metal. Furthermore, if penetration is kept low there is less of a tendency for the weld bead to pick up an excessive amount of carbon from the base metal, thereby leaving the weld bead considerably more ductile and better able to absorb shrinkage stresses.

Low-hydrogen electrodes with iron-powder coatings produce minimum penetration and ductile welds. For welding high-carbon steels in the upper carbon range, special stainless steel electrodes of the E-310-15 type (25 percent chromium and 20 percent nickel) are frequently recommended because of their high ductility.

Crater cracks. These cracks occur across the weld bead crater and result from shrinkage. See Figure 21-1. When a weld bead is deposited, solidification of the molten metal takes place from the sides and moves towards the center. The center of the crater cools rapidly because of the smaller amount of metal while the remaining bead cools more slowly, causing a concentration of stresses which eventually result in cracks. Crater cracks are particularly prevalent in thin concave fillet welds. See Figure 21-2. Most crater cracks can be prevented by proper electrode manipulation. Care should be taken to completely fill up the craters and the weld bead rounded slightly by using a shorter arc.

Root cracks. In making a groove or a fillet weld, the first pass is in the form of a narrow stringer bead along the weld seam. This is followed by one or more layers of weld beads. It is the first layer, or root bead, that is the most susceptible to cracking. See Figure 21-3. The cracking is generally due to the excessive carbon which the bead picks up from the base metal, thereby making the weld metal hard and brittle. As the weld metal cools, it shrinks, and as additional layers of beads are deposited on the root weld, tensile stresses form which develop into cracks. Root cracks can be prevented by preheating the base metal, using a more ductile weld metal, and providing sufficient space between the plates so they can move as the weld cools.

Porosity. Porosity is a common problem in welding high-carbon steel. Molten high-carbon steel readily absorbs gases such as hydrogen and carbon monoxide which are released as the metal begins to cool. If this gas does not reach the surface before the metal begins to solidify it becomes entrapped in the metal, leaving small gas pockets or blowholes. The formation of blowholes is very common in welding steel with a high sulfur content. The presence of sulfur along with phosphorous and silicon when combined with other gases always generates a degree of porosity that seriously affects the strength of the weld.

Skillful welders usually can prevent porosity by proper manipulation of the electrode. The secret of gas-free welds is to keep the surface of the deposited metal pool fluid enough so the gas is rapidly released. Welding with low-hydrogen electrodes as a rule will prevent any extensive formation of porosity.

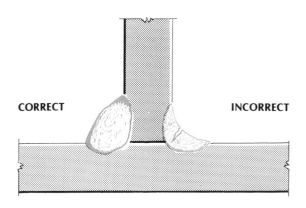

Figure 21-2. Crater cracks often occur in improperly formed concave fillet welds.

Excessive Hardening and Softening of Base Metal

As in the case of medium-carbon steel, the development of a hard layer in the weld zone is the result of too rapid cooling. The rapid cooling transforms the metal into a martensitic condition which leaves a very brittle area. The best procedure to avoid excessive hardening is to utilize a controlled system of preheating and postweld heat treatment. Preheat treatment should be between 200° and 400°F (94 to 204°C) for steels with a carbon content of 0.45 to 0.65 percent and 400° to 700°F (204 to 371°C) when the carbon content is over 0.60 percent. Postweld heat treatment should be in the 1100° to 1200°F (592° to 650°C) range.

The strength of high-carbon steel is dependent upon its hardness which is obtained by a heat-treating process. If the part is to be heat treated after welding, then excessive hardness or softness is not necessarily important. On the other hand, if the part is not to be heat treated then suitable precautions must be taken to prevent a loss of the hardness properties which are characteristic of high-carbon steel

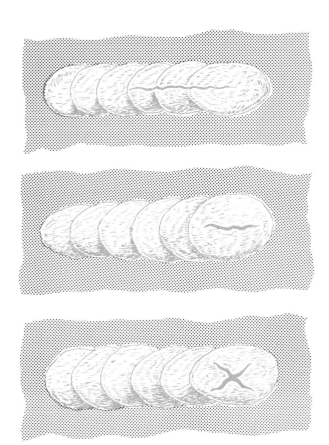

Figure 21-1. Crater cracks occur when the center of the crater cools more rapidly than the rest of the bead, resulting in a concentration of stresses.

Figure 21-3. Root cracks in the root bead can be prevented by preheating and by using the correct root opening.

after heat treatment.

Welding will often have some softening effect on hardened high-carbon steel especially near the weld. To minimize both excessive hardening and softening, the use of a high-chromium, high-nickel stainless steel electrode is recommended. A small diameter electrode and an intermittent welding procedure will also reduce excessive softening and hardening of the base metal.

Points to Remember

1 Use E-60 or E-70 series electrodes to weld low-carbon steel.

2 Use low-hydrogen electrodes when welding medium-carbon steels.

3 Preheat steels having a high-carbon content to prevent cracking.

4 The control heat welding process will often eliminate the need for preheating carbon steels.

5 For many high-carbon steels a postweld heat treatment is often advisable for stress relief.

6 Keep the cooling rate of the weld puddle as slow as possible in welding all kinds of high-carbon steel.

7 Low-hydrogen electrodes with iron powder coatings will usually minimize cracking in welding high-carbon steel.

8 Use correct electrode manipulation to allow gases to come to the surface in order to avoid gas pockets and blowholes.

QUESTIONS FOR STUDY AND DISCUSSION

1 What happens to high-carbon steel when it develops into a martensitic structure?

2 Why are low-carbon steels relatively easy to weld?

3 What type of electrodes are normally used for welding low-carbon steels?

4 When is a steel classified as a medium-carbon steel?

5 What type of electrodes are required for welding medium-carbon steel?

6 Why is a preheating treatment sometimes necessary in welding medium and high-carbon steels?

7 At what temperature should preheating be carried out?

8 What is meant by controlled heat welding?

9 What is the function of postweld heat treatment?

10 At what temperature should postweld heat treatment be done?

11 Why are high-carbon steels more difficult to weld?

12 How can weld cracks be avoided when welding high-carbon steel?

13 What are crater cracks and what causes them?

14 What are root cracks and why do they occur?

15 What causes porosity in a weld?

16 How can porosity in a weld be avoided?

17 Why must the condition of a heat-treated part be considered before welding it?

18 How can excessive softening of a heat-treated part be kept to a minimum in a welding operation?

19 What is the difference between a killed and semikilled steel?

20 What is meant by a rimmed steel?

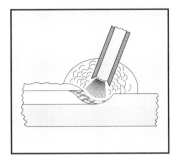

Shielded Metal-Arc Welding – SMAW

An alloy steel is a steel which is mixed with one or more additional elements, other than carbon or iron, in large enough percentages to alter the characteristics and properties of the steel. The addition of such substances as manganese, nickel, chromium, tungsten, molybdenum, or vanadium produces a steel that is greater in strength and toughness. Alloy steels are known by their main alloying elements, such as manganese steel, chromium-nickel steel, molybdenum steel, etc.

Practically all alloy steels can be welded, but as a rule the welding operation is much more difficult to perform than in welding mild steel. This is due to the fact that the characteristics of the basic alloying ingredients are sometimes destroyed as the weld is made. Furthermore, unless precautions are taken, cracks appear near the welded areas, or slag inclusions and gas pockets form in the bead, all of which weaken the weld. Many of the difficulties, however, can be avoided or minimized by using special electrodes designed specifically for welding alloy steels.

Preheating and Postweld Heat Treatment

Successful welding of many alloy steels requires a controlled rate of cooling because when heated to a high temperature and cooled rapidly, they readily harden. Welding these steels without proper heat control produces an embrittlement in the heat-affected zone parallel to the weld joint. With proper preheating and postweld heat treatment, the rate of cooling is delayed and consequently the metal near the weld zone does not harden appreciably.

The application of preheating and postweld heat treatment is also an important factor in preventing weld cracks caused by shrinkage stresses. By slowing the rate of cooling, the stresses are more readily distributed throughout the weld and released while the metal is still hot.

The preheating and postweld heat treatment temperatures will depend on the alloying content of the base metal. Some elements produce greater hardenability and therefore the cooling rate must be slower. In any event, the temperature should never exceed the original hardening and tempering temperatures unless the piece is to be heat treated again after the weld is completed.

When a preheat temperature is difficult to ascertain, the *clip test* can be used to make a rapid check. This test is not applicable for thin steels but will produce good results on heavy sections down to ⅜″ in thickness.

The test involves the welding of a piece of low-carbon steel ½″ in thickness and 2″ or 3″ square to the steel plate which is being checked for a preheat temperature. A convex contour fillet weld is made with an electrode and welding current similar to the ones used for the welding job. The weld is allowed to cool for five minutes, and then the welder, wearing safety glasses, hammers the clip until it breaks off. See Figure 22-1A. If the lug breaks through the weld after a number of blows, the test indicates that no serious underbead cracking will result when the welding is carried out in the same manner at normal room temperature. See Figure 22-1B. If the lug breaks and pulls out some of the parent metal the test shows that this particular steel must be preheated.

Identifying Low-Alloy High-Tensile Steel Electrodes

The greatest usage of alloy steels is in the group commonly referred to as low-alloy high-tensile steel

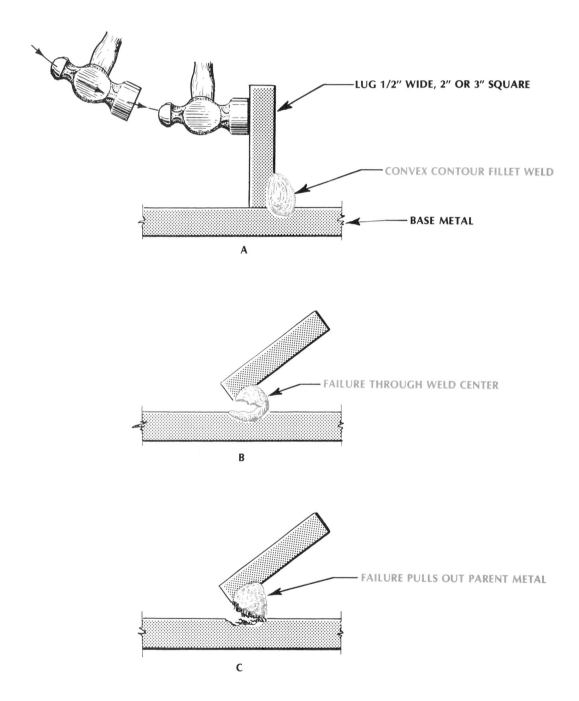

LUG 1/2" WIDE, 2" OR 3" SQUARE

CONVEX CONTOUR FILLET WELD

BASE METAL

A

FAILURE THROUGH WELD CENTER

B

FAILURE PULLS OUT PARENT METAL

C

Figure 22-1. The clip test can be used to determine if preheating is needed. if the weld fails through the center when struck with a hammer, preheating is not necessary. If the weld metal pulls out some of the parent metal, preheating is required to maintain the properties of the metal prior to welding.

TABLE 22-1. LOW-ALLOY HIGH-STRENGTH ELECTRODES.

TYPE	WELDING APPLICATION
E-7018-A1	Carbon and Molybdenum steels
E-8016-B2 E-8018-B2L	Chromium—Molybdenum steels
E-8016-C1 E-8018-C1 E-8018-C2 E-8018-C3	Nickel steels
E-9016-B3 E-9018-B3L	Chromium—Molybdenum steels
E-10016-D2	Manganese—Molybdenum steels

such as chrome molybdenum, nickel-chrome-molybdenum, and manganese-chrome nickel. Low-alloy steels contain low percentages of alloying elements. Their greatest application is where high-impact requirements must be maintained.

These steels are generally welded with special low-alloy high-tensile-strength steel electrodes of the E-70xx, E-80xx, E-90xx, and E-100xx classification. In addition to their regular classification symbols, the low-alloy high-tensile steel electrodes carry a suffix in the form of a letter and a final digit. The letters show the chemical composition of the deposited metal as follows:

A — carbon-molybdenum
B — chromium-molybdenum
C — nickel steel
D — manganese-molybdenum
G — other low-alloy electrodes with minimum elements

The final digit indicates the exact composition of these broad chemical classifications. Thus low-alloy high-tensile steel arc welding electrodes are designated as E-7010-Al, E-8016-B2, etc.

Table 22-1 includes several commonly used low-alloy high-tensile-strength steel electrodes. However, when welding any alloy steel it is always a good policy to consult with area electrode suppliers.

Welding austenitic manganese steel. Austenitic manganese steel is a tough, nonmagnetic alloy noted especially for its high strength, excellent ductility, and outstanding wear resistance. Welding this metal requires considerable attention since it is so sensitive to reheating. Any prolonged period of heating will embrittle the metal and result in loss of tensile strength and ductility. Consequently, low welding current and rapid welding rate without any extensive preheating is necessary.

Welding low-manganese steels. These steels are used for fabricating parts or structures where they must withstand impact stresses and resist wear. In welding these metals care must be taken to prevent an excessive amount of the parent metal from mixing into the weld. Slight preheating is advisable since it reduces underbead cracking.

Welding low-alloy molybdenum steels. Two of the more common molybdenum alloy steels are known as carbon-moly and chrome-moly. Carbon-moly steels have high strength especially at high temperatures. For this reason they are used extensively in piping systems where high pressures at high temperatures are encountered. See Figure 22-2.

Chrome-moly steels are widely used for highly stressed parts. Many aircraft components are fabricated from this metal, such as landing gear supports, tubular frames and engine mounts.

Low-carbon moly steels are readily weldable in much the same way as mild steel. Some preheating is advisable at temperatures of 400° to 650°F (204° to 339°C). When the thickness of the metal is over 3/8″, stress relief is often necessary after the welding is completed. Stress relief is accomplished at temperatures of 1200° to 1250°F (650° to 677°C) for periods of one hour for each inch of thickness followed by slow cooling in the furnace at a rate of 200° to 250°F (94° to 121°C) per hour. When the heat reaches about 150°F (65°C) the rest of the cooling may be done in still air.

Chrome-moly steels having a low carbon content may be welded with E-6012, E-6013, or E-6024 electrodes. These electrodes will pick up enough alloy from the parent metal to produce the necessary tensile strength in the weld. For higher carbon chrome-moly steels special low-hydrogen electrodes of the E-7018 type are used with adequate preheat and postweld heat treatment to prevent extreme brittleness of the metal at the fusion zone.

Welding low-alloy nickel steels. The addition of nickel (3 to 5 percent) to steel greatly increases the elastic properties as well as the strength, toughness, and corrosion resistance.

To weld low-alloy nickel steels where the tensile

Figure 22-2. Low-alloy high-strength steels are used in high-pressure piping systems. (*Hobart Brothers Company*)

strength must be equal to the base plate, low-alloy nickel type electrodes in the E-80xx series are generally used. On heavy sections, preheating to a dull red is generally advisable.

Welding high-manganese steels. This steel is usually found in a cast form because of its tough core and hard abrasion-resisting surface. This metal is used widely for stone crushing equipment parts, power shovel buckets, and other structures which are subjected to great wear. See Figure 22-3.

Welding high-manganese steel involves joining two pieces, repairing cracks, or building up worn surfaces. Building up worn surfaces is a surfacing operation commonly referred to as *hardfacing*. A detailed description of hardfacing is presented in Chapter 28.

To secure the best possible results in welding pieces of high-manganese steel, observe the following:

1. V the joint and clean the surfaces carefully.

2. Use the lowest possible current to prevent the formation of a brittle zone next to the weld.

3. The electrode most frequently recommended for high-manganese steel is a stainless steel 18-8 type. (E-308-15, E-308-16, E-309-15, E-309-16, E-310-15, E-310-16).

As a rule these electrodes are easier to apply and produce the most satisfactory welds. Other types of electrodes used for welding high-manganese steel are molybdenum-copper-manganese and nickel-manganese. However, more skill is needed to execute good welds with these electrodes.

4. Strike the arc ahead of the crater and continue

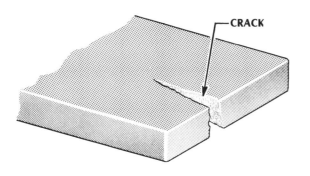

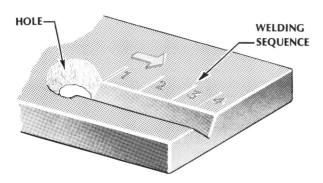

Figure 22-3. To weld a crack in manganese steel, deposit a root bead along section 1 and then weld additional layers over the root bead until the groove is full. Repeat the procedure for other sections in order.

the weld bead in the direction of travel. Keep the weld beads short (about 2″) and allow them to cool between each run to avoid concentrated and prolonged heat in any localized area of the base metal. Do not continue to weld in an area unless the temperature of the metal is below 750°F (397°C). Use tempilstiks to determine temperature by marking the base metal ⅜″ to ½″ from the weld. Actually you should be able to place your hand within 6″ to 8″ of the weld at any time. If you have trouble controlling the heat, place wet rags on areas adjacent to the weld.

5. Use a narrow weave with the electrode held at about a 45° angle in the direction of the weld bead travel.

6. Due to the high thermal expansion of manganese steel, many stresses form as the weld cools, causing cracks to develop during contraction. To reduce cracking, peen each run of weld bead when it is completed.

Repairs. To repair cracks in manganese steel follow this procedure:

1. Cut a hole at the end or ends of the crack to prevent stresses from spreading the crack further into the sound metal.

2. Gouge a U or V-groove in the crack and grind away all oxide from the weld area.

3. Deposit a root bead along section 1 as indicated in Figure 22-4. Weld additional layers over the root bead until the groove is full. Then follow the same procedure for sections 2, 3, and 4 in order.

4. When the entire groove is filled, fill the hole or holes at the end of the original crack.

Welding Stainless Steel

The first stainless steel developed was the chromium-iron type, having as the main constituents chromium and iron. Later nickel was introduced into stainless steel, producing a metal that has been popularly referred to as *18-8.* (Approximately 18 percent chromium and 8 percent nickel.) Further developments have brought into use a whole series of stainless steels designed to meet more rigid fabricating demands.

Figure 22-4. Inert gas welding is the preferred process on stainless steel to minimize heat input and retain the metal's corrosion-resistant properties.

How Stainless Steel is Classified

Stainless steel today is classified into two general AISI (American Iron and Steel Institute) series—200-300, and 400. Each series includes several different kinds of steel, all of course having some special characteristic. See Table 22-2.

The 400 series. The 400 series is further classified into two groups according to their crystalline structure. One group is known as ferritic and is non-hardenable and magnetic. The other group is referred to as martensitic, which is hardenable by heat treatment and is also magnetic.

Ferritic type stainless steels have better resistance to high temperature corrosion than the martensitic group. When subjected to the heat of welding they

will develop some degree of brittleness. This brittleness may be reduced and ductility improved by cold working the weld area by peening and annealing.

Types 405 and 430 Ti (same as regular 430 but with titanium added) have been developed especially for welding.

The martensitic chromium steels will harden when cooled from welding temperatures. Since these steels are usually subjected to further heat-treating processes, the hardening effects present no problem, especially if the annealing or tempering treatment is begun immediately after the welding operation.

The 200-300 series. These steels have an austenitic structure which make them extemely tough and ductile in the as-welded condition. Hence, they

TABLE 22-2. TYPES AND WELDING CHARACTERISTICS OF STAINLESS STEELS.

CHROMIUM—NICKEL TYPES—NON-HARDENABLE

AISI	STRUCTURE	WELDING PROPERTIES
201	Austenitic	Very good, tough welds
202	Austenitic	Very good, tough welds
301	Austenitic	Very good, tough welds
302	Austenitic	Very good, tough welds
303	Austenitic	Fusion welding not recommended
304	Austenitic	Very good, tough welds
305	Austenitic	Very good, tough welds
308	Austenitic	Good, tough welds
309	Austenitic	Good, tough welds
310	Austenitic	Good, tough welds
316	Austenitic	Very good, tough welds
321	Austenitic	Very good, tough welds
347	Austenitic	Good, tough welds

CHROMIUM TYPES—NON-HARDENABLE

405	Ferritic	Good, fairly tough welds
409	Ferritic	Good
430	Ferritic	Fair, non-ductile welds
430Ti	Ferritic	Good
434	Ferritic	Fair, non-ductile welds
436	Ferritic	Fair, non-ductile welds
442	Ferritic	Fair, non-ductile welds
446	Ferritic	Fair, non-ductile welds

CHROMIUM TYPES—HARDENABLE

410	Martensitic	Fair, preheat 400-500°F, after welding anneal at 1250°F
414	Martensitic	Fair, preheat 400-500°F, after welding anneal at 1250°F
416	Martensitic	Poor, preheat 400-500°F, after welding anneal at 1250°F
420	Martensitic	Fair, preheat 400-500°F, after welding anneal at 1250°F
431	Martensitic	Fair, preheat 400-500°F, after welding anneal at 1250°F

are ideal for welding and require no annealing after welding if used in normal atmospheric conditions or when subjected to mildly corrosive actions. If they are to encounter severe corrosive conditions, it is advisable to anneal the welded structure.

Physical Properties of Stainless Steel

The coefficient of expansion of the 400 chromium types of steel is approximately the same as that of carbon steel. Consequently, the allowances for expansion are practically the same as those for carbon steel. The chromium-nickel 200-300 series have about a 50 to 60 percent greater coefficient of expansion than carbon steel and therefore require greater consideration in expansion control.

The heat conductivity of the 400 series is approximately 50 to 65 percent of carbon steel. With the 200-300 series the heat conductivity is almost 40 to 50 percent of carbon steel. Consequently in both series the heat is not conducted away as fast as in ordinary steel and as a result, stainless steels take longer to cool. This phenomenon is particularly important to remember in welding thin gauges since there is greater danger of burning through the material.

Methods of Reducing Effects of Heat

Unfavorable effects of heat can be reduced substantially by means of chill plates. The use of chill plates such as copper pieces will help conduct the heat away. Rigid jigs and fixtures should be employed wherever possible, especially for the 200-300 series. When stainless steels are allowed to cool in a jig, warping and distortion are practically eliminated.

If jigs cannot be used, special welding procedures will be necessary to counteract expansion forces. The common practice is to resort to *skip* or *step-back* methods of welding. See Chapter 3.

Weldability of Stainless Steel

It is generally conceded that stainless steels in the 200-300 series have better welding qualities than those in the 400 series. However, this does not mean that the 400 series stainless steels are not weldable. Greater precautions simply have to be taken, especially in applying proper heat-treating methods after the weld is completed.

All methods of welding may be used in joining stainless steel. Oxyacetylene welding is sometimes used for welding 20 gauge and lighter stainless steel

sheets, and the metallic arc for heavier plates. See Figure 22-5.

Today the inert-gas shielded arc is utilized a great deal more for welding stainless steel of all types because of the ease with which welds can be made and since with this process there is less danger of destroying the corrosion-resistant properties in the steel.

Choosing the Correct Joint Design

For thin metal, the flange-type joint is probably the most satisfactory design. Slightly heavier sheets, up to ⅛" in thickness, may be butted together. For plates heavier than ⅛", the edges should be beveled to provide a V so fusion can be obtained entirely to the bottom of the weld.

Since stainless steels have a much higher coefficient of expansion with lower thermal conductivity than mild steels, there are greater possibilities for distortion and warping. Therefore, whenever possible, clamps and jigs should be used to keep the pieces in line until they have cooled.

Stainless Steel Electrodes

In arc welding, flux-coated electrodes should always be used. The flux shields the molten metal from air, preventing oxidation of the chromium and producing strong, corrosion-resistant welds. Also, the flux has a tendency to act as a stabilizing agent helping to maintain a steady arc with an even metal flow into the shielded area.

The slag formed by flux-coated electrodes will flow to the surface where it should be brushed off before subsequent beads are laid on the weld. To produce a good strong weld, the electrode should ordinarily be as low in carbon as possible. It is also desirable for the electrode coatings to be free from undesirable elements such as carbon.

The alloy content of electrodes normally should be higher than or the same as that of the base metal to compensate for expected alloy loss. Moreover, a columbium-bearing electrode must be used for both the columbium (Type 347) and the titanium (Type 321) stabilized grades. Chromium-nickel electrodes are often used in welding chromium grades due to the fact that they provide a ductile weld metal. See Table 22-3.

Stainless steel electrodes are identified in a different way than mild steel. For example, a standard 18-8 electrode for AC-DC current is designated as E-308-16. The prefix *E* indicates a metallic arc electrode. The next three digits are the AISI (American

Figure 22-5. The gas tungsten-arc welding process is used to weld stainless steel pipe. (*Hobart Brothers Company*)

Iron and Steel Institute) symbols for a particular type of metal. Thus 308 represents a metal containing 18 percent chromium and 8 percent nickel. The last two digits following the dash may be either 15 or 16; the 1 indicates an all-position welding, and the 5 or 6 specifies the type of covering and applicable welding current. The 5 designates a lime-coated electrode for DC reverse polarity. The 6 is an electrode designed for AC and DC reverse polarity and has a titanium type covering.

Electrode Selection[1]

The selection of the proper electrode for stainless steel application is in most cases a more critical choice than with mild steel because of the number of types of and grades of stainless steel and the varying degrees of severity of heat, corrosion media, etc., to which the weldment will be subjected. Selecting the right electrode for most satisfactory results is a matter of analyzing all the conditions applying to the particular job. To determine the right type and size of electrode best suited to a given set of conditions the following factors must be considered:

1. The analysis of the base metal to be welded.

2. Dimensions of the section to be welded.
3. Type of welding current available.
4. Welding position or positions to be used.
5. The fit-up of the section to be welded.
6. Specific properties of the weld deposit.

TABLE 22-3. ELECTRODES FOR WELDING STAINLESS STEEL.

AUSTENITIC	ELECTRODE
201, 202	308-15, 16
301, 302, 304	
305, 308	308-15, 16
309	309-15, 16
310	310-15, 16
316	316-15, 16
321, 347	347-15, 16
FERRITIC	
405, 409	410-15, 16
430, 434	
436, 442	430-15, 16
446	446-15, 16
MARTENSITIC	
410, 414	
416, 420	410-15, 16
431	430-15, 16

---

1. Airco

7. Requirements of a specific code, standard or specification.

8. Selection of the electrode must be exercised carefully because of the high cost of the material to be welded.

9. Not only must the stainless steel weld metal have sufficient tensile strength and ductility but it must also have corrosion resistance equivalent to the parent metal. Hence, an electrode having an analysis comparable to the base metal should be used.

Welding Current

Both direct and alternating current are used in the arc welding of stainless steel. Reversed polarity will produce deeper weld penetration and more consistent fusion when welding stainless steel sheets and light plates with direct current.

Since stainless steel has a lower melting point than mild steel, at least 20 percent less current is recommended than would ordinarily be used for mild steel. Also, the low thermal conductivity of stainless localizes the heat from the arc along the weld, again lowering current requirements.

Welding Procedure

To produce good welds, all butting edges should be squared. Stainless steel sheets 18 gauge and lighter are fitted with no gap. Heavier gauge sheets and plates are set up with gaps.

It is necessary to have the material to be welded free from scale, grease, and dirt in order to prevent weld contamination.

To start arc welding, the arc is struck by touching the metal electrode on the work and quickly withdrawing it a short distance (enough to maintain the proper arc). The tendency for the electrode to stick, or freeze, to the work may be overcome by using a striking motion similar to that used when striking a match. Any coating on the electrode tip must be removed before an arc can be struck.

To maintain the arc, the electrode should be fed continuously into the arc to compensate for metal deposited, and also moved rapidly in a continuous movement in the direction of the welding.

To finish the weld or break the arc, the electrode should be held close to the work, thus shortening the arc, then moved quickly back over the finished bead.

In order to reduce weld oxidation and porosity, the arc should be as short as possible. Too long an arc is inefficient and increases spattering. Vertical and overhead welding require a short arc together with

smaller diameter electrodes than are used in horizontal welding.

After welding, all slag, scale, and discoloration should be completely removed from the weld bead and adjacent base metal. See Figure 22-6. Light weld discoloration may be removed electrolytically. Scale or oxide is best removed by grinding, pickling, or sandblasting. When grinding, refinish with progressively finer grits. The smoother and cleaner the surface of any stainless part, the better the corrosion resistance.

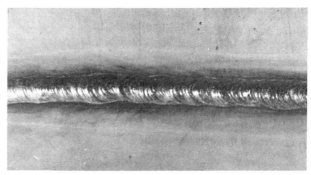

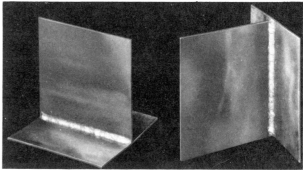

Figure 22-6. The weld made on stainless steel should have a smooth bead with little heat discoloration.

Flat position welding[1]. In welding butt joints in the flat position, the current selected should be high enough to insure ample penetration with good wash-up on the sides. When several beads are required it is advisable to use a number of small beads rather than to try and fill up the groove with one or two passes. A fairly short arc should be maintained, and any weaving should be limited to 2½ times the electrode diameter. In general, it is good practice to hold the electrode vertical or very slightly tilted in the direction of travel. The latter case should only be used with small diameter electrodes. For lack of a hard and fast rule it may be said that the correct position is one that gives a clean pool of metal and which solidifies uniformly as the work progresses.

---

1. Airco

The movement of the electrode across the pool controls the flow of the metal and slag. Any weave technique employed should be in the form of a U for best results.

Horizontal fillet welding[1]. Horizontal fillet welds and lap welds require a machine setting high enough to give good penetration into the root of the joint and a well-shaped bead. Too low a current is easily recognized, since difficulty will be experienced in controlling or concentrating the arc in the joint, and a convex bead with poor fusion will result.

When two legs of equal thickness are being welded, the electrode should be held at equal distance from each face and tilted slightly forward in the direction of travel. If one leg is lighter than the other, the electrode should be pointed towards the heavier leg. Undercutting on the vertical is caused by dwelling too long on that leg or by too high a current.

Vertical welding[1]. The welding of butt joints in the vertical position progressing upward can be accomplished with a reduced current from that used in the flat positions for a given electrode diameter. Oscillation or whipping is not recommended but instead a motion in the form of a V may be used for the first pass. The point of the V is the root of the joint, hesitating momentarily at this point to assure adequate penetration and to bring the slag to the surface. The arc is then brought out on one side of the point about ⅛″ and immediately returned to the root of the joint.

After the momentary pause at the root the procedure is repeated on the other side of the weld. Electrodes of ³⁄₁₆″ in diameter may be used on sections heavy enough to give rapid dissipation of the heat, but ⁵⁄₃₂″ diameter electrodes are the generally accepted maximum size for less massive sections. Welding should progress from the bottom upward except for single-pass corner welds with ³⁄₃₂″ and smaller diameter electrodes, which may be used from the top downward. In the usual vertical fillet weld the electrode is inclined slightly below the horizontal position (holder end lower than arc end) and the weave motion should be rapid across the center of the bead.

Overhead welding[1]. Stringer beads are recommended when welding in the overhead position, since attempts to carry a large puddle of molten metal will result in an irregular convex bead. To assure best results, a short arc should be maintained and the machine should be set properly thereby providing good penetration of the base metal.

---

1. Airco

Welding Clad Steels

Clad steels are those which have a mild steel core and one or both sides covered with a thin layer of some other metal. The covering may consist of zinc, aluminum, chromium, or various forms of nickel-copper alloys or chromium-nickel alloys.

Light-gauge materials are usually welded with the gas tungsten-arc welding process. Shielded metal-arc or gas metal-arc welding is preferred for thicker materials. In making any clad weld there should be a good joint fitting. With a poor fit-up the molten clad material is likely to penetrate the steel base, thereby forming hard, brittle welds with low ductility. Care also must be taken to use correct type electrodes. Remember, the purpose of cladding is to impart a corrosive resistance coating to the base metal. Unless the electrodes correspond as much as possible in composition to the cladding material, the benefit of the cladding is lost.

Groove welds. A ¹⁄₁₆″ root face should be provided above the cladding in a groove weld. See Figure 22-7. The lip acts as a barrier to prevent the cladding material from flowing into the mild steel. The weld should be made on the steel side with the root bead fused to the base of the steel lip but not penetrating into the cladding. Normally a larger bevel than used for regular steel welding and smaller diameter electrodes are necessary to control the depth of penetration of the root bead.

Welding electrodes for the mild steel side should be E-6020 or E-6027 when welding in the flat position. In vertical or overhead positions E-6010 or E-6011 electrodes are recommended. After sufficient passes are made with mild steel electrodes the final pass on the clad side should be with a correct alloy electrode.

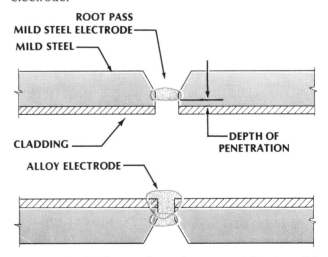

Figure 22-7. Different electrodes are used for the mild steel and the alloy cladding when welding clad steel.

Points to Remember

1 Some preheating is required when welding most alloy steels.

2 Before welding any alloy steel check with a supplier for the recommended electrode.

3 Use the lowest possible current when welding high-manganese steel.

4 Best results are obtained with E-308, E-309, or E-310 electrodes for high-manganese steel welding.

5 Peen each bead to minimize cracking when welding manganese steel.

6 When welding stainless steel, use a short arc with only slight weaving motion.

7 Be sure to use the right kind of electrodes for the type of stainless steel to be welded.

8 In flat position welding of stainless steel hold the electrode vertical or tilted only slightly in the direction of travel.

9 When making vertical welds on stainless steel, avoid any whipping action of the electrode. Instead use a motion in the form of a V.

10 Use E-70XX type electrodes to weld low-carbon moly steels.

11 Keep preheating temperatures between 400° and 650°F (204° to 339°C) for low-carbon moly steels.

QUESTIONS FOR STUDY AND DISCUSSION

1 Why is some form of preheating recommended when arc welding alloy steels?

2 What is the purpose of a clip test and how is it conducted?

3 What are some of the basic characteristics of austenitic manganese steel?

4 Why must the lowest possible current be used when welding manganese steel?

5 What type of electrode produces the most satisfactory results on manganese steel?

6 What should be done to prevent cracks when welding high-manganese steel?

7 How are stainless steels classified?

8 What are the qualities of stainless steel that make this metal so valuable?

9 Why are chill plates frequently used when welding stainless steel?

10 How does the symbol classification of stainless steel electrodes differ from mild steel electrodes?

11 Why is less current required in welding stainless steel?

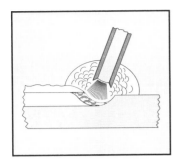

NONFERROUS METALS

Shielded Metal-Arc Welding – SMAW

Nonferrous metals are those which do not contain iron—such as aluminum, copper, brass, bronze, Monel, and Inconel. Continued experimentation and development have now made possible successful welding of these metals. Actually, with a little practice, you can master the art of welding nonferrous metals as easily as you can weld most steels.

Identification of Nonferrous Electrodes

Nonferrous electrodes are identified by a classification number such as E-Al-43. The E stands for electrode, Al for aluminum, and 43 for the alloy composition which in this case is 5 percent silicon.

Copper and copper alloy electrodes use chemical symbols of the principal alloying ingredients. For example, in ECU-Al-A2, the CU represents copper, Al aluminum, and A2 specifies the composition.

Welding Aluminum

Aluminum plays an important role in the manufacture of every conceivable type of equipment ranging from railroad cars, trucks, and buildings, to cooking utensils. See Figure 23-1. In most cases, welding is used in fabricating these aluminum products.

Aluminum weighs only about one-third as much as other commonly used metals; it has a high strength-to-weight ratio, is highly resistant to corrosion, possesses great electrical conductivity, and permits unusual ease in fabrication.

Most types of aluminum used in making commercial products are readily weldable by oxyacetylene, shielded metal-arc, or gas shielded-arc processes. The gas shielded-arc processes, gas tungsten arc and gas metal arc, are used most often for welding aluminum. But caution is necessary when welding on aluminum. Aluminum does not change colors when heated.

The welding technique employed depends on such factors as experience of the welder and kind of work to be done. In some instances it is more convenient to weld with the oxyacetylene flame, while for other jobs the task is simplified if the welding is performed with the shielded metal-arc or gas shielded-arc process.

Classification of Aluminum

Aluminum is classified into three main groups: commercially pure aluminum, wrought alloys, and casting alloys. *Commercially pure aluminum* has a purity of at least 99 percent, with the remaining 1 percent consisting of iron and silicon. Since it lacks alloying ingredients, this aluminum does not possess a very high tensile strength. One of its chief qualities is its ductility, which makes the metal especially adaptable for pressing and forming operations.

Wrought alloys are those which contain one or more alloying elements and possess a much higher tensile strength. The main alloying elements are copper, manganese, magnesium, silicon, chromium, zinc, and nickel. The wrought aluminums are either non-heat-treatable or heat-treatable. The nonheat-treatable types are those which are not hardenable by any forms of heat treatment; their varying degrees of hardness are controlled by cold working. The heat-treatable alloys are those in which hardness and strength are further improved by subjecting them to heat-treating processes.

Casting alloys are used to produce aluminum castings, the metal being poured into a sand or permanent metal mold. A great many of the castings are weldable, but extreme care must be exercised when welding the heat-treated types to prevent any loss of the characteristics achieved by the heat-treating process.

Method of Designating Aluminum and Aluminum Alloys

Wrought aluminum and wrought aluminum alloys are designated by a four digit system. The first digit shows the alloy groups. See Table 23-1. Thus 1xxx indicates aluminum of at least 99.00 percent purity, 2xxx indicates an aluminum alloy in which copper is the major alloying element, and 3xxx an aluminum alloy with manganese as the major alloying element.

Pure aluminum. In the 1xxx groups for alumi-

Figure 23-1. Welding is used in fabricating many aluminum products, from jet airplanes to kitchen equipment. *(Boeing Commercial Airplane Group [top]; Alcoa [bottom])*

num of at least 99.00 percent purity, the last two of the four digits in the designation indicate the degree of aluminum purity expressed in percent.

The second digit in the designation indicates modifications in impurity limits. If the second digit in the designation is zero, it indicates that there is no special control on individual impurities; while integers 1 through 9, which are assigned consecutively as needed, indicate special control of one or more in-

dividual impurities. Thus 1030 indicates 99.30 percent minimum aluminum without special control on individual impurities and 1130, 1230, 1330, etc. indicate the same purity with special control on one or more impurities. Likewise 1075, 1175, etc. indicate 99.75 percent minimum aluminum; and 1097, 1197, etc. indicate 99.97 percent.

Aluminum alloys. In the 2xxx through 7xxx alloy groups the last two of the four digits in the desig-

TABLE 23-1. DESIGNATIONS FOR ALUMINUM.

Aluminum, 99.00 percent minimum and greater 1xxx

	MAJOR ALLOY	AA NUMBER
	Copper	2xxx
Aluminum	Manganese	3xxx
alloys	Silicon	4xxx
grouped	Magnesium	5xxx
by major	Magnesium and Silicon	6xxx
alloying	Zinc	7xxx
elements	Other element	8xxx

nation have no special significance but serve only to identify the different alloys in the group.

The second digit in the alloy designation indicates alloy modifications. If the second digit in the designation is zero, it indicates the original alloy, while integers 1 through 9, which are assigned consecutively, indicate the various alloy modifications.

Temper designations. Temper designation is based on the sequence of basic treatments used to produce the various tempers. The temper designation follows the alloy designation and is separated from it by a dash. Basic temper designations consist of letters. Subdivisions of the basic tempers, where required, are indicated by digits following the letter, and they specify sequences of basic treatments.

Basic temper designations:

F = as fabricated, no effort is made to control the mechanical properties
O = fully annealed—applies to wrought products only
H = strength improved by strain-hardening—applies to wrought products only. Digits which follow the letter indicate the manner and extent to which strain-hardening is achieved
W = Solution heat-treated—applies to alloys which age at room temperature after solution heat-treatment
T = Thermally treated to produce stable tempers. The T is always followed by one or more digits

Subdivisions of H temper. The first digit which follows the H-temper aluminum indicates specific operations, as follows:
H1 = Strain-hardened only

H2 = Strain-hardened and partially annealed
H3 = Strain-hardened and stabilized

The second digit following the H1, H2, and H3 classifications indicates the degree of strain-hardening. These numbers run from 1 to 8, with 8 representing a temper having the highest ultimate tensile strength.

Examples:

1100-0—fully annealed
3003-H18—cold worked, hardest temper
3003-H24—cold worked and partially annealed
5052-H32—cold worked and stabilized

Subdivisions of T temper. The thermally treated aluminums are designated by temper numbers of 1 to 10 and have the following meanings. (The temper designation follows the alloy designation and is separated from it by a dash.)

T1 = Strength is increased by room temperature aging
T2 = Annealed to improve ductility and dimensional stability, cast products only
T3 = Solution heat-treated and then cold worked
T4 = Solution heat-treated and naturally aged to a stable condition
T5 = Artificially aged
T6 = Solution heat-treated and then artificially aged
T7 = Solution heat-treated and then stabilized
T8 = Solution heat-treated, cold-worked, and then artificially aged
T9 = Solution heat-treated, artificially aged, and then cold-worked
T10 = Artificially aged and then cold worked

Examples:

2017-T4
6070-T2
7075-T8

Cast aluminum designation system. Cast aluminum alloys have the same four digit numerical designation as used to identify aluminum and aluminum alloys. The first digit indicates the alloy group as shown in Table 23-2.

The second two digits identify the aluminum alloy or indicate the aluminum purity. The last digit, which is separated from the others by a decimal point, indicates the product form: that is, casting or ingot. A zero is used to designate a casting and 1 or 2 as an ingot. Numbers 1 and 2 simply specify the chemical composition limits of the ingot.

TABLE 23-2. DESIGNATIONS FOR CAST ALUMINUM.

Aluminum, 99, percent of greater purity	1xx.x
Aluminum alloys	
Copper	2xx.x
Silicon, with added Copper and/or Magnesium	3xx.x
Silicon	4xx.x
Magnesium	5xx.x
Zinc	7xx.x
Tin	8xx.x
Other elements	9xx.x

Metallurgical Aspects[1]

In high-purity form aluminum is soft and ductile. Most commercial uses, however, require greater strength than pure aluminum affords. This is achieved in aluminum first by the addition of other elements to produce various alloys which singly or in combination impart strength to the metal. Further strengthening is possible by means which classify the alloys roughly into two categories, nonheat-treatable and heat-treatable.

Nonheat-treatable alloys. The beginning or initial strength of alloys in this group depends upon the hardening effect of elements such as manganese, silicon, iron and magnesium, singly or in various combinations. The nonheat-treatable alloys are usually designated, therefore, in the 1000, 3000, 4000, or 5000 series. Since these alloys are work-hardenable, further strengthening is made possible by various degrees of cold working, denoted by the H series of tempers. Alloys containing appreciable amounts of magnesium when supplied in strain-hardened tempers are usually given a final elevated-temperature treatment called stabilizing to insure stability of properties.

Heat-treatable alloys. The initial strength of alloys in this group is enhanced by the addition of alloying elements such as copper, magnesium, zinc, and silicon. Since these elements singly or in various combinations show increasing solid solubility in aluminum with increasing temperature, it is possible to subject them to thermal treatments which will impart pronounced strengthening.

The first step, called heat treatment or solution

1. Aluminum Company of America, Aluminum Association.

heat treatment, is an elevated-temperature process designed to put the soluble element or elements in solid solution. This is followed by rapid quenching, usually in water, which momentarily freezes the structure and for a short time renders the alloy very workable. It is at this stage that some fabricators retain this more workable structure by storing the alloys at below freezing temperatures until they are ready to form them. At room or elevated temperatures the alloys are not stable after quenching, and precipitation of the constituents from the supersaturated solution begins.

After a period of several days at room temperature, a process termed aging or room-temperature precipitation begins and the alloy becomes considerably stronger. Many alloys approach a stable condition at room temperature, but some alloys, particularly those containing magnesium and silicon or magnesium and zinc, continue to age-harden for long periods of time at ambient or ordinary room temperature.

By heating for a controlled time at slightly elevated temperatures, even further strengthening is possible and properties are stabilized. This process is called artificial aging or precipitation hardening. By the proper combination of solution heat treatment, quenching, cold working and artificial aging, the highest strengths are obtained.

Clad alloys. The heat-treatable alloys in which copper or zinc are major alloying constituents, are less resistant to corrosive attack than the majority of nonheat-treatable alloys. To increase the corrosion resistance of these alloys in sheet and plate form they are often clad with high-purity aluminum, a low magnesium-silicon alloy, or an alloy containing 1 percent zinc. The cladding, usually from 2½ to 5 percent of the total thickness on each side, not only protects the composite due to its own inherently excellent corrosion resistance, but also exerts a galvanic effect which further protects the core material.

Special composite may be obtained such as clad nonheat-treatable alloys for extra corrosion protection, for brazing purposes, or for special surface finishes. Some alloys in wire and tubular form are clad for similar reasons and experimental extrusions have also been clad.

WELDING CHARACTERISTICS OF ALUMINUM

Nonheat-treatable wrought aluminum alloys in the 1000, 3000, 4000 and 5000 series are weldable. The heat of the welding may remove some of the material's strength developed by cold working, but the strength will never be below that of the material in its fully annealed condition. Nevertheless, the welding procedure should confine the concentration of heat in the narrowest zone possible.

The heat-treatable alloys in the 2000, 6000, and 7000 series can be welded except that the 2000 and 7000 are not generally recommended for oxyacetylene welding, nor should the 7000 types be welded with the shielded metal arc. See Table 23-3.

TABLE 23-3. WELDABILITY OF ALUMINUM.

| TYPE | WELDING PROCESS | | |
	GAS	METAL-ARC	GAS SHIELDED-ARC
1060	A	A	A
1100	A	A	A
3003	A	A	A
3004	A	A	A
5005	A	A	A
5050	A	A	A
5052	A	A	A
2014	X	C	C
2017	X	C	C
2024	X	C	C
6061	A	A	A
6063	A	A	A
6070	C	B	A
6071	A	A	A
7070	X	X	A
7072	X	X	A
7075	X	X	C

A—Readily weldable
B—Weldable in most applications—may require special technique
C—Limited weldability
X—Not recommended

To penetrate these alloys, higher welding temperatures and speeds are needed than those possible with an oxyacetylene flame. In most instances resistance welding is preferred for high-strength alloys.

Welding Procedure

The following paragraphs describe the general procedure for shielded metal-arc welding aluminum alloys.

Type of electrode. A heavy shielded electrode containing 95 percent aluminum and 5 percent silicon is recommended for welding most alloys. See Table 23-4 for correct diameter of electrode.

A common aluminum electrode is the E-Al-43 (AWS classification). This electrode has a coating that readily dissolves the aluminum oxide and provides a smooth operating arc. It is an all-position electrode using DC current with reverse polarity. The electrode can be used to weld aluminum sheets, plates or castings. It produces homogeneous, dense welds, free from cracks.

Position of the weld. While welding can be performed in any position, the task is simplified and the quality of the completed joint is more satisfactory if the welding is done in a flat position. See Figure 23-2. Welding aluminum in an overhead position is particularly difficult and should be avoided whenever possible.

Current setting. For the best results, use reversed polarity. See Table 23-4 for the correct setting for various metal thicknesses.

Type of joint. Due to the difficulty of controlling the arc at low currents, it is not very practical to use the shielded metal-arc process to weld material less than 1/16" in thickness. Since it is easy to secure good penetration in aluminum, no edge preparation is necessary on material 1/4" or less in thickness. On heavier metal, it is better to bevel the edges. See Figure 23-3. For some butt welds, beveling can be dispensed with on stock 3/16" to 3/8" in thickness by running a bead on both sides. See Figure 23-4. Copper back-up blocks should be used whenever possible, especially on plates 1/8" or less in thickness to avoid distortion of the weld pieces.

Striking the arc. Scratch the end of the electrode across the surface as in striking a match. During arc starting, the electrode melts very fast, and the metal solidifies rapidly when the arc is extinguished. If you try to just touch the plate, the electrode will freeze.

Manipulating the electrode. Keep the arc as short as possible, with the electrode coating almost touching the molten pool of metal. Hold the electrode with a work angle of 90° at all times. Direct the arc so both edges of the joint are uniformly heated. Move the electrode either forward or backward along the seam and advance it at such a rate as to produce a uniform bead. See Figure 23-5. Avoid weaving the electrode as in welding mild steel. Just keep it moving in a straight line. Before starting a new electrode, remove the slag from the crater at approximately one inch back of the crater. Then quickly move the electrode back about 1/2" over the finished weld and proceed forward as soon as the crater is completely remelted.

When making a lap weld or fillet weld, hold the electrode so that the work angle between the electrode and the plate is approximately 45°. Manipulate

Figure 23-2. Welding brackets to the head of a tank is simplified when performed in a flat position. (*Alcoa*)

the electrode with a small rotary motion, playing the arc first on the vertical member and then on the horizontal plate.

Cleaning the finished weld. After a weld is completed, it is important that you remove the slag over the bead. This slag is solidified flux from the coating on the electrode. If it is not removed, it will attack the aluminum, especially when moisture is present. To remove the flux, chip it off with a chipping tool and then give it a vigorous brushing with a wire brush followed by a warm water rinse.

Welding Aluminum Castings

Most aluminum alloy castings can be welded by the shielded metal-arc process as shown in Figure 23-6. However, considerable precautions must be taken to prevent the formation of cracks during contraction on cooling and the loss of mechanical properties through excessive heat. To weld a casting successfully proceed as follows:

1. Thoroughly remove oil, grease, and dirt from the weld area with a suitable solvent.

2. Chip out a 45° bevel on sections ³⁄₁₆″ thick or heavier.

3. Clamp the pieces to be welded to hold them in correct alignment.

4. Preheat the entire casting to between 500° and 800°F (260° to 427°C) in a furnace or with a gas torch. Be careful not to overheat the casting.

5. Use an electrode having approximately the same composition as the base metal. The E-Al-43 electrodes usually will produce a good weld.

Welding Copper

Copper is a soft, tough, and ductile metal. It cannot

TABLE 23-4. ELECTRODE SIZE, CURRENT, AND NUMBER OF PASSES FOR ARC WELDING ALUMINUM.

METAL THICKNESS (INCHES)	ELECTRODE (INCH) DIA	APPROXIMATE CURRENT (AMPERAGE)	NUMBER OF PASSES (BUTT, LAP, FILLET)
0.081	1/8	60	1
.102	1/8	70	1
.125	1/8	80	1
.156	1/8	100	1
.188	5/32	125	1
.250	3/16	160	1
.375	3/16 for laps and fillets 1/4 for butts	200	2
.500	3/16 for laps and fillets 1/4 for butts	300	3
1.000	5/16	450	3

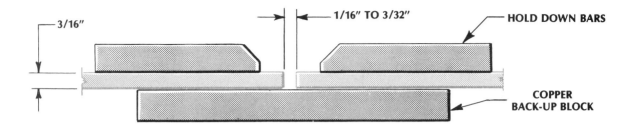

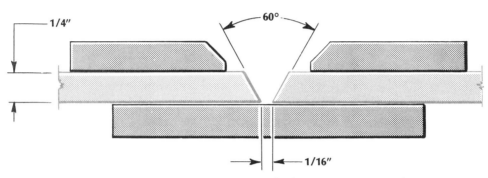

Figure 23-3 When welding aluminum plate, root opening and edge preparation are determined by the thickness of the pieces. Back-up blocks are used to prevent distortion.

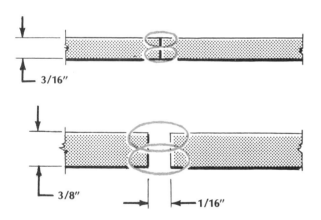

Figure 23-4. In a butt joint on aluminum, beveling is not necessary if a bead is deposited on both sides.

Figure 23-5. When welding aluminum, move the electrode in a straight line at a rate that will produce a uniform bead.

be heat-treated but will harden when cold-worked. Commercially available coppers are divided into two groups: oxygen-bearing copper and oxygen-free copper.

Oxygen-bearing copper. This electrolytic tough pitch copper is practically 99.9 percent pure and is

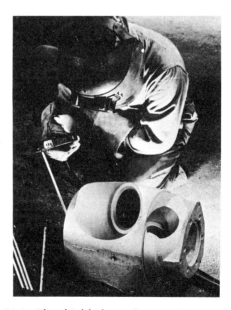

Figure 23-6. The shielded metal-arc welding process can be used to weld most aluminum alloy castings if precautions are taken as in welding cast iron.

considered to be the best conductor of heat and electricity. A small amount of oxygen in the form of copper oxide is uniformly distributed throughout the metal, but it is insufficient to affect the ductility of the copper. However, if heated above 1680°F (965°C) for prolonged periods the copper oxide tends to migrate to the grain boundaries, causing a reduction in strength and ductility. Also when exposed to this temperature the copper will absorb carbon monoxide and hydrogen which react with the copper oxide and release carbon dioxide or water vapor. Since these gases are not soluble in copper they exert pressure between the grains and produce internal cracking and embrittlement.

Oxygen-free copper. This group of copper contains a small percentage of phosphorus or some other deoxidizer, thereby leaving the metal free of oxygen and consequently with no copper oxide. The absence of copper oxide gives the metal superior fatigue resistance qualities and better cold-working properties over the oxygen-bearing copper.

Weldability. Oxygen-bearing copper is not recommended for gas welding because it causes embrittlement. Some welds can be made with the shielded metal-arc process in situations where the tensile strength requirements are extremely low (19,000 psi or less), providing a high welding current and high travel speeds are used. The high current and speed will not allow embrittlement to develop.

Deoxidized copper is the most widely used type for fabrication by welding. A properly made weld will have a tensile strength of about 30,000 psi. This copper can be welded with all standard welding processes including oxyacetylene, shielded metal arc, gas tungsten arc, and gas metal arc. Since copper has a very high coefficient of expansion, contraction precautions must be taken. Provide suitable jigging and clamping to prevent movement while cooking. Contraction forces will often cause cracking during the cooling temperature range.

Special shielded metal-arc electrodes have been developed to weld sheet copper. The most common are phosphor bronze (E-CUSN-A) and aluminum bronze (E-CUAL-A).

For thicker copper sheets, the shielded metal-arc process is actually the fastest process to use. With this method it is possible to concentrate the heat and bring about instant fusion of the parent metal and electrode.

The procedure for welding copper is much the same as for welding other metals using the shielded metal-arc process. The gas tungsten arc process is best used for thinner gauges of copper and copper alloys.

Joint design must include relatively large root openings and groove angles. Tight joints should be avoided to prevent buckling, poor penetration, slag inclusions, undercuttings, and porosity. Copper backing strips are often advisable.

Welding Brass

Brass is a copper-zinc alloy. Since application of heat tends to vaporize the zinc, arc welding this metal is somewhat difficult. When the zinc volatilizes, the zinc fumes and oxides often obscure vision and make welding hard to perform. Furthermore, the formation of oxides produces a dirty surface which ruins the wetting properties of the molten metal.

To arc weld brass, use heavily coated phosphor-bronze electrodes. Best results will be obtained in welding brass if small deposits of metal are made at a time.

WARNING: Be sure there is an abundance of fresh air circulating around the area to remove the harmful zinc oxide fumes.

Welding Bronze

Bronze is a copper-tin alloy. It possesses higher mechanical properties than either brass or copper. Actually it has about the same mechanical properties as mild steel, but with the corrosion resistance of copper. Hence, bronze is often used in fabricating products requiring superior fatigue-resisting qualities.

Since the thermal conductivity of bronze is near that of steel, this metal can easily be welded. When using the shielded metal-arc process, a heavily coated phosphor-bronze electrode is necessary and the current set in reversed polarity. It is very important that the metal be absolutely clean to get sound welds.

Welding Monel and Inconel

Monel metal is an alloy containing approximately 67 percent nickel, 30 percent copper, and small quantities of other ingredients such as iron, aluminum, and manganese.

Inconel is an alloy having about 80 percent nickel, 15 percent chromium, and 5 percent iron.

Both Monel and Inconel are not hardened by regular heat-treating processes. Their high strength is obtained by cold-working such as rolling or drawing. These metals are extensively used as corrosion-resisting linings in tanks and liquid-carrying vessels.

Monel and Inconel can be welded with good results by the shielded metal-arc process. The operation is performed with almost as much ease as in welding mild steel. Although these metals may be welded in any position, better results are obtained if welded in a flat position. In general, shielded metal-arc welding should not be attempted on sheets lighter than 0.050" (18 gauge); the gas tungsten arc welding process is best used on thinner gauges.

The procedure for arc welding Monel or Inconel is as follows:

1. Remove the thin, darkly colored oxide film around the area to be welded. The oxide can be removed by grinding, sandblasting, rubbing with emery cloth, or pickling.

2. No preheating is necessary to arc weld these metals.

3. Use a heavy coated electrode especially designed for welding Monel and Inconel. Reverse polarity will produce the best results.

4. For welds in a flat position, hold the electrode at a travel angle of about 20° from the vertical, ahead of the puddle. In this position it is easier to control the molten flux and to estimate slag trappings. To make welds in other positions, hold the electrode approximately at right angles to the plate.

5. Whenever it is necessary to withdraw the electrode, draw the arc slowly from the crater. Such a procedure permits a blanket of flame to cover the crater, protecting it from oxidation while the metal solidifies.

6. Avoid depositing wide beads. Hold the arc weaving motion to a minimum.

Points to Remember

Welding Aluminum

1 Use a heavily coated electrode containing about 95 percent aluminum and 5 percent silicon such as E-Al-43. Electrodes are identified by E standing for electrode, AL for aluminum, and 43 for the alloy composition.

2 Try to place the weld in a flat position.

3 Use a low current in reverse polarity.

4 Bevel the edges of metal ¼" or more in thickness.

5 Use a scratching motion to start the electrode arc.

6 Keep the arc as short as possible and avoid any weaving motion.

7 Clean the bead after completing the weld.

8 To weld aluminum casting, remove all oil, grease, and dirt from the weld area, plus any oxide coating of aluminum.

9 Chip out a groove in the area to be welded.

10 Preheat casting to between 500° and 800°F (260 to 427°C).

11 Be careful when welding aluminum as there is no color change when the metal is heated.

Welding Copper

1 Use the shielded metal-arc process on thicker sheets.

2 Use phosphor-bronze heavily coated electrodes.

3 Use the gas tungsten arc process on thinner gauges.

Welding Brass

1 Use heavily coated phosphor-bronze electrodes.

2 Make small deposits of beads at a time.

3 Do not weld in a confined area. Have plenty of fresh air circulating to remove harmful fumes. If necessary, a respirator should be used.

Welding Bronze

1 Use heavily coated phosphor-bronze electrodes.

2 Set current in reversed polarity.

3 Clean edges to be welded.

4 Follow the same techniques as in welding mild steel.

Welding Monel and Inconel

1 Place metal in flat position if at all possible.

Points to Remember, cont.

2 Use the gas tungsten – arc welding process on sheets lighter than 18 gauge.

3 Remove the oxide film from the surface to be welded.

4 Do not preheat.

5 Use heavily coated electrodes especially designed for these metals.

6 Use reverse polarity current.

7 Withdraw the arc slowly from the crater to prevent oxidation.

8 Deposit narrow beads.

QUESTIONS FOR STUDY AND DISCUSSION

1 What is meant by a nonferrous metal?

2 What are some of the outstanding properties of aluminum?

3 What is the difference between nonheat-treatable and heat-treatable aluminum?

4 Why is it difficult to arc weld brass?

5 What temper designations are used for nonheat-treatable aluminum?

6 Why are some heat-treatable aluminum alloys difficult to weld?

7 What kind of an electrode should be used to weld nonheat-treatable aluminum?

8 When should the edges of aluminum be beveled for welding?

9 Why should a scratching motion be used in starting an aluminum electrode?

10 Why should the slag be removed from a weld made on aluminum?

11 What type of copper is the easiest to weld?

12 Why is the shielded metal-arc process better for welding thicker copper sheet?

13 In the four digit system used to identify aluminum, what does each digit represent?

14 Brass is an alloy consisting of what elements?

15 What are the principle alloying ingredients in bronze?

16 What is the advantage of bronze over brass or copper?

17 What are the chief ingredients in Monel and Inconel?

18 What must be done to the surface of Monel and Inconel before welding?

19 When welding Monel or Inconel what type of current should be used?

20 In what position should the electrode be held when welding Monel and Inconel?

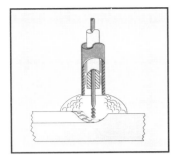

GAS TUNGSTEN-ARC WELDING—GTAW

Gas Shielded-Arc Welding

The primary consideration in any welding operation is to produce a weld that has the same properties as the base metal. Such a weld can only be made if the molten puddle is completely protected from the atmosphere during the welding process. Otherwise atmospheric oxygen and nitrogen will be absorbed in the molten puddle, and the weld will be porous and weak. In gas shielded-arc welding, a gas is used as a covering shield around the arc to prevent the atmosphere from contaminating the weld.

Originally gas shielded-arc welding was developed to weld corrosion-resistant and other difficult-to-weld metals. Today the various gas shielded-arc processes are being applied to all types of metals. Gas shielded-arc welding will eventually displace much of the shielded metal-arc and oxyacetylene production welding due to the superiority of the weld, greater ease of operation, and increased welding speed. In addition to manual welding, the process can be automated, and in either case can be used for both light and heavy gauge ferrous and nonferrous metals. See Figure 24-1.

Figure 24-1. The gas tungsten-arc welding process is used extensively in industrial applications. (*Miller Electric Manufacturing Company*)

SPECIFIC ADVANTAGES OF GAS-SHIELDED ARC

Since the shielding gas excludes the atmosphere from the molten puddle, welded joints are stronger, more ductile, and more corrosion resistant than welds made by most other welding processes. The gas-shielded arc particularly simplifies the welding of nonferrous metals since no flux is required. Whenever a flux is needed, there is the problem of removing traces of the flux after welding. Furthermore, with the use of flux there is always the possibility that slag inclusions and gas pockets will develop.

Another advantage of the gas-shielded arc is that a neater and sounder weld can be made because there is very little smoke, fumes, or sparks to contend with. Since the shielding gas around the arc is transparent, the welder can clearly observe the weld as it is being made. More important, the completed weld is clean and free of complications often encountered in shielded metal-arc welding.

Welding can be done in all positions with a minimum of weld spatter. Inasmuch as the weld surface is smooth, there is a substantial saving in production cost because little or no metal finishing is required. Also, there is less distortion of the metal near the weld.

Types of Gas Shielded-Arc Processes

There are two general types of gas shielded-arc welding: gas tungsten-arc welding — GTAW (sometimes referred to as TIG for tungsten inert gas) and gas metal-arc welding — GMAW (sometimes referred to as MIG for metal inert gas). Each has certain distinct advantages; however both produce welds that are deep penetrating and relatively free from atmospheric contamination. Most industrial metals can be welded easily with either the gas tungsten-arc or gas metal-arc process. These include such metals as aluminum, magnesium, low-alloy steel, carbon steel, stainless steel, copper, nickel, Monel, Inconel, titanium, and others.

Both welding processes can be applied by any of the four basic processes: manual, semiautomatic, machine, and automatic. In the manual process the operation is done by hand. In the semiautomatic process the operator controls the speed of travel and direction. With the machine process the weld size, weld length, rate of travel, and starting and stopping are controlled by equipment under observation and control of the operator. The automatic process performs the welding without constant observation and adjustment of the controls by an operator.

Gas Tungsten-Arc Welding — GTAW

In the gas tungsten-arc process, a virtually non-consumable tungsten electrode is used to provide the arc for welding. During the welding cycle a shield of inert gas expels the air from the welding area and prevents oxidation of the electrode, weld puddle, and surrounding heat-affected zone. See Figure 24-2.

In gas tungsten-arc welding, the electrode is used only to create the arc. It is not consumed in the weld. In this way it differs from the shielded metal-arc process, where the stick electrode is consumed in the weld. For joints where additional weld metal is needed, a filler rod is fed into the puddle in a manner similar to welding with the oxyacetylene flame process. See Figure 24-3.

WELDING MACHINES

Any standard DC or AC arc welding machine can be used to supply the current for gas tungsten-arc welding. However, it is important that the generator or transformer have good current control in the low range. This is necessary in order to maintain a stable arc. It is especially necessary when welding thin gauge materials. If an old DC machine which has poor low-range current control is to be used, then it is advisable to install a resistor in the ground line between the generator and the work bench. Such a resistor will enable the electrical system to provide a very low, stable arc current.

Specially designed machines with all of the necessary controls are available for gas tungsten-arc welding. Many power supply units are made to produce both AC and DC current. See Figures 24-4 and 24-5.

The choice of an AC or DC machine depends on what weld characteristics may be required. Some metals are joined more easily with AC current, while with others better results are obtained when DC current is used. See Table 24-1. To understand the effects of the two different currents an explanation of their behavior in a welding process is necessary.

Direct Current Reverse Polarity (DCRP)

With direct current the welding circuit may be either straight or reverse polarity. When the machine is set for straight polarity the flow of electrons from the electrode to the plate creates considerable heat in the plate. In reverse polarity, the flow of electrons is from the plate to the electrode, thus causing a greater concentration of heat at the electrode. See

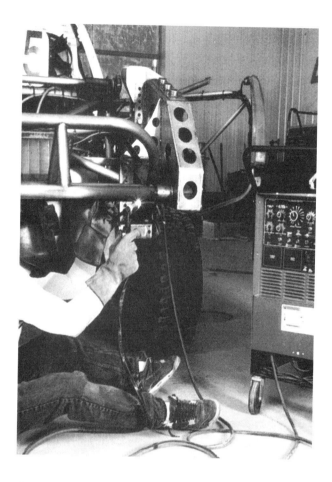

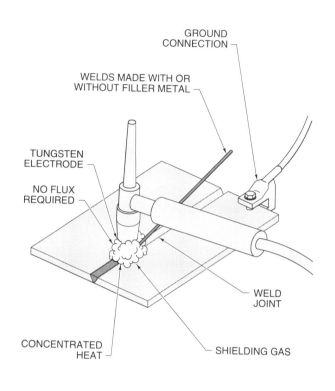

GROUND
CONNECTION

WELDS MADE WITH OR
WITHOUT FILLER METAL

TUNGSTEN
ELECTRODE

NO FLUX
REQUIRED

CONCENTRATED
HEAT

WELD
JOINT

SHIELDING GAS

Figure 24-2. In gas tungsten-arc welding, a nonconsumable tungsten electrode is surrounded by a shield of inert gas. *(Miller Electric Manufacturing Company)*

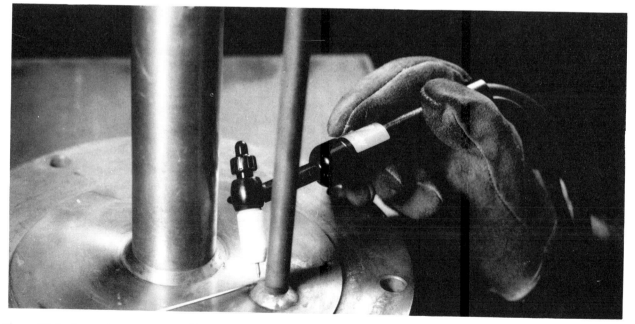

Figure 24-3. In gas tungsten-arc welding, a filler rod is fed into the weld puddle when additional weld metal is needed. *(Airco)*

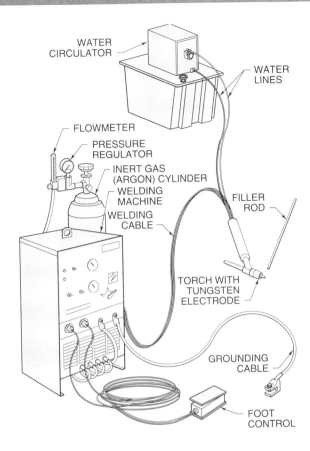

Figure 24-4. Machine and equipment typically necessary for gas tungsten-arc welding include the welding machine, torch, gas cylinder, gas regulator, and foot control.

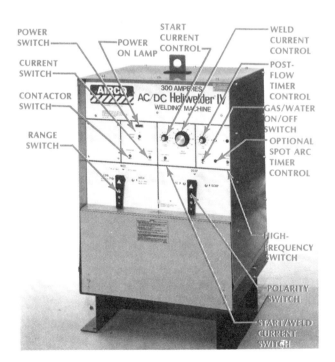

Figure 24-5. The power supply unit specially designed for gas tungsten-arc welding has all the necessary controls to produce both AC and DC current. (*Airco*)

TABLE 24-1. CURRENT SELECTION FOR GAS TUNGSTEN-ARC WELDING.

METAL	AC CURRENT with High Frequency Stabilization	DC CURRENT Straight Polarity	Reverse Polarity
Magnesium up to 1/8" thick	1	NR	2
Magnesium above 3/16" thick	1	NR	NR
Magnesium castings	1	NR	2
Aluminum	1	NR	2
Aluminum castings	1	NR	NR
Stainless steel up to 0.050"	1	2	NR
Stainless steel 0.050"	2	1	NR
Brass alloys	2	1	NR
Silver	2	1	NR
Hastelloy alloys	2	1	NR
Silver cladding	1	NR	NR
Hard-facing	1	2	NR
Cast iron	2	1	NR
Low carbon steel 0.015" to 0.030"	2	1	NR
Low carbon steel 0.030" to 0.125"	NR	1	NR
High carbon steel 0.015" to 0.030"	2	1	NR
High carbon steel 0.030" and up	2	1	NR
Deoxidized copper up to 0.090"	NR	1	NR

Key:

1. Excellent operation — best recommendation.
2. Good operation — second recommendation.
NR = Not recommended.

Figure 24-6. The intense heat at the electrode tends to melt off the end of the electrode and may contaminate the weld. Hence, for any given current, DCRP requires a larger diameter electrode than DCSP. For example, a 1/16" diameter tungsten electrode normally can handle about 125 amperes in a straight-polarity circuit. However, if reverse polarity is used with this amount of current the tip of the electrode will melt off. Consequently, a 1/4" diameter electrode will be required to handle 125 amperes of welding current.

Polarity also affects the shape of the weld. DCSP produces a narrow deep weld whereas DCRP with its larger diameter electrode and lower current forms

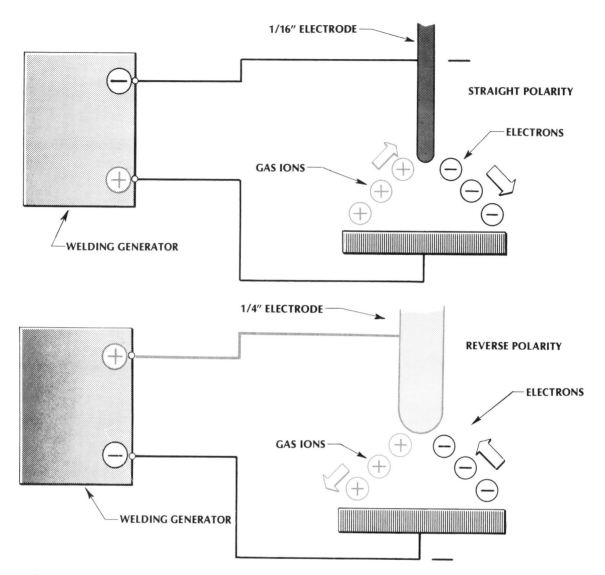

Figure 24-6. In direct current straight polarity, electron flow from the electrode to the plate creates heat at the plate. In direct current reverse polarity, electron flow from the plate to the electrode creates heat at the electrode.

a wide and shallow weld. See Figure 24-7. For this reason *DCRP is never used in gas tungsten-arc welding except occasionally for welding aluminum and magnesium.* These metals have a heavy oxide coating which is more readily removed by the greater current cleaning action of DCRP.

The same cleaning action is present in the reverse-polarity half of the AC welding cycle. No other metals require the kind of cleaning action that is normally needed on aluminum and magnesium. The cleaning action develops because of a bombardment of positive-charged gas ions that are attracted to the negative-charged workpiece. These gas ions when striking the metal have sufficient power to break the oxide and dislodge it from the surface. Generally speaking, better results are obtained in welding aluminum and

magnesium with alternating current.

Direct Current Straight Polarity (DCSP)

Direct current straight polarity is used for welding most metals because better welds are achieved. With the heat concentrated at the plate the welding process is more rapid, there is less distortion of the base metal, and the weld puddle is deeper and narrower than with DCRP. Since more heat is directed at the puddle, smaller diameter electrodes can be used.

Alternating Current High Frequency (ACHF)

AC welding is actually a combination of DCSP and DCRP. See Figure 24-8A. Each half of the complete

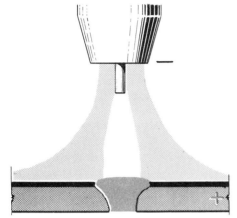

WELDING WITH DCSP PRODUCES DEEP PENETRATION
BECAUSE IT CONCENTRATES HEAT AT THE JOINT

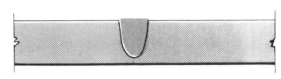

DC STRAIGHT POLARITY
DEEP PENETRATION—NARROW WELD

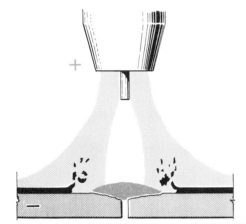

WELDING WITH DCRP PRODUCES GOOD CLEANING
ACTION BUT WELD PENETRATION IS SHALLOW

DC REVERSE POLARITY
SHALLOW PENETRATION—WIDE WELD

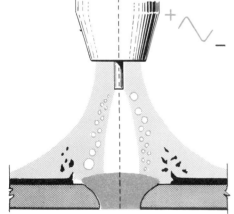

WELDING WITH ACHF COMBINES THE DESIRED CLEANING
ACTION, ON THE POSITIVE HALF OF EACH CYCLE, WITH
THE HEATING REQUIRED FOR GOOD WELD PENETRATION,
ON THE NEGATIVE SWING

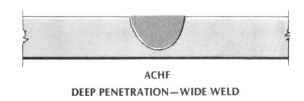

ACHF
DEEP PENETRATION—WIDE WELD

Figure 24-7. The different types of operating current directly affect weld penetration, contour, and metal transfer. (*Linde Company*)

AC cycle is DCSP and the other half is DCRP.

Unfortunately, oxides, scale, and moisture on the workpiece often tend to prevent the full flow of current in the reverse-polarity direction. If no current whatsoever flowed in the reverse-polarity direction during a welding operation, the partial or complete stoppage of current flow (rectification) would cause the arc to be unstable and sometimes go out. See Figure 24-8B. To prevent such rectification, AC welding machines incorporate a high-frequency current flow unit. The high-frequency current is able to jump the gap between the electrode and the workpiece, piercing the oxide film and forming a path for the welding current to follow.

WELDING EQUIPMENT

Torches

Manually operated welding torches are constructed to conduct both the welding current and the inert gas to the weld zone. These torches are either air or water-cooled. See Figures 24-9 and 24-10. Air-cooled torches are designed for welding light gauge materials where low current values are used. Water-cooled torches are recommended when the welding requires amperages over 200 amps. A circulating stream of water flows around the torch to keep it from overheating.

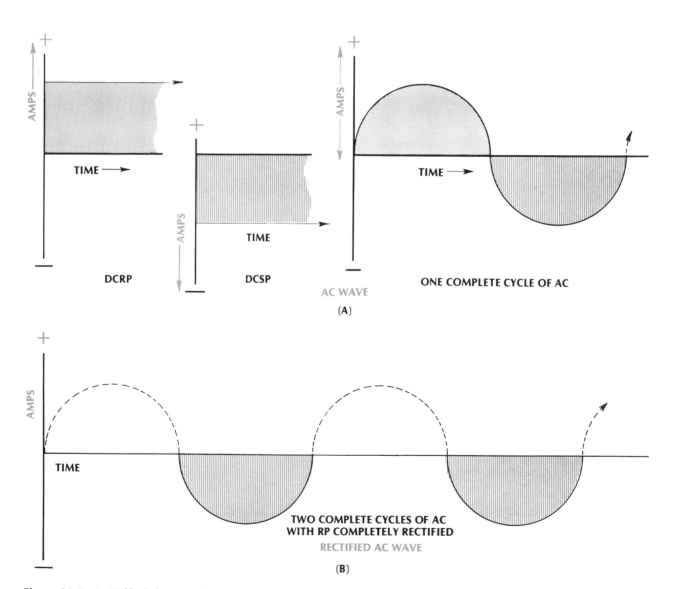

Figure 24-8. A: Half of the complete AC cycle is DCSP and the other half is DCRP. B: If no current flowed in the reverse-polarity direction during welding, rectification would cause the arc to be unstable and possibly to go out.

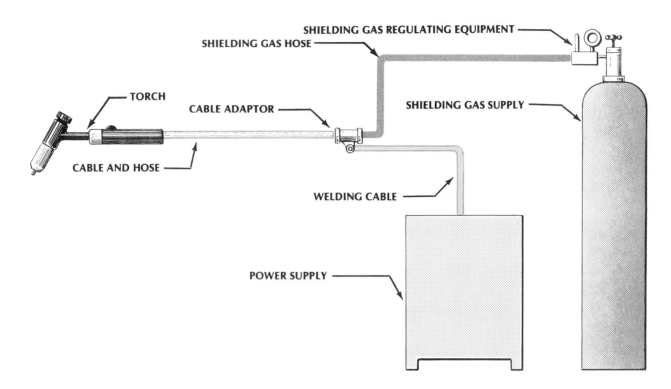

Figure 24-9. A gas tungsten-arc welding unit with an air-cooled torch is designed for welding light gauge materials.

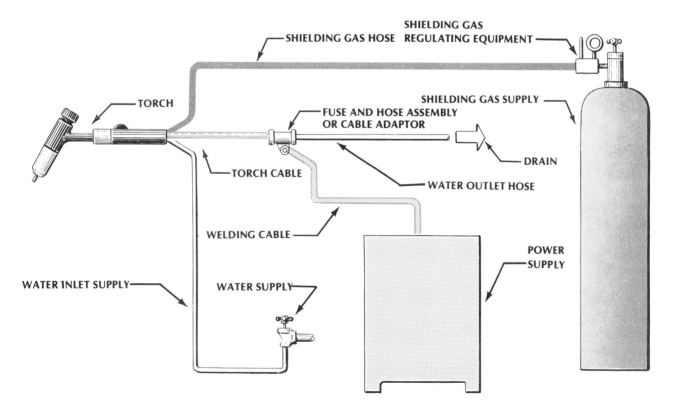

Figure 24-10. A gas tungsten-arc welding unit with a water-cooled torch is recommended when welding requires over 200 amps.

The tungsten electrode which supplies the welding current is held rigidly in the torch by means of a collet that screws into the body of the torch. A variety of collet sizes are available so different diameter electrodes can be used. Gas is fed to the weld zone through a nozzle which consists of a ceramic cup. Gas cups are threaded into the torch head to provide directional and distributional control of the shielding gas. The cups are interchangeable to accommodate a variety of gas flow rates.

Some torches are equipped with a gas lens to eliminate turbulence of the gas stream which tends to pull in air and cause weld contamination. Gas lenses have a permeable barrier of concentric fine-mesh stainless steel screens that fit into the nozzle. See Figure 24-11.

WITHOUT LENS

WITH LENS

Figure 24-11. A gas lens in the nozzle eliminates turbulence of the shielding gas stream.

Pressing a control switch on the torch starts the flow of both current and gas. A timer that maintains gas flow after the weld current is stopped (post purge) is used to protect the weld from atmospheric contamination. On some equipment the flow of current and gas is energized by a foot control. The advantage of the foot control is that the variable current flow can be utilized as the end of the weld is reached. By gradually decreasing the current it is less likely for a cavity to remain in the end of the weld puddle and less danger of cutting short the shielding gas.

Gas cups vary in size. The size to be used depends upon the type and size of torch and the diameter of the electrode. See Table 24-2.

TABLE 24-2. APPROXIMATE CUP ORIFICE FOR GAS TUNGSTEN-ARC WELDING.

TUNGSTEN ELECTRODE diameter (inches)	CUP ORIFICE diameter (inches)
1/16	1/4 -3/8
3/32	3/8 -7/16
1/8	7/16-1/2
3/16	1/2 -3/4

Electrodes

Basic diameters of nonconsumable electrodes are .040″, ¹⁄₁₆″, ³⁄₃₂″, and ¹⁄₈″. They are either pure tungsten or alloyed tungsten. The alloyed tungsten electrodes usually have one to two percent thorium or zirconium. The addition of thorium increases the current capacity and electron emission, keeps the tip cooler at a given level of current, minimizes movement of the arc around the electrode tip, permits easier arc starting, and the electrode is not as easily contaminated by accidental contact with the workpiece. The two percent thoria electrodes normally maintain their formed point for a greater period than the one percent type. The higher thoria electrodes are used primarily for critical sheet metal weldments in aircraft and missile industries. They have little advantage over the lower thoria electrode for most steel welds. The introduction of the striped electrode combines the advantage of the pure, low, and high thoriated tungsten electrodes. This electrode has a solid stripe of two percent thoria inserted in a wedge the full length of the electrode.

The diameter of the electrode selected for a welding operation is governed by the welding current to be used. Larger diameter tungsten electrodes are required with reversed polarity than with straight polarity. (See Tables 24-4 through 24-9 for recommended sizes of electrodes, current, and material thickness for gas tungsten-arc welding.)

Electrode shapes. To produce good welds the tungsten electrode must be shaped correctly. The general practice is to use a pointed electrode with DC welding, and a spherical end with AC welding. See Figure 24-12.

It is also important that the electrode be straight; otherwise the gas flow will be off-center from the arc.

Shielding Gas

Shielding gas for gas tungsten-arc welding can be argon, helium, or a mixture of argon and helium.

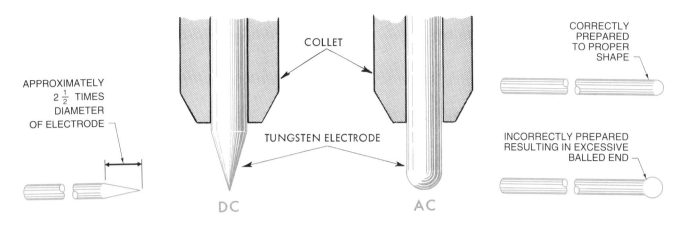

Figure 24-12. Welding with DC current requires an electrode with a pointed tip. Welding with AC current requires a spherical electrode tip.

Argon is the most popular shielding gas used in the gas tungsten-arc process. Helium is rarely used because of its higher cost as compared to argon. In addition, since argon is heavier than air, it provides a better blanket over the weld. A mixture of argon and helium is sometimes used in welding metals that require a higher heat input. See Table 24-3 for the recommended selection of gases.

Argon is supplied in steel cylinders containing approximately 330 cubic feet at a pressure of 2000 psi. Either a single or two-stage regulator may be used to control the gas flow or a specially designed regulator containing a flowmeter is used. See Figure 24-13. The advantage of the flowmeter is that it provides better gas flow control. The flowmeter is calibrated either to show the flow of gas in cubic feet per hour (cfh) or liters per minute (lpm).

The correct flow of argon to the torch is controlled by turning the adjusting screw on the flowmeter. The rate of flow required depends on the thickness of the metal to be welded.

Filler Rod

When gas tungsten-arc welding heavy gauge metals a filler rod is required. Normally filler metal is not necessary on light gauge materials since they can be made to readily flow together. Occasionally filler metal is added on thin pieces when it is essential to reinforce the joint.

Filler metal must be of the same composition as the base metal. Thus mild steel rods are used to weld low-carbon steel, aluminum rods for welding alu-

minum, copper rods for joining copper, and so on. Sometimes strips of the parent metal will serve as satisfactory filler metal.

Special filler rods are available for gas tungsten-arc welding. These rods are similar in classification to the filler wires used for gas metal-arc welding. See Table 25-2 in Chapter 25. The copper-coated mild steel rods used for oxyacetylene welding are not recommended for gas tungsten-arc welding since they tend to contaminate the tungsten electrode. The special gas tungsten-arc filler rods contain a greater amount of deoxidizers, thereby producing less spattering in the weld and sounder weld joints.

In general, the diameter of the filler rod should be about the same as the thickness of the metal to be welded.

Protective Equipment

A helmet like the one used in shielded metal-arc welding is required to protect the welder from arc radiation. The shade of the lens to be used depends upon the intensity of the arc.

Besides the welding helmet, protective clothing such as an apron and gloves must be worn whenever welding with gas tungsten arc.

Joint Preparation

Regardless of the type of joint used, proper cleaning of the metal is essential. All oxidation, scale, oil, grease, dirt, and other foreign matter must be removed by physical or chemical means.

TABLE 24-3. SELECTION OF GASES.

METAL	TYPE	GAS	RESULT
Al	Manual welding	Argon	Better arc starting, cleaning action and weld quality; lower gas consumption
	Machine welding	Helium	High welding speeds possible
		Argon-Helium	Better weld quality, lower gas flow than required with straight helium
Mg	0-1/16″	Helium	Controlled penetration
	0-1/16″ +	Argon	Excellent cleaning, ease of puddle manipulation, low gas flows
Mild Steel	0-1/8″	Argon	Ease of manipulation, freedom from overheating
	0-1/8″ +		(Mig process preferred)
	Spot welding	Argon	Generally preferred for longer electrode life
			Better weld nugget contour
			Ease of starting, lower gas flow
		Argon-Helium	Helium addition improves penetration on heavy gauge metal
	Manual welding	Argon	Better puddle control, especially for position welding
SS	Machine welding	Argon	Permits controlled penetration on thin gauge material (up to 14 gauge)
		Argon-Helium	Higher heat input, higher welding speeds possible on heavier gauges
		Argon-Hydrogen (65%-35%)	Prevents undercutting, produces desirable weld contour at low current levels, requires lower gas flows
		Helium	Provides highest heat input and deepest penetration
Cu & Ni Cu-Ni Alloys (Monel & Inconel)		Argon	Ease of obtaining puddle control, penetration, and bead contour on thin gauge metal
		Argon-Helium	Higher heat input to offset high heat conductivity of heavier gauges
		Helium	Highest heat input for high welding speed on heavy metal sections
Ti		Argon	Low gas flow rate minimize turbulence and air contamination of weld; improved metal transfer; improved heat affected zone
		Helium	Better penetration for manual welding of thick sections (inert gas backing required to shield back of weld against contamination)
Si Bronze		Argon	Reduces cracking of this 'hot short' metal
Al Bronze		Argon	Less penetration of base metal

The following are the most common joints designed for gas tungsten-arc welding:

Butt joint. For light materials the *square butt joint*, as shown in Figure 24-14, is the easiest to prepare and can be welded with or without filler rod.

If the weld is to be made without filler rod, extreme care must be taken to avoid burning through the metal.

The *single-V butt joint* is preferable on material ranging in thickness from ⅜″ to ½″ in order to secure

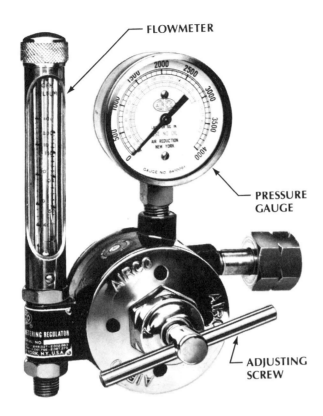

Figure 24-13. A flow meter controls the flow of shielding gas in cubic feet per hour to the torch. (*Airco*)

Figure 24-14. The square butt joint is the easiest to prepare and can be welded with or without filler rod.

complete penetration. The included angle of the V should be approximately 60° with a root face of about ⅛″ to ¼″. See Figure 24-15.

Double-V butt joint is needed when the metal ex-

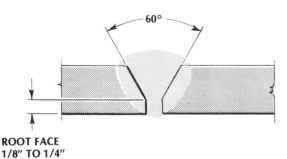

ROOT FACE
1/8″ TO 1/4″

Figure 24-15. The single-V butt joint has an included angle of approximately 60° and a root face of 1/8″ to 1/4″.

ceeds ½″ in thickness and the design is such that the weld can be made on both sides. With a double-V there is greater assurance that penetration will be complete. See Figure 24-16.

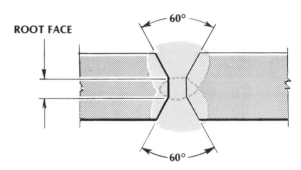

Figure 24-16. A double-V butt joint can be welded on both sides, assuring complete penetration.

Lap joint. The only special requirement for making a good weld is to have the pieces in close contact along the entire length of the joint. See Figure 24-17. On metal ¼″ or less in thickness, the weld can be made with or without filler rod. As a rule, the lap joint is not recommended for material exceeding ¼″ in thickness.

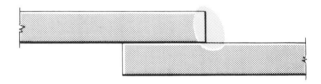

Figure 24-17. A lap joint must have the pieces in close contact along the entire length of the joint.

Corner joint. On light material up to ⅛″ in thickness, no filler rod is required for a corner joint. See Figure 24-18A and 24-18B. With heavier metal the use of a filler rod is advisable. If the metal exceeds ¼″, one edge of the joint should be beveled. See Figure 24-18C. The number of passes will depend on the size of the V and the thickness of the metal.

T-joint. Filler rod is necessary to weld T-joints regardless of the thickness of the metal. As a rule, a weld should be made on both sides of the joint. See Figure 24-19. The number of passes over the seam will depend on the thickness of the material and the size of the weld to be made.

Edge joint. The edge joint is suitable only on

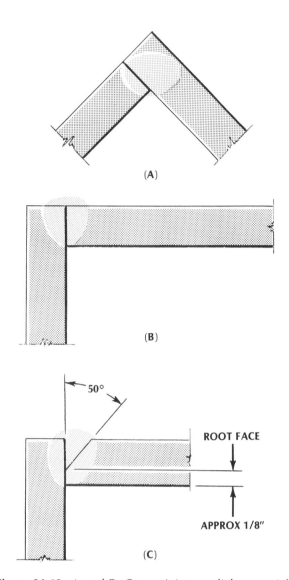

Figure 24-18. A and B: Corner joints on light materials usually do not require beveling. C: One edge of the joint should be beveled if the metal exceeds 1/4".

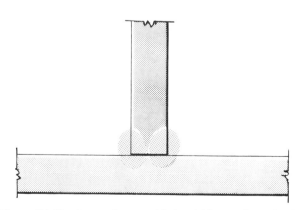

Figure 24-19. As a rule, a weld should be made on both sides of the T-joint. The number of passes over the joint depends on the thickness of the metal.

very light material. No filler rod is needed to make this weld. See Figure 24-20.

Backing the Weld

For many welding jobs, some suitable backing is necessary. On light gauge metals, backing is used to protect the underside of the weld from atmospheric contamination and burning through. On heavier stock, back-up bars draw some of the heat generated by the intense arc.

The type of material used for back-up bars depends on the metal to be welded. Copper bars are suitable for stainless steel. When welding aluminum or magnesium, steel or stainless steel back-up bars are needed.

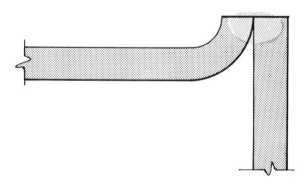

Figure 24-20. The edge joint, usually welded with no filler rod, is used on very light materials.

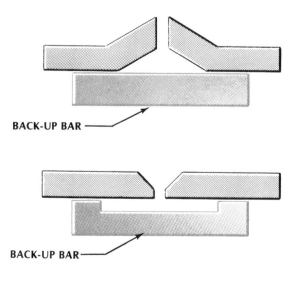

Figure 24-21. A back-up bar should be designed so that it does not touch the weld zone.

The back-up bar should be designed so it does not actually touch the weld zone. See Figure 24-21.

Welding Travel Direction

Either a backhand or forehand travel direction may be used in gas tungsten-arc welding. In backhand welding the direction of the travel is from left to right. In forehand welding travel is from right to left. See Figure 25-19.

WELDING PROCEDURE

Preliminary Steps

Before starting to weld, follow these steps:

1. Check all electrical circuit connections to make sure they are tight.

2. Check for the proper diameter electrode and cup size. (Follow manufacturer's recommendations.)

3. Adjust the electrode stickout so it extends about ⅛″ to ³⁄₁₆″ beyond the end of the gas cup for butt welding and approximately ¼″ to ⅜″ for fillet welding. See Figure 24-22.

4. Check the electrode to be certain that it is firmly held in the collet. If the electrode moves in the nozzle, tighten the collet holder or gas cup. Be careful not to overtighten the gas cup because this will strip the threads very easily.

5. Set the machine for the correct welding amperage. (See Tables 24-4 to 24-9.)

6. If a water-cooled torch is to be used, turn on the water.

7. Turn on the inert gas and set to the correct flow. Set the afterflow (post purge) timer.

Starting the arc. If you are using an AC machine, the electrode should not touch the metal to start the arc. To strike the arc, first turn on the welding current and hold the torch in a horizontal position about 2″ above the work. See Figure 24-23. Angle the end of the torch toward the workpiece so the end of the electrode is ⅛″ above the plate. See Figure 24-24.

Figure 24-23. To start the AC arc, swing the electrode from 2″ to 1/8″ away from the work.

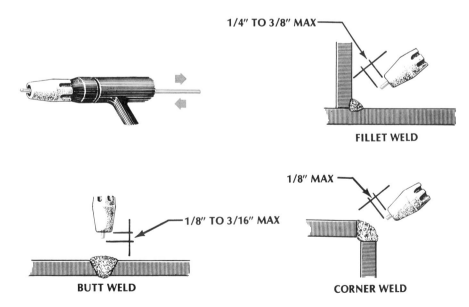

Figure 24-22. Adjust the electrode so that it extends beyond the edge of the gas cup the appropriate distance for the particular joint being welded.

The high-frequency current will jump the gap between the electrode and the plate, establishing the arc. *Be sure the downward motion is made rapidly* to provide the maximum amount of gas protection to the weld zone.

If a DC machine is used, hold the torch in the same position, but in this case the electrode can touch the plate to start the arc. See Figure 24-25. When the arc is struck, withdraw the electrode so it is about 1/8″ above the workpiece.

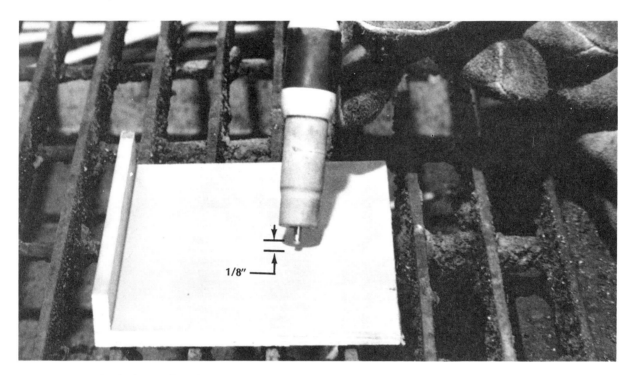

Figure 24-24. Establish the arc length at 1/8″.

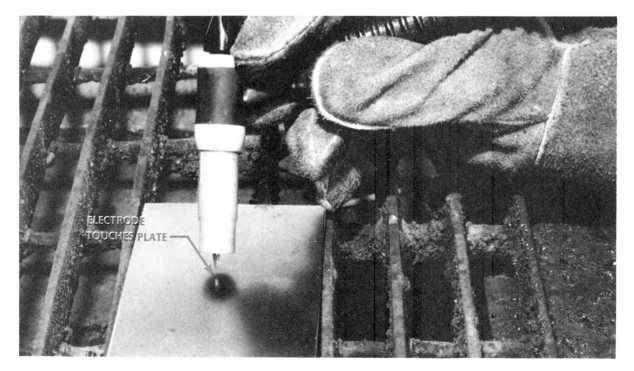

Figure 24-25. When starting the arc using DC current, the electrode must touch the plate.

To stop the arc on the AC or DC machine, swing the electrode back to the horizontal position. Make this movement rapidly to avoid marring or damaging the weld surface.

Some machines are equipped with a foot pedal to permit a gradual decrease of current. With such control it is easier to fill the crater completely and prevent crater cracks.

CAUTION: If you are using a water-cooled cup, do not allow the cup to come in contact with the work when the current is on. The hot gases may cause the arc to jump from the electrode to the cup instead of the plate, thereby damaging the cup. Be sure that the water flow is set according to the manufacturer's recommendations.

Welding a butt joint. Preheat the starting point of the weld by moving the torch in small circles. See Figure 24-26. As soon as the puddle becomes bright and fluid, move the torch slowly and steadily along the joint to form a uniform bead. No circular motion of the torch is necessary. Hold the torch at a 15° travel angle to the surface of the work. Because the electrode is pointing toward the filler metal, or pushing it, this is referred to as push angle. See Figure 24-27.

When filler rod is to be added, hold the rod about 15° from the work. As the puddle becomes fluid, move the arc to the rear of the puddle and add the rod by touching the leading edge of the puddle. Remove the rod and bring the arc back to the leading

edge of the puddle. Repeat this sequence for the entire length of the seam. See Figure 24-28.

Welding a lap and T-joint. To weld a lap or T-joint without filler rod, first form a puddle on the bottom piece. After the puddle is formed, shorten the arc to about 1/16″. Then rotate the torch directly over the joint until the pieces are joined. After the welding is started, no further torch rotation is necessary. Move the torch along the joint with the end of the electrode just above the edge of the top sheet.

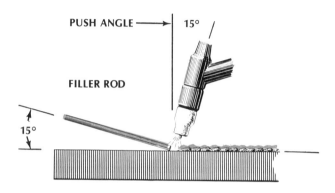

Figure 24-27. Hold the torch at a 15° push angle and the filler rod at a 15° angle.

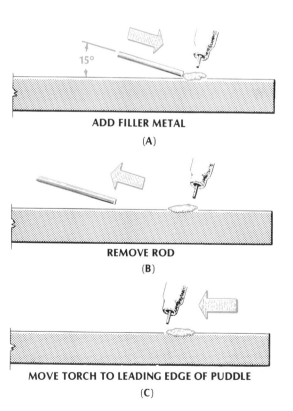

Figure 24-28. Filler rod is added by touching the lead edge of the weld puddle.

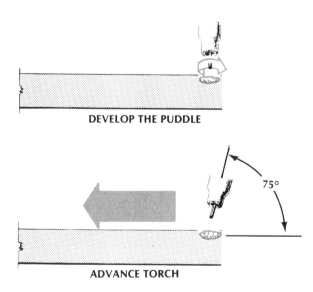

Figure 24-26. Start the puddle by moving the torch in a circular motion. Once the puddle is formed, no circular motion of the torch is necessary when moving along the joint.

In welding a lap joint, you will find that the puddle forms a V shape. The center of the V is called a *notch*, and the speed at which this notch travels determines how fast the torch should be moved. Do not get ahead of it. See Figure 24-29. Make certain that this notch is completely filled for the entire length of the seam. Otherwise there will be insufficient fusion and penetration. Dip the end of the filler rod in and out of the puddle about every ¼" travel of the puddle. See Figure 24-30. Watch carefully to avoid laying bits of filler rod on the cold, unfused base metal. If you add just the right amount of rod at the correct moment, you will get a uniform bead of the proper proportions.

Welding a corner joint.   A corner joint does not need any filler rod. Start the puddle at the beginning edge and move the torch straight along the seam. If you find that the molten metal has a tendency to roll off the edge, your speed is too slow. On the other hand, if the completed portion of the weld is rough and uneven, then your speed is too fast.

Vertical welding.   Vertical gas tungsten-arc welding on thin material is usually done in a downward position to achieve an adequate weld without burning through the metal. When filler rod is to be used, add it from the bottom or leading edge of the puddle. See Figure 24-31.

On heavier materials, an upward welding technique is preferred since deeper penetration can be achieved. Upward welding generally requires a filler rod. See Figure 24-32.

Horizontal welding.   Start the arc about ½" from the right of the joint. Once the arc is started, move it to the beginning of the joint. Hold the torch at a

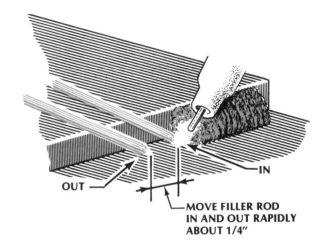

Figure 24-30. Move the filler rod in and out of the puddle consistently when welding.

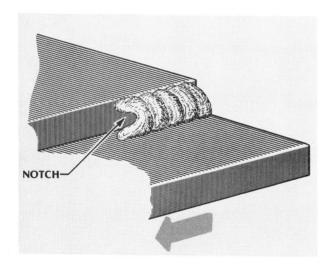

Figure 24-29. Advance the torch so that the notch in the weld bead continues to form and move forward.

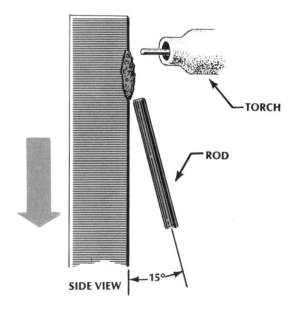

Figure 24-31. When welding vertical down, hold the filler rod at a 15° angle and the torch at a 90° angle. Add filler rod from the bottom or leading edge of the puddle.

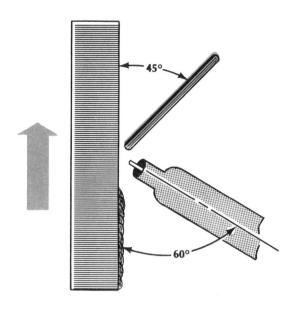

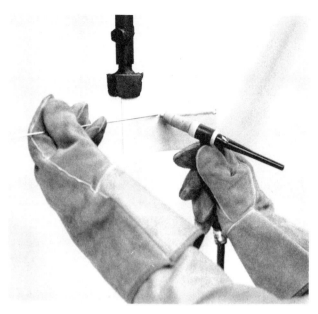

Figure 24-32. When welding vertical up, hold the filler rod at a 45° angle and the torch at a 60° angle.

Figure 24-33. Dip the filler rod into the high side at the front of the puddle when welding a horizontal butt joint.

work angle of 15° and a push angle of 15°. Dip the filler rod into the front of the puddle and preferably on the high side as the torch is moved along the joint. See Figure 24-33. While the rod is dipped into the puddle, withdraw the torch slightly. This allows the molten metal to solidify and prevents it from sagging. Keep the arc length as close as possible to the electrode diameter. Good arc length and correct speed eliminates undercutting and permits complete penetration.

Overhead welding. When gas tungsten-arc welding in an overhead position the current should be reduced between 5 to 10 percent to what is normally used for flat position welding. Slightly reduced current will give better control of the weld puddle. Both the torch and rod should be held somewhat as in flat welding. See Figure 24-34. A smaller bead is advisable since it is less affected by the pull of gravity. Dip the filler rod in and out as in other welding positions. By pulling the arc back a little further the filler metal will solidify faster.

MECHANIZED GAS TUNGSTEN-ARC WELDING

Although mechanized gas tungsten-arc welding is not as common as mechanized gas metal-arc welding, it does serve as a rapid and efficient welding process in many fabrication situations. In a gas tungsten-arc mechanized system the filler metal is automatically fed from a wire feeder that runs to a wire

torch mounted behind the gas tungsten-arc welding torch. See Figure 24-35.

Linde has developed what is known as a gas tungsten-arc hot-wire welding process which produces quality welds at the speed of gas metal-arc welding. The filler wire is preheated to a molten state as it enters the weld puddle, leaving the gas tungsten arc completely free to concentrate on the weld and not the wire. Wire melting is accomplished by passing an AC current through the wire by means of an AC unit. The power unit is regulated so the wire reaches the melting point as it enters the weld puddle.

By attaching the hot-wire torch behind the welding torch, the operator is given an unobstructed view

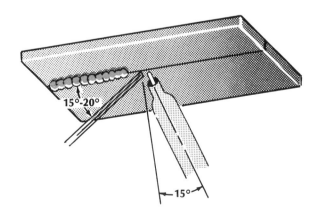

Figure 24-34. The position of the torch and filler rod in overhead welding is similar to the position in flat welding.

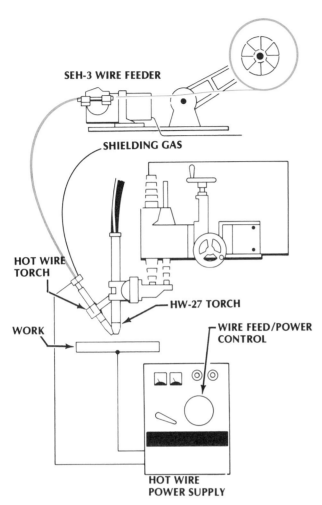

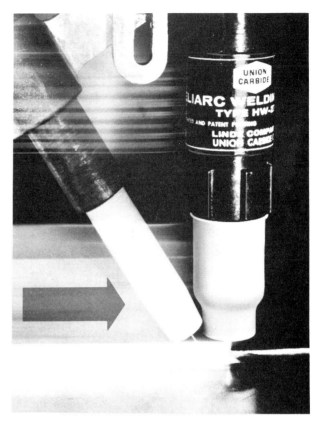

Figure 24-36. The hot-wire welding torch is mounted behind the gas tungsten-arc welding torch to give the operator an unobstructed view of the weld. (*Linde Company*)

Figure 24-35. The gas tungsten-arc hot-wire system preheats the filler wire to a molten state by passing an AC current through the wire. (*Linde Company*)

of the actual weld. See Figure 24-36. By preheating the filler wire, weld porosity is eliminated and welds are made with greater quality and speed.

PULSED-CURRENT GAS TUNGSTEN-ARC WELDING

A pulsed-current process used with a gas tungsten arc has been developed by Hobart Brothers Company whereby two levels of welding current, high and low, are used instead of the ordinary single level current. This high-and-low level generates a pulsating current or arc which produces overlapping spot welds. See Figure 24-37. The arc spot welds are actually formed by the high-level current. Then when the current switches to a low level the welds are allowed to cool and partially solidify between each spot. The welding controls are set so the arc spot overlaps will produce a continuous seam.

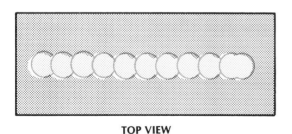

TOP VIEW

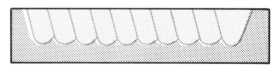

LONGITUDINAL CROSS SECTION

Figure 24-37. The pulsed-current GTAW process produces overlapping spot welds that form a continuous seam. (*Hobart Brothers Company*)

Pulsed-current gas tungsten-arc welding can be manual or automatic and with or without filler rod. The process is particularly appropriate for welding very thin materials where critical control of metallurgical factors is necessary. A pulsed current permits more tolerance of edge misalignment, greater variations in back-up bar and fixturing, better root penetration and less distortion.

Another distinct advantage of the pulsed arc is in welding curved seams or pipes. Normally with other welding techniques a change in current or travel speed must be made when changing positions around curved edges to insure uniform weld appearance. The pulsed current is more tolerant of welding position and allows continuous welding without having to vary travel speed, voltage or current. See Figure 24-38.

GAS TUNGSTEN-ARC WELDING COMMON METALS

The actual technique of gas tungsten-arc welding common metals such as carbon steels, aluminum, stainless steel, magnesium, copper, and low-alloy

steels is virtually the same. In general these metals can be gas tungsten-arc welded more easily and with better results than by the oxyacetylene or shielded metal-arc process. Several specific welding characteristics for each of these metals are described in the paragraphs that follow.

Carbon steels. Gas tungsten arc is being used more extensively in welding low and medium-carbon and low-alloy steels because of the ease with which the welding can be accomplished and the greater protection from atmospheric contamination. For economic reasons gas tungsten-arc welding is limited to materials under ¼″ in thickness. When the gas tungsten-arc process is used without filler rod there may be evidence of some pitting in the weld. This porosity can be eliminated by lightly brushing the seam with a mixture of aluminum powder and methyl alcohol. Ordinarily when filler rods are used they should contain sufficient deoxidizers to prevent porosity. Remember oxyacetylene rods are not suitable for gas tungsten-arc welding. See Exercises 1-6.

High-carbon steels are weldable, but preheating, special welding techniques, and postweld heat treatment stress relief are required. Steels with a high amount of carbon are subject to rapid grain growth and unless precautions are taken, the welded area loses its toughness, strength, and ductility. Quite often postweld heat treatment assures proper grain refinement of the steel, but this process adds to the problems and economics of fabrication.

Very high-carbon steels are rarely recommended for gas tungsten-arc welding as the required welding temperature tends to destroy their mechanical properties. The usual practice in repairing broken parts made with such steels is to use a brazing process where the heat is not sufficient to affect their metallurgical structure. Table 24-4 includes specific welding requirements for gas tungsten-arc welding plain carbon and low-alloy steels.

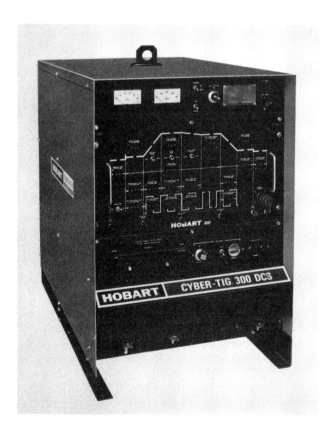

Figure 24-38. The welding machine for pulsed-current gas tungsten-arc welding can be used for manual, semiautomatic, and machine process applications depending on the equipment used. (*Hobart Brothers Company*)

TABLE 24-4. GTAW PLAIN CARBON AND LOW ALLOY STEELS.

STOCK THICKNESS	DC CURRENT STRAIGHT POLARITY (amperes)	FILLER ROD dia (inches)	ARGON FLOW (psi) lpm	chf
0.035	100	1/16	4-5	8-10
.049	100-125	1/16	4-5	8-10
.060	125-140	1/16	4-5	8-10
.089	140-170	1/16	4-5	8-10

psi—pounds per square inch
lpm—liters per minute
cfh—cubic feet per hour

Running Beads on Mild Steel Using the GTAW Process   Exercise 1

1 Use a 3/32 1% thoriated tungsten electrode (tip pointed).

2 Adjust the stickout (electrode extension) from 1/8″ to 3/16″ beyond the end of the gas cup.

3 Set the machine for DCSP (direct current straight polarity). High frequency should be set for start only. Weld current remote should be off and contactor control on.

4 Argon is used for a shielding gas at 20 cfm with a post purge of 15 seconds.

5 Set the amperage at 50-60 amps.

6 Obtain a piece of 16 gauge 4″ x 6″ mild steel.

7 Activate the current flow by pushing the foot pedal or turning on the switch at the torch.

8 To start the arc, position the torch nozzle at a 45° angle with the electrode 1/8″ from the plate.

9 When the arc starts, raise the torch to a 90° work angle and a 20° push angle. Push angle refers to the travel angle when the torch is pointing forward while welding. If the torch is pointing backward, the travel angle is referred to as the drag angle.

10 Hold a close arc. Maintain a puddle of approximately 1/8″ wide while forming a consistent bead across the plate.

11 Run a series of straight consistent beads on the plate approximately 3/8″ apart.

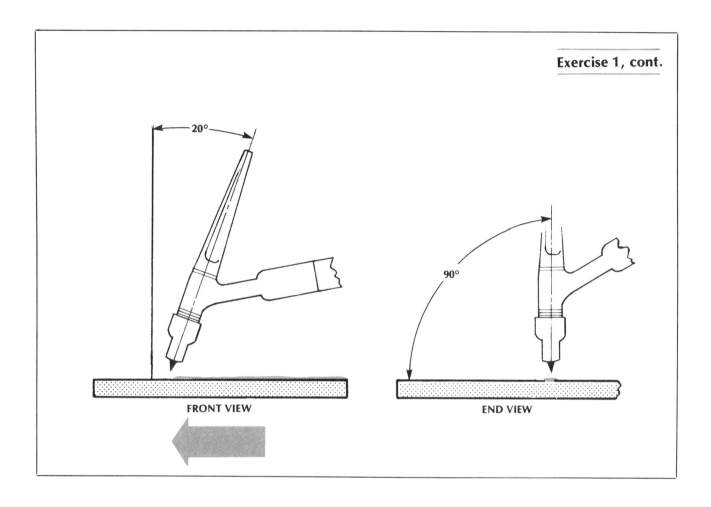

FRONT VIEW

END VIEW

Running Beads with Filler Rod on Mild Steel Using the GTAW Process | **Exercise 2**

1 Refer to Exercise 1 for equipment set-up and adjustment.

2 Obtain a piece of 16 gauge 4″ x 6″ mild steel and recommended filler rod for mild steel.

3 Establish a weld puddle. While maintaining the arc, hold the filler rod at 20° and dip into the leading edge of the puddle. Use an in and out motion. Do *not* touch the filler rod to the tungsten electrode.

4 Use a small rotary motion with the torch and maintain a bead width approximately ³⁄₁₆″ wide.

5 Complete the plate with a series of straight consistent beads ³⁄₈″ apart.

Welding a Butt Joint on Mild Steel in Flat Position Using the GTAW Process — Exercise 3

1. Refer to Exercise 1 for equipment set-up and adjustment.

2. Obtain two pieces of 16 gauge 1½″ x 6″ mild steel and recommended filler rod for mild steel.

3. Tack weld the two pieces of steel to form a butt joint with no root opening.

4. Use the same procedure for running beads with filler rod. Deposit a bead using the joint as the center of the weld across the piece.

5. There should be complete penetration with a bead width of ³⁄₁₆″.

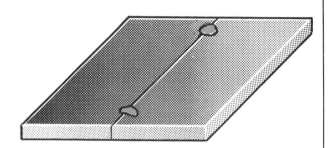

Welding a Lap Joint on Mild Steel in Horizontal Position Using the GTAW Process — Exercise 4

1. Refer to Exercise 1 for equipment set-up and adjustment.

2. Obtain two pieces of 16 gauge 1½″ x 6″ mild steel and recommended filler rod.

3. Position the pieces and tack weld ends to form a lap joint.

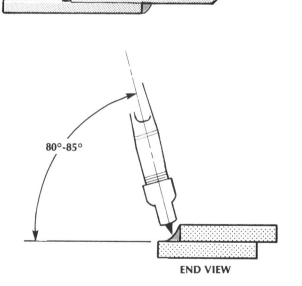

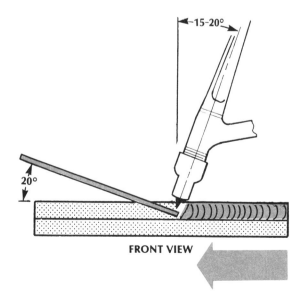

15-20°

20°

FRONT VIEW

80°-85°

END VIEW

Exercise 4, cont.

4 Hold the torch at an 80° to 85° work angle and a 15° to 20° push angle. Position the filler rod at a 20° angle.

5 Melt the top edge of the piece and add filler metal using an in and out motion to the leading edge of the puddle.

6 Carefully observe the formation of the weld and maintain a consistent bead across the piece.

Welding a T-Joint on Mild Steel in the Horizontal Position Using the GTAW Process

Exercise 5

1 Refer to Exercise 1 for equipment set-up and adjustment.

2 Obtain two pieces of 16 gauge 2″ x 6″ mild steel and recommended filler rod.

3 Position and tack weld the pieces to form a T-joint.

4 Hold the torch at a 45° work angle and a 15° push angle.

5 Hold the filler rod at a 20° angle from the bottom plate and a 20° angle from the vertical plate.

6 Form a puddle at the center of the joint. Weave the torch slightly while adding filler rod with an in and out motion at the lead edge of the puddle.

7 Avoid excess heat build-up on the vertical piece.

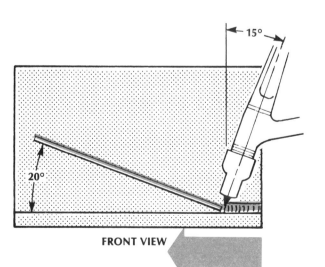

FRONT VIEW

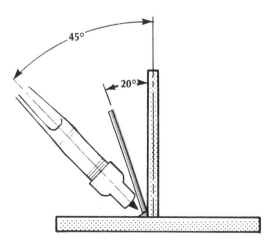

Welding a T-Joint on Mild Steel in the Vertical Position Using the GTAW Process

Exercise 6

1 Refer to Exercise 1 for equipment set-up and adjustment.

2 Obtain two pieces of 16 gauge 2″ x 6″ mild steel and recommended filler rod.

3 Position and tack weld the pieces to form a T-joint.

4 Hold the torch at a 45° work angle and a 20° push angle. Deposit a fillet weld from the bottom to the top of the piece.

5 Angle the filler rod 20° centered in the joint. Use an in and out motion while adding filler rod to the leading edge of the puddle.

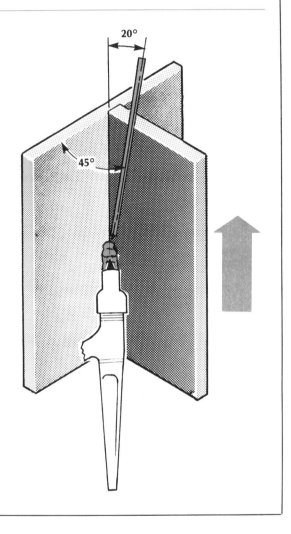

Aluminum. Nonheat-treatable wrought aluminum alloys in the 1000, 3000 and 5000 series are readily weldable. The heat-treatable alloys in the 2000, 6000 and 7000 series can be welded but higher welding temperatures and speeds are needed. Elimination of weld cracking in these alloys can often be achieved with a rod having a higher alloy content than that of the base metal.

While welding can be performed in any position the task is simplified and the quality of the completed joint is more satisfactory if the weld is done in a flat position. Copper back-up blocks should be used wherever possible especially on plates ⅛″ or less in thickness. In most cases, the torch should be moved in a straight line without a weaving motion.

Best results are obtained by using ACHF current with argon as a shielding gas. See Table 24-5 for recommended welding requirements of gas flow, current, etc. See Exercises 7 and 8.

TABLE 24-5. GTAW WELDING—ALUMINUM.

STOCK THICKNESS (inches)	TYPE OF JOINT	AMPERES, AC CURRENT			ELECTRODE (inches) dia	ARGON FLOW 20 psi		FILLER ROD (inches) dia
		FLAT	HORIZONTAL & VERTICAL	OVERHEAD		lpm	cfh	
1/16	Butt	60-80	60-80	60-80	1/16	7	15	1/16
	Lap	70-90	55-75	60-80	1/16	7	15	1/16
	Corner	60-80	60-80	60-80	1/16	7	15	1/16
	Fillet	70-90	70-90	70-90	1/16	7	15	1/16
1/8	Butt	125-145	115-135	120-140	3/32	8	17	1/8
	Lap	140-160	125-145	130-160	3/32	8	17	1/8
	Corner	125-145	115-135	130-150	3/32	8	17	1/8
	Fillet	140-160	115-135	140-160	3/32	8	17	1/8
3/16	Butt	190-220	190-220	180-210	1/8	10	21	5/32
	Lap	210-240	190-220	180-210	1/8	10	21	5/32
	Corner	190-220	180-210	180-210	1/8	10	21	5/32
	Fillet	210-240	190-220	180-210	1/8	10	21	5/32
1/4	Butt	260-300	220-260	210-250	3/16	12	25	3/16
	Lap	290-340	220-260	210-250	3/16	12	25	3/16
	Corner	280-320	220-260	210-250	3/16	12	25	3/16
	Fillet	280-320	220-260	210-250	3/16	12	25	3/16

Running Beads on Aluminum in Flat Position Using the GTAW Process

Exercise 7

1 Use a ³⁄₃₂″ pure tungsten electrode (tip spherical).

2 Adjust the electrode extension about ⅛″ to ³⁄₁₆″ beyond the end of the gas cup.

3 Set the machine for AC (alternating current). High frequency should be set for continuous. Weld current remote and contactor control should be on.

4 Argon is used for a shielding gas at 20 cfm with a post purge of 15 seconds.

5 Set the amperage at 140-150 amps, continuous high frequency.

6 Obtain a ⅛″ x 4″ x 6″ piece of pure aluminum.

7 Clean the aluminum piece with a stainless steel wire brush.

8 To shape the tungsten electrode switch the machine to DCRP (direct current reverse polarity). Using the procedure for mild steel, strike the arc on a copper plate and position the torch at a 90° angle. A small ball will form on the end of the electrode.

9 Start the arc by using the procedure for mild steel. The foot control can be used to vary the amperage.

10 The position of the torch should be a 90° work angle and a 20° push angle.

11 Melt the base metal and add filler metal to form a ¼″ bead. Use an in and out motion to add filler rod to the leading edge of the puddle. A slight weaving motion can be used.

12 To fill the crater at the end of the weld reduce the amperage with the foot control and continue to add filler rod.

13 Run a series of beads on the piece approximately ⅜″ apart.

```
┌─────────────────────────────────────────────────┐
│ ┌───────────────────────────────────────────┐   │
│ │ Welding Joints on Aluminum                 │   │
│ │ Using the GTAW Process    Exercise 8       │   │
│ └───────────────────────────────────────────┘   │
│                                                   │
│  ▇   Refer to Exercise 7 for equipment set-up    │
│  1   and adjustment.                              │
│                                                   │
│  ▇   Obtain ⅛″ aluminum plates.                   │
│  2                                                │
│                                                   │
│  ▇   Using the procedures presented for           │
│  3   welding mild steel, complete the follow-     │
│      ing joints: butt, lap, and T-joint.          │
│                                                   │
└─────────────────────────────────────────────────┘
```

steels. If a filler rod is used it should have a slightly higher chromium content. Danger of cracking is greatly reduced if the metal is preheated to a temperature of 300° to 500°F(145° to 260°C). See Table 24-6 for specific welding requirements.

Copper and copper alloys. Deoxidized copper is the most widely used type for gas tungsten-arc welding.

Copper alloys such as brass, bronze, and copper alloys of nickel, aluminum, silicon and beryllium are also readily welded with the gas tungsten-arc process. DCSP is generally used for welding these metals. However ACHF or DCRP is often recommended

TABLE 24-6. GTAW WELDING—STAINLESS STEEL.

STOCK THICKNESS (inches)	TYPE OF JOINT	DC CURRENT STRAIGHT POLARITY amperes			ELECTRODE (inches) dia	ARGON FLOW 20 psi		FILLER ROD (inches) dia
		flat	horizontal & vertical	overhead		lpm	cfh	
1/16	Butt	80-100	70-90	70-90	1/16	5	11	1/16
	Lap	100-120	80-100	80-100	1/16	5	11	1/16
	Corner	80-100	70-90	70-90	1/16	5	11	1/16
	Fillet	90-110	80-100	80-100	1/16	5	11	1/16
3/32	Butt	100-120	90-110	90-110	1/16	5	11	1/16
	Lap	110-130	100-120	100-120	1/16	5	11	1/16
	Corner	100-120	90-110	90-110	1/16	5	11	1/16
	Fillet	110-130	100-120	100-120	1/16	5	11	1/16
1/8	Butt	120-140	110-130	105-125	1/16	5	11	3/32
	Lap	130-150	120-140	120-140	1/16	5	11	3/32
	Corner	120-140	110-130	115-135	1/16	5	11	3/32
	Fillet	130-150	115-135	120-140	1/16	5	11	3/32
3/16	Butt	200-250	150-200	150-200	3/32	6	13	1/8
	Lap	225-275	175-225	175-225	3/32	6	13	1/8
	Corner	200-250	150-200	150-200	3/32	6	13	1/8
	Fillet	225-275	175-225	175-225	3/32	6	13	1/8
1/4	Butt	275-350	200-250	200-250	1/8	6	13	3/16
	Lap	300-375	225-275	225-275	1/8	6	13	3/16
	Corner	275-350	200-250	200-250	1/8	6	13	3/16
	Fillet	300-375	225-275	225-275	1/8	6	13	3/16

Stainless steel. Stainless steels, especially those in the 300 series, are very easy to weld using the gas tungsten-arc process. Either direct current straight polarity or alternating current with high-frequency stabilization can be used. The gas tungsten-arc process is particularly adaptable for welding light gauge stainless steel.

The procedure for welding martensitic stainless steels is the same as for welding austenitic stainless

for beryllium copper or for copper alloys less than 0.040″ thick. Workpieces thicker than ¼″ should be preheated to approximately 300-500°F (145° to 260°C) prior to welding. A forehand welding technique will usually produce the best results. See Table 24-7 and 24-8 for specific welding requirements.

Always be sure that there is good ventilation when welding copper or copper alloys. The fumes of these metals are highly toxic; therefore, a high-velocity

TABLE 24-7.   GTAW WELDING—DEOXIDIZED COPPER.

STOCK THICKNESS (inches)	TYPE OF JOINT	DC CURRENT STRAIGHT POLARITY AMPERES FLAT POSITION	ELECTRODE (inches) dia	ARGON FLOW 20 psi lpm	ARGON FLOW 20 psi cfh	FILLER ROD (inches) dia
1/16	Butt	110-140	1/16	7	15	1/16
	Lap	130-150	1/16	7	15	1/16
	Corner	110-140	1/16	7	15	1/16
	Fillet	130-150	1/16	7	15	1/16
1/8	Butt	175-225	3/32	7	15	3/32
	Lap	200-250	3/32	7	15	3/32
	Corner	175-225	3/32	7	15	3/32
	Fillet	200-250	3/32	7	15	3/32
3/16	Butt	250-300	1/8	7	15	1/8
	Lap	275-325	1/8	7	15	1/8
	Corner	250-300	1/8	7	15	1/8
	Fillet	275-325	1/8	7	15	1/8
1/4	Butt	300-350	1/8	7	15	1/8
	Lap	325-375	1/8	7	15	1/8
	Corner	300-350	1/8	7	15	1/8
	Fillet	325-375	1/8	7	15	1/8

Linde Co.

TABLE 24-8.   GTAW WELDING—COPPER ALLOYS.

STOCK THICKNESS (inches)	TYPE OF JOINT	DC CURRENT—STRAIGHT POLARITY AMPERES FLAT	HORIZONTAL & VERTICAL	OVERHEAD	ELECTRODE (inches) dia	ARGON FLOW 20 psi lpm	ARGON FLOW 20 psi cfh	FILLER ROD (inches) dia
1/16	Butt	100-120	90-110	90-110	1/16	6	13	1/16
	Lap	110-130	100-120	100-120	1/16	6	13	1/16
	Corner	100-130	90-110	90-110	1/16	6	13	1/16
	Fillet	110-130	100-120	100-120	1/16	6	13	1/16
1/8	Butt	130-150	120-140	120-140	1/16	7	15	3/32
	Lap	140-160	130-150	130-150	1/16, 3/32	7	15	3/32
	Corner	130-150	120-140	120-140	1/16	7	15	3/32
	Fillet	140-160	130-150	130-150	1/16, 3/32	7	15	3/32
3/16	Butt	150-200	—	—	3/32	8	17	1/8
	Lap	175-225	—	—	3/32	8	17	1/8
	Corner	150-200	—	—	3/32	8	17	1/8
	Fillet	175-225	—	—	3/32	8	17	1/8
1/4	Butt	150-200	—	—	3/32	9	19	1/8, 3/16
	Lap	250-300	—	—	1/8	9	19	1/8, 3/16
	Corner	175-225	—	—	3/32	9	19	1/8, 3/16
	Fillet	175-225	—	—	3/32	9	19	1/8, 3/16

ventilating system is absolutely necessary.

Magnesium. The welding characteristics of magnesium are somewhat comparable to those of aluminum. Both, for example, have high heat conductivity, low melting point, high thermal expansion, and oxidize rapidly.

With gas tungsten arc several current variations are possible. DC reverse polarity with helium gas produces wider weld deposits, higher heat, larger heat-affected zone, and shallower penetration. AC current

TABLE 24-9. GTAW WELDING— MAGNESIUM

STOCK THICKNESS (inch)	TYPE OF JOINT	AMPERES AC CURRENT FLAT POSITION	WELDING ROD (inches) dia	ARGON FLOW 15 psi lpm	cfh	REMARKS
0.040	Butt	45	3/32, 1/8	6	13	Back-up
.040	Butt	25	3/32, 1/8	6	13	No backing
.040	Fillet	45	3/32, 1/8	6	13	
.064	Butt	60	3/32, 1/8	6	13	Back-up
.064	Butt and corner	35	3/32, 1/8	6	13	No backing
.064	Fillet	60	3/32, 1/8	6	13	
.081	Butt	80	1/8	6	13	Back-up
.081	Butt, corner and edge	50	1/8	6	13	No backing
.081	Fillet	80	1/8	6	13	
.102	Butt	100	1/8	9	19	Back-up
.102	Butt, corner and edge	70	1/8	9	19	No backing
.102	Fillet	100	1/8	9	19	
.128	Butt	115	1/8, 5/32	9	19	Back-up
.128	Butt, corner and edge	85	1/8, 5/32	9	19	No backing
.128	Fillet	115	1/8, 5/32	9	19	
3/16	Butt	120	1/8, 5/32	9	19	1 pass
3/16	Butt	75	1/8, 5/32	9	19	2 passes
1/4	Butt	130	5/32, 3/16	9	19	1 pass
1/4	Butt	85	5/32	9	19	2 passes

with superimposed high frequency and helium, argon, or a mixture of these gases, will join material ranging in thickness from 0.20″ to over 0.25″. Both DCRP and AC current provide excellent cleaning action of the base metal surface. Direct current straight polarity with helium as a shielding gas produces a deep penetrating arc but no surface cleaning. This technique is used for mechanized butt welding of sheets up to ¼″ in thickness without beveling. See Table 24-9 for recommended welding requirements.

Points to Remember

1 Gas tungsten-arc welding can be used for joining practically all metals and alloys in various thicknesses and types of joints.

2 Be sure to use a cup of the correct size. Nozzles having too small an orifice tend to overheat and either crack or deteriorate very rapidly.

3 A water-cooled torch is recommended for welding currents that are above 200 or 250 amperes.

4 Argon is generally the inert gas recommended for gas tungsten-arc welding.

5 The power source can be either a DC or AC machine. With a DC machine, better penetration is usually obtained with straight polarity. For some metals better welds are made with an AC machine having a high-frequency voltage than with a DC machine.

6 The diameter of the electrode depends on the kind and thickness of the metal to be welded. Make certain the tip is properly shaped for the type of current used.

7 When welding light gauge metals, it is often necessary to use back-up bars.

Points to Remember, cont.

8 Before starting to weld, always check to make sure the electrode extends the correct distance.

9 Follow the recommendations for the correct gas flow; otherwise the shielding gas will not be effective.

10 If filler rod is to be used, be sure it is of the right diameter.

11 If using a water-cooled torch, always make sure water is flowing before attempting to weld.

12 Never attempt to adjust the tungsten electrode without first shutting down the power supply machine.

QUESTIONS FOR STUDY AND DISCUSSION

1 What are some of the advantages of gas shielded-arc welding compared to other welding processes?

2 What is meant by gas tungsten-arc welding?

3 What kind of metal can be welded with the gas tungsten-arc welding process?

4 When is a filler rod used in gas tungsten-arc welding?

5 In gas tungsten-arc welding, what type of power supply unit may be used?

6 Why should an AC machine be of the high-frequency type?

7 What polarity is used in gas tungsten-arc welding?

8 What determines whether an air-cooled or water-cooled torch is used?

9 Why is it important to use a cup of the correct size?

10 What is meant by an inert gas?

11 What is the function of a flowmeter in a gas regulator assembly?

12 In gas tungsten-arc welding, what results can be expected when DCRP or DCSP current is used with respect to heat distribution?

13 What determines the size of the tungsten electrode to be used for welding?

14 What should be the shape of the tungsten electrode for DC and AC welding?

15 How is the arc started and stopped in gas tungsten-arc welding?

16 How far should the electrode extend beyond the end of the gas cup on the torch?

17 What is the proper torch angle for welding a butt joint?

18 What precaution should be observed when using a water-cooled torch?

19 When using a filler rod, how should it be manipulated?

20 What is gas tungsten-arc hot-wire welding?

21 How are welds produced by the gas tungsten-arc pulsed current process?

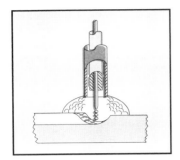

Gas Shielded-Arc Welding

The gas metal-arc welding process (GMAW), sometimes referred to as MIG, uses a continuous consumable wire electrode. The molten weld puddle is completely covered with a shield of gas. The wire electrode is fed through the torch at preset controlled speeds. The shielding gas is also fed through the torch. See Figures 25-1 and 25-2.

The weld can be applied by the semiautomatic, machine, or automatic process. When semiautomatic, the wire feed, power setting and gas flow are preset, but the torch is manually operated. The welder directs the torch over the weld seam, holding the correct wire stickout distance and speed. See Figures 25-3 and 25-4.

Gas metal-arc welding is sometimes referred to by the tradename of the manufacturer such as *Micro-*

wire Welding (Hobart), *Aircomatic Welding* (Airco), *Sigma Welding* (Linde), and *Millermatic Welding* (Miller).

SPECIFIC ADVANTAGES OF GAS METAL-ARC WELDING

The following are considered to be some of the more important advantages of gas metal-arc welding:

1. Since there is no flux or slag and very little spatter to remove, there is a considerable saving in total welding cost. By minimizing weld clean-up time, money is saved.

2. Less time is required to train an operator. Welding operators who are proficient in other welding processes can usually master the technique of gas metal-arc welding in a matter of hours. The main concern of the operator is to watch the angle of the welding gun and speed of travel, and wire stickout.

3. The welding process is faster, especially when compared with shielded metal-arc welding. There is no need to start and stop in order to change electrodes, thereby eliminating one common cause of weld failures. Starting and stopping of welding often is the cause of slag inclusions, cold lapping and other problems like crater cracking.

4. Because of the high speed of the gas metal-arc process, better metallurgical benefits are imparted to the weld area. With faster travel there is a narrower heat-affected zone and consequently less molecular disarrangement, less grain growth, less heat transfer in the parent metal, and greatly reduced distortion.

5. Although originally gas tungsten-arc welding was considered more practical for welding thin sheet because of its lower current, the development of the short-circuiting transfer technique now makes it possible to weld thin stock equally as effectively and more economically with the gas metal-arc process.

6. Since gas metal-arc welding has deep penetrating characteristics, narrower beveled joint designs can be used. The size of fillet welds is reduced by comparison to other welding methods.

Welding current. Different welding currents have a large effect on the results obtained in gas metal-arc welding. Optimum efficiency is achieved with

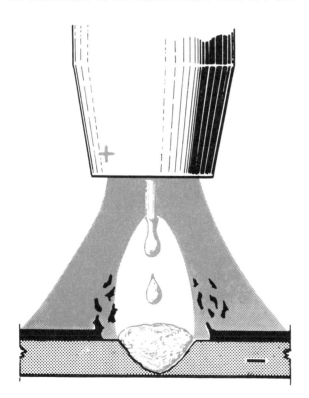

Figure 25-1. In the gas metal-arc welding process, a high-temperature electric arc melts the advancing wire electrode into a globule of liquid metal. A shield of inert gas completely covers the molten weld puddle. (*Linde Company*)

Figure 25-2. In the GMAW process, a consumable wire electrode is fed into a weld puddle protected by shielding gas. (*Miller Electric Manufacturing Company*)

Figure 25-3. Gas metal-arc welding, when applied by the semiautomatic process, is manually operated; but the wire feed, power setting, and gas flow are preset. *(Miller Electric Manufacturing Company)*

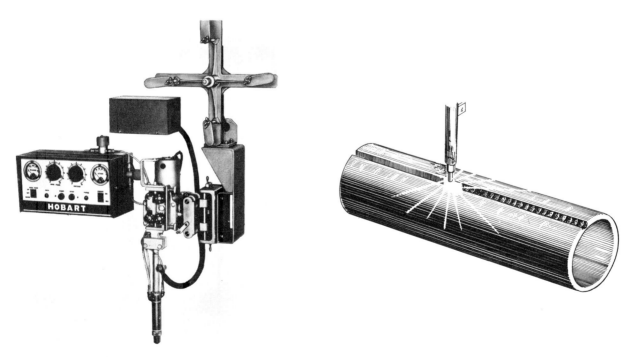

Figure 25-4. Machine-controlled gas metal-arc welding machines deposit the weld under the supervision of an operator. *(Hobart Brothers Company)*

direct current reverse polarity (DCRP). See Figure 25-5. The heat in this instance is concentrated at the weld puddle and therefore provides deeper penetration at the weld. Furthermore, with DCRP there is greater surface cleaning action which is important in welding metals having heavy surface oxides such as aluminum and magnesium.

Straight polarity (DCSP) is impractical with gas metal-arc welding because weld penetration is wide and shallow, spatter is excessive, and there is no surface cleaning action. The ineffectiveness of straight polarity largely results from the pattern of metal transfer from the electrode to the weld puddle. Whereas in reverse polarity the transfer is in the form of a fine spray, with straight polarity the transfer is largely of the erratic globular type. The use of AC current is not recommended since the burn-offs are unequal on each half-cycle.

TYPES OF METAL TRANSFER

When welding with consumable wire electrodes, the transfer of metal is achieved by three methods: spray transfer, globular transfer, and short-circuiting transfer. The type of metal transfer that occurs will depend on electrode wire size, shielding gas, arc voltage, and welding current.

Spray transfer. In spray transfer very fine droplets or particles of the electrode wire are rapidly projected through the arc plasma from the end of the electrode to the workpiece in the direction in which the electrode is pointed. The droplets are equal to or smaller than the diameter of the electrode. While in the process of transferring through the welding arc, the metal particles do not interrupt the flow of current and there is a nearly constant spray of metal.

Spray transfer requires a high current density. With the higher current, the arc becomes a steady quiet column having a well-defined narrow incandescent cone-shape core within which metal transfer takes place. See Figure 25-6A. The use of argon or a mixture of argon and oxygen is also necessary for spray transfer. Argon produces a pinching effect on the molten tip of the electrode, permitting only small droplets to form and transfer during the welding process. This makes the spray transfer process useful when welding aluminum.

With high heat input, heavy wire electrodes will melt readily and deep weld penetration becomes possible. Since the individual drops are small the arc is stable and can be directed where required. The fact that the metal transfer is produced by directional force which is stronger than gravity makes spray transfer effective for out-of-position welding. *It is particularly appropriate for welding heavy gauge metal.* It is not too practical for welding light gauge metal because of the resulting burn-through.

Globular transfer. This type of transfer occurs

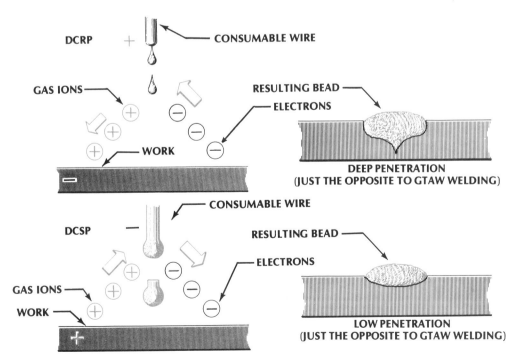

Figure 25-5. In gas metal-arc welding, direct current reverse polarity is used as straight polarity results in low penetration and excessive spatter.

when the welding current is low or below what is known as the transition current. The transition range extends from the minimum value where the heat melts the electrode to the point where the high current value induces spray transfer. Only a few drops are transferred per second at low current values, whereas many small drops are transferred when high current values are used.

In globular transfer the molten ball at the tip of the electrode tends to grow in size until its diameter is two or three times the diameter of the wire before it separates from the electrode and transfers across the arc to the workpiece. See Figure 25-6B. As the globule moves across the arc it assumes an irregular shape and rotary motion because of the physical forces of the arc. This frequently causes the globule to reconnect with the electrode and workpiece, causing the arc to go out and then reignite. The result is poor arc stability, poor penetration, and excessive spatter.

Because of these characteristics a globular type transfer is usually not very effective for most gas metal-arc welding operations. Its use is generally restricted to where a low heat input is desired and for welding thin sections.

Short-circuiting transfer (short arc). The short-circuiting transfer permits welding thinner sections with greater ease. It is extremely practical for welding in all positions, especially for vertical, horizontal and overhead welding where normally puddle control is a little more difficult.

With this process a shallow weld penetration is obtained. See Figure 25-7 A-E. It is generally considered to be the most practical at current levels below 200 amperes with fine wire of 0.045″ or less in diameter. The use of fine wire produces weld pools that remain relatively small and are easily managed, making all-position welding possible.

As the molten wire is transferred to the weld, each drop touches the weld puddle before it has broken away from the advancing electrode wire. The circuit is shorted, and the arc is then extinguished.

Electromagnetic pinch force squeezes the drop from the wire. The short circuit is broken and the arc reignites. Shorting occurs from 20 to 200 times a second according to preset controls. Shorting of the arc pinpoints the effective heat. The result is a small, relatively cool weld puddle which reduces burn-through. Intricate welds are possible in most all of the positions, as indicated in Figure 25-7 C and D.

In *short-arc welding*, the shielding gas mixture consists of 75 percent carbon dioxide, which provides increased heat for higher speeds, and 25 per-

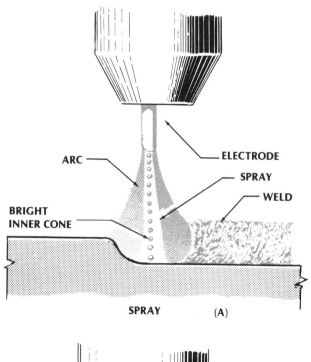

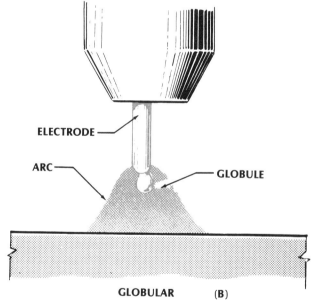

Figure 25-6. Different types of metal transfer occur by varying the current and shielding gas. (*Airco*)

cent argon which controls spatter. However, considerable usage is now being made of straight CO_2 where bead contour is not particularly important but good penetration is very essential.

GAS METAL-ARC WELDING EQUIPMENT

The gas metal-arc welding equipment consists of

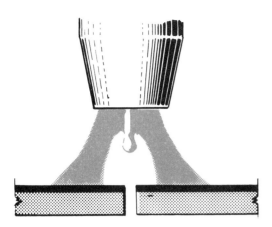

Figure 25-7A. At the start of the short circuit transfer, a high-temperature electric arc melts the advancing wire electrode into a drop of liquid metal. Arc heat is regulated by the power supply, and the wire is fed mechanically through the welding torch. (*Linde Company*)

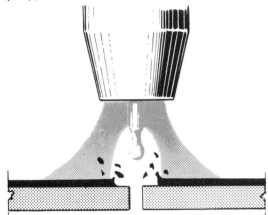

Figure 25-7B. As the molten electrode moves toward the workpiece, the argon gas mixture shields the molten wire and seam, insuring regular arc ignition and preventing spatter and weld contamination. (*Linde Company*)

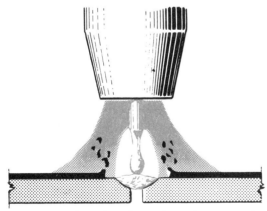

Figure 25-7C. When the electrode makes contact with the workpiece, it creates a short circuit. The arc is extinguished allowing it to cool. Depending on job requirements, the frequency of arc extinction can vary from 20 to 200 times per second. (*Linde Company*)

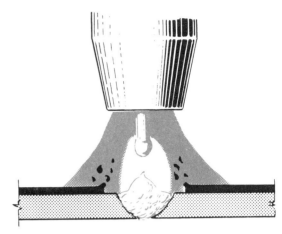

Figure 25-7D. A drop of molten wire breaks contact with the electrode, causing the arc to reignite. Pinch force, a squeezing power common to all current carriers, is what causes the electrode to break. The amount and suddenness of pinch is controlled by the power supply. (*Linde Company*)

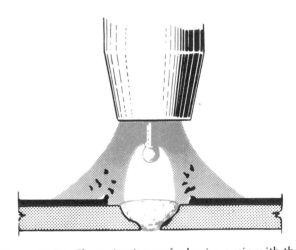

Figure 25-7E. Short circuit transfer begins again with the reignition of the arc. Because of its precision control of arc characteristics and its cool, uniform operation, the short circuit transfer produces perfect welds on metals as thin as 0.030". (*Linde Company*)

four major units: power supply, wire feeding mechanism, welding gun, and gas supply. See Figure 25-8.

Power Supply

The recommended machine for gas metal-arc welding is a rectifier or motor generator supplying direct current with normal limits of 200 to 250 amperes for all position welding. Direct current reverse polarity (DCRP) is used for optimum efficiency. DCRP contributes to better melting, deeper penetration, and excellent cleaning action.

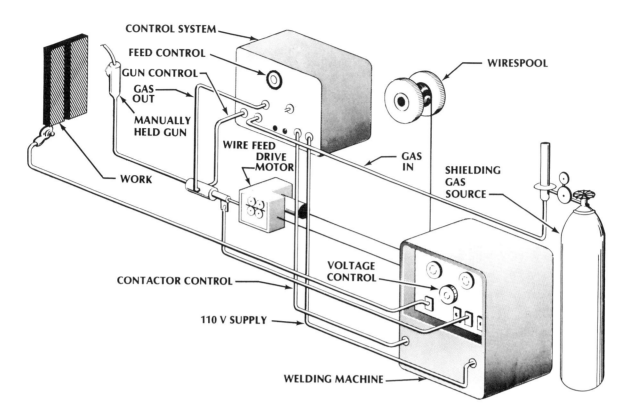

CONTROL SYSTEM

FEED CONTROL

GUN CONTROL

GAS OUT

MANUALLY HELD GUN

WIRE FEED DRIVE MOTOR

WORK

CONTACTOR CONTROL

110 V SUPPLY

WELDING MACHINE

VOLTAGE CONTROL

GAS IN

SHIELDING GAS SOURCE

WIRESPOOL

Figure 25-8. The four major components of a gas metal-arc welding unit are the power supply, wire feeding mechanism, welding gun, and gas supply.

Constant current versus constant potential power supply. In gas metal-arc welding, heat is generated by the flow of current through the gap between the end of the wire electrode and the workpiece. A voltage forms across this gap which varies with the length of the arc. To produce a uniform weld, the welding voltage and arc length must be maintained at a constant value. This can be accomplished by (1) feeding the wire into the weld zone at the same rate at which it melts, or (2) melting the wire at the same rate it is fed into the weld zone.

With the conventional constant current welding machine used for many years in shielded metal-arc (stick) welding, the power source produces a constant current over a range of welding voltages. The current has a steep drooping volt-ampere characteristic.

When a conventional power source is used for gas metal-arc welding, the wire feed speed must be adjusted to narrow limits to prevent the wire from burning back to the nozzle or plunging into the weld plate. Whenever the nozzle-to-work distance changes, the arc length (voltage) changes. If the nozzle-to-work distance increases, the arc length increases, resulting in a nonuniform weld.

With the need for better arc control, the *constant voltage (potential) power supply* was developed. Constant potential welding power supply has a nearly flat volt-ampere characteristic. See Figure 25-9. This means that the preset voltage level can be held throughout its range. Although its static voltage potential at open circuit is lower than a machine with drooping characteristic, it maintains approximately the same voltage regardless of the amount of current drawn. Accordingly there is unlimited amperage to melt the consumable wire electrode.

The power supply becomes self-correcting with respect to arc length. The operator can change the wire feed speed over a considerable range without affecting the stubbing or burning back of the wire. In other words, the arc length can be set on the power supply and any variations in nozzle-to-work distance will not produce changes in the arc length. For example, if the arc length becomes shorter than the preselected value, there is an automatic increase of current and the wire speed automatically adjusts itself to maintain a constant arc length. Similarly, if the arc becomes too long, the current decreases and the wire begins to feed faster.

Stated in another way, when the wire is fed into

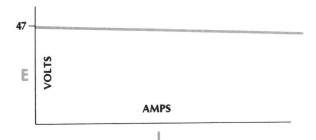

Figure 25-9. The constant potential welding power supply has a nearly flat volt-ampere curve. (*Airco*)

trode wire is fed faster, the current increases; if it is fed slower the current decreases.

Because of this self-correcting feature, less operator skill is necessary to achieve good welds. There are only two basic controls: a rheostat on the welding machine to regulate the voltage and a rheostat on the wire feed mechanism to control the speed of the wire feed motor.

There is no current control on a constant voltage type machine: the welding current output is determined by the wire feeder.

Slope control. Some power supply units designed for gas metal-arc welding have provisions for controlling the slope. The incorporation of slope control gives the machine greater versatility. Thus by altering the flat shape of the slope it is possible to control the pinch force on the consumable wire which is particularly important in the short circuiting transfer method of welding. With better control of the short circuit, the weld puddle can be kept more fluid with better resulting welds. Slope control also helps to decrease the sudden current surge when the electrode makes its initial contact with the workpiece. By slowing down the rate of current rise, the amount of spatter can be reduced.

Wire Feeding Mechanism

The wire feeding mechanism automatically drives the electrode wire from the wire spool to the gun and arc. See Figures 25-10 and 25-11. Control on the panel can be adjusted to vary the wire feeding speed. In addition, the control panel usually includes a welding power contactor and a solenoid to energize the gas flow.

The wire feeder can be mounted on the power

Figure 25-10. The wire feeding unit for a gas metal-arc welding system automatically drives the electrode wire from the wire spool to the gun and arc. (*Miller Electric Manufacturing Company*)

the arc at a specific rate, a proportionate amount of current is automatically drawn. The constant potential welder therefore provides the necessary current required by the load imposed on it. When the elec-

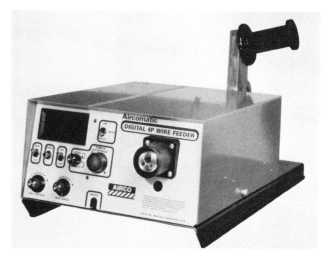

Figure 25-11. The control panel of the wire feeder permits adjustment of wire speed for welding. (*Left: Hobart Brothers Company; Right: Airco*)

supply machine or it can be separate from the welding machine and mounted elsewhere to facilitate welding over a large area. See Figure 25-12.

Figure 25-12. The wire feeder can be mounted on the power supply, or it can be positioned elsewhere for welding convenience. (*Miller Electric Manufacturing Company*)

Welding Gun

The function of the welding gun is to deliver the wire, shielding gas, and welding current to the arc area. Guns are either of the push or pull type. The pull gun has drive rolls that pull the welding wire from the wire feeder, and the push gun has the wire pushed to it by drive rolls in the wire feeder itself. The pull gun handles small diameter wires; the push gun moves heavier diameter wires. The pull type is also used to weld with soft wires such as aluminum and magnesium while the push gun is considered more suitable for welding with hard wires such as carbon and stainless steels and where currents are often in excess of 250 amperes.

Both guns have a trigger switch that controls the wire feed and arc as well as the shielding gas. When the trigger is released the wire feed, arc, and shielding gas stop immediately. With some equipment a timer is included to permit the shielding gas to flow for a predetermined time to protect the weld until it solidifies.

Guns are available with a straight or curved nozzle. See Figure 25-13. The curved nozzle provides easy access to intricate joints and difficult-to-weld patterns.

Shielding Gas[1]

In any gas shielded-arc welding process, the shielding gas can have a large effect upon the properties of a weld deposit. In gas metal-arc welding, gases are supplied from an external source. See Table 25-1.

The air in the arc area is displaced by the shielding

---

1. Hobart Brothers Company

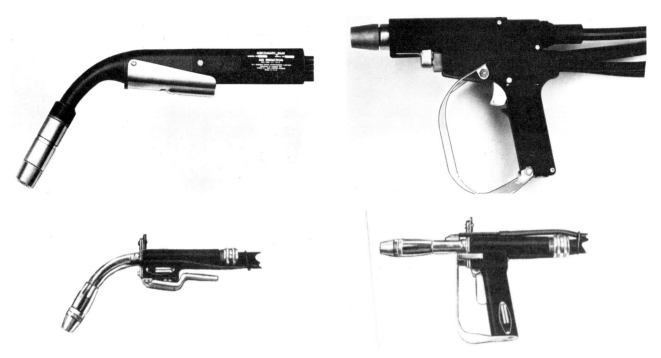

Figure 25-13. Guns used in gas metal-arc welding are available with straight or curved nozzles designed for easy access to the weld area.

gas. The arc is then struck under the blanket of shielding gas and the welding is accomplished. Since the molten weld metal is exposed only to the shielding gas it is not contaminated and strong dense weld deposits are obtained. The reason for shielding the arc area is to prohibit air from coming in contact with the molten metal.

By volume, air is made up of 21 percent oxygen, 78 percent nitrogen, 0.94 percent argon, and 0.04 percent other gases (primarily carbon dioxide). The atmosphere will also contain a certain amount of water depending upon its humidity. Of all of the elements that are in the air, the three which cause the most difficulty as far as welding is concerned are oxygen, nitrogen, and hydrogen.

Oxygen is a highly reactive element and combines readily with other elements in the metal or alloy to form undesirable oxides and gases. The oxide-forming aspect of the oxygen can be overcome with the use of deoxidizers in the steel weld metal.

The deoxidizers, such as manganese and silicon, combine with the oxygen and form a light slag which floats to the top of the weld pool. If the deoxidizers are not provided, the oxygen will combine with the iron and form compounds which can lead to inclusions in the weld material, and lower its mechanical properties. On cooling, the free oxygen in the arc area combines with the carbon of the alloy material and forms carbon monoxide. If this gas is trapped in

the weld material as it cools, it collects in pockets which cause pore or hollow spaces in the weld deposit.

Of all of the elements in the air, nitrogen causes the most serious problems in welding steel materials. When iron is molten, it is able to take a relatively large amount of nitrogen into solution. At room temperature, however, the solubility of nitrogen in iron is very low. Therefore, in cooling, the nitrogen precipitates or comes out of the iron as nitrites. These nitrites cause a high yield and tensile strength, and increased hardiness, but a pronounced decrease in the ductility and impact resistance of the steel materials. The loss of ductility often leads to cracking in and near the weld metal. Since air contains approximately 78 percent nitrogen by volume, therefore, if the weld metal is not protected from the air during welding, very pronounced decreases in weld quality will occur. In excessive amounts, nitrogen can also lead to gross porosity in the weld deposit.

Hydrogen is also harmful to welding. Very small amounts of hydrogen in the atmosphere produce an erratic arc. Of more importance is the effect that hydrogen has on the properties of the weld deposit. As in the case of nitrogen, iron can hold a relatively large amount of hydrogen when it is molten but upon cooling it has a low solubility for hydrogen. As the metal starts to solidify, it rejects the hydrogen. Hydrogen that becomes entrapped in the solidifying

TABLE 25-1. SHIELDING GASES FOR GAS METAL-ARC WELDING.

MATERIAL	PREFERRED GAS	REMARKS
Aluminum alloys	Argon	With DC reverse polarity removes oxide surface on work piece
Magnesium aluminum alloys	75% He 25% A	Greater heat input reduces porosity tendencies. Also cleans oxide surface
Stainless steels	Argon + 1% O_2	Oxygen eliminates under-cutting when DC reverse polarity is used
	(Argon + 5% O_2)	When DC straight polarity is used 5% O_2 improves arc stability
Magnesium	Argon	With DC straight polarity removes oxide surface on work piece
Copper (deoxidized)	75% He, 25% A (Argon)	Good wetting and increased heat input to counteract high thermal conductivity. Light gauges
Low-carbon steel	Argon + 2% O_2	Oxygen eliminates under-cutting tendencies also removes oxidation
Low-carbon steel	Carbon dioxide (spray transfer)	High quality low current out of position welding low spatter
	Carbon dioxide (buried arc)	High speed low cost welding accompanied by spatter loss
Nickel	Argon	Good wetting decreases fluidity of weld metal
Monel	Argon	Good wetting decreases fluidity of weld metal
Inconel	Argon	Good wetting decreases fluidity of weld metal
Titanium	Argon	Reduces heat-affected zone, improves metal transfer
Silicon bronze	Argon	Reduces crack sensitivity of this hot short material
Aluminum bronze	Argon	Less penetration of base metal Commonly used as a surfacing material

Note: ( ) = Second choice

metal collects at certain points and causes large pressures or stresses to occur. These pressures lead to minute cracks in the weld metal which can later develop into large cracks. Hydrogen also causes defects known as fish eyes and underbead cracking.

The effects of oxygen, nitrogen, and hydrogen make it essential that they be excluded from the weld area during welding. This is done by using inert gases for shielding. The inert gases consist of atoms which are very stable and do not react readily with other atoms. In nature there are only six elements possessing this stability and each of these elements exists as a gas.

The six inert gases are helium, neon, argon, krypton, xenon and radon. Since the inert gases do not readily form compounds with other elements, they are very useful as shielding atmospheres for arc welding. Of the six inert gases only helium and argon are important to the welding industry. This is because they are the only two which can be obtained in quantities at an economical price.

Carbon dioxide gas can also be used for shielding the weld area. Although it is not an inert gas, compensations can be made for its oxidizing tendencies and it can readily be employed for shielding the weld. Characteristics of this gas will be explained in detail later.

Argon. Argon gas has been used for many years as a shielding medium for fusion welding. Argon is obtained by the liquification and distillation of air. Air contains approximately 0.94 percent argon by volume or 1.3 percent by weight. This seems like a small quantity, but calculations show that the amount of air covering one square mile of the earth's surface contains approximately 800,000 pounds (364 metric tonnes) of argon.

In manufacturing argon, air is put under great pressure and refrigerated to very low temperatures. Then, the various elements in air are boiled off by raising the temperature of the liquid. Argon boils off from the liquid at a temperature of $-302.4°F$ ($-185.9°C$). For welding the purity of the argon is approximately 99.995 percent. When greater purity is required, the gas can be chemically cleaned to the purity of 99.999 percent.

Argon has a relatively low ionization potential. This means that the welding arc tends to be more stable when argon is used in the shielding gas. For this reason argon is often used in conjunction with other gases for arc shielding. The argon gives a quiet arc and thereby reduces spatter. Since argon has a low ionization potential, the arc voltage is reduced when argon is added to the shielding gas. This results in lower power in the arc and therefore lower penetration. The combination of lower penetration and reduced spatter makes the use of argon desirable when welding sheet metal.

Straight argon is seldom used for arc shielding except in welding such metals as aluminum, copper, nickel, and titanium. When welding steel the use of straight argon gas leads to undercutting and poor bead contour. Also, the penetration with straight argon is shallow at the bead edges and deep at the center of the weld. This can lead to lack of fusion at the root of the weld.

Argon plus oxygen. In order to reduce the poor bead contour and penetration pattern obtained with argon gas when welding on mild steel, it has been found that the addition of oxygen to the shielding gas is desirable. Small amounts of oxygen added to the argon produce significant changes. Normally, the oxygen is added in amounts of 1, 2, or 5 percent. Using gas metal-arc welding wires, the amount of oxygen which can be employed is limited to 5 percent. Additional oxygen might lead to the formation of porosity in the weld deposit.

Oxygen improves the penetration pattern by broadening the deep penetration finger at the center of the weld bead. It also improves bead contour and eliminates the undercut at the edge of the weld that is obtained with pure argon. Argon-oxygen mixtures are very common for welding alloy steels, carbon steels and stainless steel.

Carbon dioxide. Unlike argon or helium gases which are made up of single atoms, the carbon dioxide gas is made up of molecules. Each molecule contains one carbon atom and two oxygen atoms. The chemical formula for the carbon dioxide molecule is CO_2. Often, carbon dioxide is referred to simply as C-O-TWO.

At normal temperatures, carbon dioxide is essentially an inert gas. However, when subjected to high temperatures, carbon dioxide will disassociate into carbon monoxide and oxygen. In the high temperature of the welding arc this disassociation takes place to the extent that 20 to 30 percent of the gases in the arc area are oxygen (O_2). Because of this oxidizing characteristic of the CO_2 gas, the wires used with this gas must contain deoxidizing elements. The deoxidizing elements have a great affinity for the oxygen and readily combine with it. This prevents the oxygen atoms from combining with carbon or iron in the weld metal and producing low quality welds. The most common deoxidizers used in wire electrodes are manganese, silicon, aluminum, titanium, and vanadium.

Carbon dioxide is manufactured in most plants from flue gases which are given off by the burning of natural gas, fuel oil, or coke. It is also obtained as a byproduct of calcining operations of lime kilns, from the manufacturing of ammonia, and from the fermentation of alcohol. The carbon dioxide given off by the manufacturing of ammonia and the fermentation of alcohol is almost 100 percent pure.

The purity of carbon dioxide gas can vary considerably depending upon the process used to manufacture it. However, standards have been set up for the purity that must be obtained if it is to be used for arc welding. The purity specified for welding grade CO_2 is a minimum dew point of minus 40°F. This means that gas of this purity will contain approximately 0.0066 percent moisture by weight.

Carbon dioxide gas eliminates many of the undesirable characteristics that are obtained when using argon for arc shielding. With the carbon dioxide a broad, deep penetration pattern is obtained. This makes it easier for the operator to eliminate weld defects such as lack of penetration and lack of fusion. Bead contour is good and there is no tendency towards undercutting. Another advantage is its relatively low cost compared to other shielding gases.

The chief drawback of the CO_2 gas is the tendency for the arc to be somewhat violent. This can lead to spatter problems when welding on thin materials where appearance is of particular importance. However, for most applications this is not a major problem and the advantages of CO_2 shielding far outweigh its disadvantages.

CO_2 is used primarily for mild steel welding, although it has some application in the formulation of other shielding gas mixtures.

Helium. Helium is an inert gas and may be compared to argon in that respect. There the similarity ends. Helium has an ionization potential of 24.5 volts. It is lighter than air and has high thermal conductivity. The helium arc plasma will expand under heat (thermal ionization) reducing the arc density.

With helium there is a simultaneous change in arc voltage where the voltage gradient of the arc length is increased by the discharge of heat from the arc stream or core. This means that more arc energy is lost in the arc itself and is not transmitted to the work. The result is that, with helium, there will be a broader weld bead with relatively shallower penetration than with argon. (For gas tungsten-arc welding, the opposite is true.) This also accounts for the higher arc voltage, for the same arc length, that is obtained with helium as opposed to argon.

Helium is derived from natural gas. The process by which it is obtained is similar to that of argon. First the natural gas is compressed and cooled. The hydrocarbons are drawn off, then nitrogen, and finally the helium. This is a process of liquifying the various gases, each at its separate distillation temperature, until at $-452°F$ ($-269°C$), the helium is produced.

Helium has sometimes been in short supply due to governmental restrictions and, therefore, has not been used as much as it might have been for welding purposes. It is difficult to initiate an arc in a helium atmosphere with the gas tungsten-arc process. The problem is less acute with the gas metal-arc process.

Because of its high cost, helium is used primarily for special welding tasks and for the nonferrous metals such as aluminum, magnesium and copper. It is also used in combination with other shielding gases.

Argon-CO₂. For some applications of mild steel welding, welding grade CO_2 does not provide the arc characteristics needed for the job. This will usually manifest itself, where surface appearance is a factor, in the form of intolerable spatter in the weld area. In such cases a mixture of argon-CO_2 has usually eliminated the problem. Some welding authorities believe that the mixture should not exceed 25 percent CO_2. Others feel that mixtures with up to 80 percent CO_2 are practical.

The reason for wanting to use as much CO_2 as possible in the mixtures is cost. By using a cylinder of each type of gas, argon and CO_2, the mixture percentages may be varied by the use of flowmeters. This method precludes the possibility of gas separation such as may occur in premixed cylinders. When it is considered that premixed argon-CO_2 gas is sold at the price of pure argon then it makes good sense to mix your own. The price of CO_2 is approximately 15 percent that of argon in most areas of the country.

Argon-CO_2 shielding gas mixtures are employed for welding mild steel, low-alloy steel and, in some cases, for stainless steels.

Argon-Helium-CO₂. This mixture of shielding gases is used primarily for welding austenitic stainless steel. The combination of gases provides a unique characteristic to the weld. It is possible to make a weld with very little build-up of the top bead profile. The result is excellent for those applications where a high crowned weld is detrimental rather than a help. This gas mixture has found considerable use in the welding of stainless steel pipe.

Gas Flow and Regulation

For most welding conditions, the gas flow rate will approximate 35 cubic feet per hour. This flow rate may be increased or decreased depending upon the particular welding application. See Tables 25-5 through 25-10.

The data presented in these tables are not intended as absolute settings but only as a point in making the starting settings. Final adjustments must often be made on a trial and error basis. Actually the correct settings will be governed by the type and thickness of metal to be welded, position of the weld, kind of shielding gas used, diameter of electrode and type of joint.

The proper amount of gas shielding usually results in a rapidly crackling or sizzling arc sound. Inadequate gas shielding will produce a popping arc sound with resultant weld discoloration, porosity, and spatter.

Gas drift may occur with high weld travel speeds or from unusually drafty or windy conditions in the weld area. Since one or more of these factors may cause the gas to drift away from the arc, the result is inadequate gas shielding. See Figure 25-14. The gas nozzle should be adjusted for proper shielding and outside influences should be eliminated by proper wind-breakers or shields. See Figure 25-15.

Correct positioning of the nozzle with respect to the work will be determined by the nature of the weld. The gas nozzle may usually be placed up to 2″ from the work. Too much space between nozzle and work reduces the effectiveness of a gas shield while too little space may result in excessive weld spatter which collects on the nozzle and shortens its life.

Wire for Gas Metal-Arc Welding

Filler wire for gas metal-arc welding should be similar in composition to the base metal. Several common wires and their suitability for a particular type of welding are listed in Table 25-2. These designations are based on the AWS classification system. Thus for mild steel wires, the *E* identifies it as an electrode, the next two digits show the tensile strength in psi per thousand, the *S* indicates a solid bare wire, and the final symbols specify a particular classification based on chemical composition of the wire.

Wires are usually available in spools of several different sizes as well as in 36″ rod lengths for gas tungsten-arc welding.

Best results are obtained by using the proper diameter wire for the thickness of the metal to be welded and the position in which the welding is to be done. See Tables 25-5 to 25-10.

Basic wire diameters are 0.020″, 0.030″, 0.035″, 0.045″, ¹⁄₁₆″ and ⅛″. Generally wires of 0.020″, 0.030″ or 0.035″ are best for welding thin metals, although they also can be used to weld low and medium-carbon steels, and low-alloy/high-strength steels of medium thickness. See Table 25-3. Medium thickness metals normally require 0.045″ or ¹⁄₁₆″ diameter electrodes. For thick metals, ⅛″ electrodes are usually recommended. However, the position of welding is a factor which must be considered in electrode selection. Thus for vertical or overhead welding, smaller diameter electrodes will be more satisfactory than larger diameter wires.

Wire feed. The amperage of the welding current used limits the speed of the wire feed to a definite range. However, it is possible to make adjustments of the wire feed within the range. For a specific amperage setting, a high speed of wire feed will result in a short arc. A low speed contributes to a long arc. Also a higher speed must be used for overhead weld-

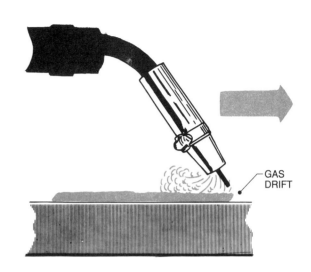

Figure 25-14. High welding travel speeds or unusually drafty conditions in the weld area can cause the gas shielding to drift away from the arc, resulting in inadequate gas shielding. (*Hobart Brothers Company*)

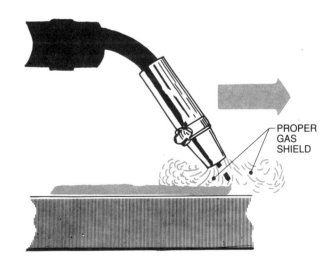

Figure 25-15. Shielding gas drift can be controlled with proper wind-breakers or nozzle adjustment. (*Hobart Brothers Company*)

ing than speeds for flat position welding. See Tables 25-5 to 25-10.

Wire stickout. Wire stickout refers to the distance the wire projects from the nozzle of the gun. See Figure 25-16. Wire stickout influences the welding current since it changes the preheating in the wire. When the stickout increases the preheating increases, which means that the power source does not have to furnish as much welding current to melt the wire at a given feed rate. Since the power source is self-regulating, the current output is automatically decreased. Conversely, if the stickout decreases, the power source is forced to furnish more current to burn off the wire at the required rate.

For most gas metal-arc welding applications, the wire stickout should measure from ⅜″ to ¾″. On

TABLE 25-2. FILLER WIRES FOR GAS METAL—ARC WELDING.

MILD STEEL WIRES	
E-60S-1	Silicon deoxidized wire for low and medium-carbon steels. Can be used either with CO_2, argon, or argon-CO_2 mixtures. Performs best on killed steels.
E-60S-2	Premium quality wire containing Al, Zr, and Ti in addition to silicon and manganese deoxidizers. Can be used with CO_2 or argon-CO_2 or argon-O_2. Recommended for pipe welding and heavy vessel construction.
E-60S-3	Used for higher quality welding either with CO_2, argon-O_2, or argon-CO_2 mixtures. Produces medium quality welds in rimmed steels and high quality welds in semi-killed steels.
E-70S-1B	Low-alloy wire for carbon steels, low-alloy steels, and high strength low-alloy steels.
E-70S-3	General purpose welding of low to medium-carbon steels. Has a silicon content high enough to permit its use in either CO_2, argon-oxygen mixtures or mixture of the two.
E-70S-6	Contains higher manganese and silicon levels and has more powerful deoxidizing characteristics for welding over rust and scale or where stringent cleaning practices cannot be followed.
E-70S-5	Contains aluminum and is designed for single or multipass welding of rimmed, semi-killed, or killed mild steels. Suitable to weld steels having rusty or dirty surfaces and normally used with CO_2 gas.
ALUMINUM WIRES	
ER-1100 ER-4043 ER-5183 ER-5554, 5556 ER-5654	To weld aluminum of similar composition
STAINLESS STEEL WIRES	
ER-308L	For welding types 304, 308, 321, 347
ER-308L-Si	For welding types 301, 304
ER-309	For welding types 309 and straight chromium grades when heat treatment is not possible. Also for 304-clad
ER-310	For welding types 310, 304-clad and hardenable steels
ER-316	For welding 316
ER-347	For welding types 321, and 347 where maximum corrosion resistance is required
COPPER AND COPPER-BASE ALLOY WIRES	
E-CuSi (Silicon Bronze) E-CuAl-A1 (Aluminum Bronze) E-Cu (Deoxidized Copper) E-CuAl-A2 (Aluminum Bronze) E-CuAl-B (Aluminum Bronze)	Special wires for welding copper and copper-based alloys.

TABLE 25-3. WIRE DIAMETERS FOR GAS METAL-ARC WELDING.
(Manual travel, single pass, flat fillet welds)

MATERIAL THICKNESS (inches)	ELECTRODE SIZE	WELDING CONDITIONS DCRP (arc volts)	(amperes)	GAS FLOW (cfh)	TRAVEL SPEED (ipm)
0.025	0.030	15-17	30-50	15-20	15-20
.031	.030	15-17	40-60	15-20	18-22
.037	.035	15-17	65-85	15-20	35-40
.050	.035	17-19	80-100	15-20	35-40
.062	.035	17-19	90-110	20-25	30-35
.078	.035	18-20	110-130	20-25	25-30
.125	.035	19-21	140-160	20-25	20-25
.125	.045	20-23	180-200	20-25	27-32
.187	.035	19-21	140-160	20-25	14-19
.187	.045	20-23	180-200	20-25	18-22
.250	.035	19-21	140-160	20-25	10-15
.250	.045	20-23	180-200	20-25	12-18

Shielding gas: CO_2, welding grade
Wire stick-out—1/4″ to 3/8″
Hobart Brothers Co.

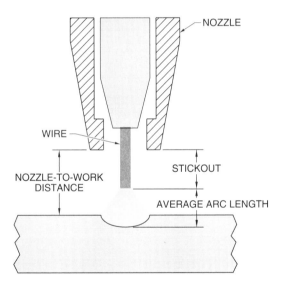

Figure 25-16. Correct wire stickout (electrode extension) is necessary to control the characteristics of the wire electrode in the weld puddle.

micro wires a shorter wire stickout ranging from ¼″ to ⅜″ is recommended. An excessive amount of wire stickout results in increased wire preheating; this tends to increase the deposit rate. Too much wire stickout may also produce a ropy appearance in the weld bead. Too little stickout will cause the wire to fuse to the nozzle tip, which decreases the life of the tip. As the amount of wire stickout increases, it may become increasingly difficult to follow the weld seam, particularly with a small diameter wire. The tip should be either flush with the gas nozzle or recessed in the nozzle.

The wire, in a near-plastic state between the tip and arc, tends to move (whip) around, describing a somewhat circular pattern. Decreasing the amount of wire stickout and straightening the welding wire tend to decrease the amount of wire whip.

Welding Current

A wide range of current values can be used with each wire diameter. This permits welding various thicknesses of metal without having to change wire diameter. The correct current to use for a particular joint must often be determined by trial. The current selected should be high enough to secure the desired penetration without cold lapping (cold shuts) but low enough to avoid undercutting and burn through. See Tables 25-5 to 25-10.

Joint Edge Preparation and Weld Backing

Preparation of the edge of each member to be joined is recommended to aid in the penetration and control of weld reinforcement. For gas metal-arc welding, beveling the edges is usually desirable for butt joints thicker than ¼″ if complete root penetration is desired. For thinner sections, a square butt joint is best.

To a considerable extent the same conventional joint design recommended for other arc welding processes can be used for gas metal-arc welding. However, some joint modifications are often incorporated to compensate for the operating characteristics of gas metal-arc welding. Thus the arc in gas metal-arc welding is more penetrating and narrower than the arc in shielded metal-arc welding. See Figure 25-17. Consequently, groove joints can have smaller root faces and root openings. Also, since the nozzle does not have to be placed within the groove, a narrower included angle can be provided. By reducing the joint area less weld metal is required; this lowers material and labor costs.

In the gas metal-arc process, weld backing is helpful in obtaining a sound weld at the roots. Backing prevents molten metal from running through the joint being welded, especially when complete weld penetration is desired.

There are several types of material used for backing: steel and copper blocks, strips and bars; carbon blocks, plastics, or the use of fire clay. Some of these serve to conduct heat away from the joint and also to form a mold or dam for the metal. The most commonly used backing for gas metal-arc welding is copper or steel.

Positioning Work and Electrode

The proper position of the welding torch and weldment is important. In gas metal-arc welding, the flat position is preferred for most joints because this position improves the molten metal flow, bead contour, and gives better gas protection. However, on gauged material, it is sometimes necessary or advantageous to weld with the work inclined 10° to 20°. The welding is done in the downhill position. This has a tendency to flatten the bead and increase the travel speed.

The alignment of the welding wire in relation to the joint is very important. The welding wire should be on the center line of the joint for most butt joints, if the pieces to be joined are of equal thickness. If the pieces are unequal in thickness, the wire may be moved toward the thicker piece.

Correct work and travel angles are necessary for correct bead formation. See Figure 25-18. The travel angle refers to the angle the gas metal-arc welding gun makes with a reference line perpendicular to the axis of the weld in the plane of the weld axis. The travel angle may be a push angle or a drag angle depending upon the position of the gun. If the gun is angled back toward the beginning of the weld, the travel angle is called a drag angle. If the gun is pointing ahead toward the end of the weld, the travel angle is called a push angle. See Figure 25-19.

When the gun is ahead of the weld, it is referred to as pulling the weld metal. If the gun is behind the weld, it is said to be pushing the metal.

Generally the penetration of beads deposited with a pulling technique is greater than with a pushing technique. Furthermore, since the welder can see the weld crater easier in a pulling action he can pro-

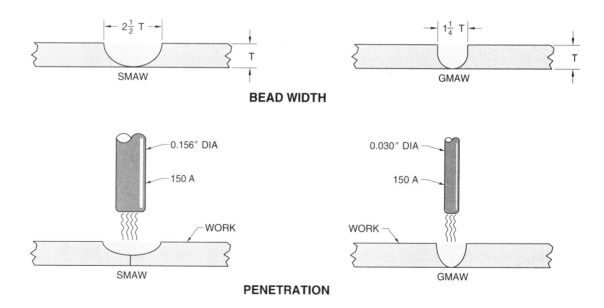

Figure 25-17. The gas metal-arc welding process produces deeper penetration and narrower beads than the shielded metal-arc process.

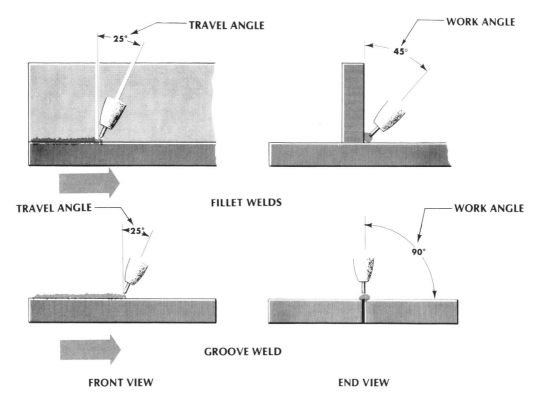

TRAVEL ANGLE

25°

WORK ANGLE

45°

FILLET WELDS

TRAVEL ANGLE

25°

WORK ANGLE

90°

GROOVE WELD

FRONT VIEW   END VIEW

Figure 25-18. The correct travel angle and work angle are necessary for correct bead formation.

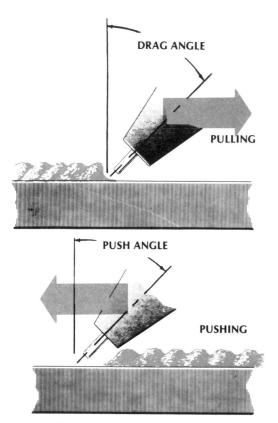

DRAG ANGLE

PULLING

PUSH ANGLE

PUSHING

Figure 25-19. The pulling (drag) technique is preferred when welding heavy materials, while the pushing technique is better for welding light gauge metals.

technique. Furthermore, since the welder can see the weld crater easier in a pulling action he can produce high quality welds more consistently. On the other hand, pushing permits the use of higher welding speeds and produces less penetrating and wider welds.

Preliminary Checks for Gas Metal-Arc Welding

Before starting to weld, it is always a good practice to check the following:

1. All electric power controls are in the OFF position.

2. All hose and cable connections from the gun to the feeder are in good condition, properly insulated, and connections have been correctly made and secured.

3. Correct nozzle for the diameter wire.

4. Wire is properly threaded through gun.

5. Orifices of contact tube and nozzle are clean. Blow out gun occasionally because sometimes the gun becomes loaded with dust which restricts proper wire feed and flow of protective gases.

6. Wire speed and feed have been predetermined and adjusted on the feeder control.

7. Shielding gas and water coolant sources are on and adjusted on the feeder control.

8. Wire stickout is correct.

9. Contact tip is in proper shape. Tips eventuallly wear out, especially under high usage, and must be replaced for good welds.

WELDING PROCEDURE

In general, gas metal-arc welding procedure follows a definite sequence regardless of the kind of welding that is being done. Basically the following steps are involved:

1. Set the voltage, wire feed, and gas flow to the prescribed conditions for the required type of welding. See Tables 25-7 to 25-10. During welding the wire speed rate may have to be varied to correct for too much or too little heat input.

2. Adjust the wire for the proper amount of stickout.

3. Start the arc and move the gun along the joint at a uniform speed, keeping the gun at the correct angle. If the arc is not started properly the filler wire may stick to the work or actually freeze to the tip. Should this happen, shut off the machine and free the wire.

4. Move the gun along the joint with a pushing or pulling motion. As the gun is moved keep the wire at the leading edge of the puddle. Also be sure the wire is centered in the gas pattern to insure adequate shielding. A slight weaving motion is helpful for complete penetration.

5. Release the trigger when reaching the end of the weld. This stops the wire feed and interrupts the welding current. However, always keep the gun over the weld until the gas stops flowing in order to protect the puddle until it solidifies.

6. Shut down the welding unit when welding is completed. Follow this sequence:

a. Turn OFF wire speed control.

b. Shut OFF gas flow at cylinders.

c. Squeeze the welding gun trigger to bleed the lines.

d. Hang up the welding gun.

e. Shut OFF welding machine.

During any welding operation certain welding conditions may have to be changed. Some of the more specific welding variables with their required changes are shown in Table 25-4.

Arc Starting

Starting an electrical arc for a welding process involves three major factors: electrical contact, arc voltage, and time. See Figure 25-20. To assure good arc starts, it is necessary for the electrode wire to make good electrical contact with the work. The electrode must exert sufficient force on the workpiece to penetrate impurities.

Arc initiation becomes increasingly more difficult as wire stickout increases. A reasonable balance of volts and amperes must be maintained in order to assure the proper arc and to deposit the metal at the best electrode melting rate. Once the arc is started, the gun is held at the correct angles and moved at a uniform speed.

Possible Weld Defects

Gas metal-arc welding like any other form of welding must be controlled properly to produce consistently high quality welds. The beginner, in practicing gas metal-arc welding, should analyze each completed weld to avoid repeated weld defects. The following are a few of the more common defects which may be encountered during the early stages of the learning process.

Cold lap. Cold laps usually occur when the arc does not melt the base metal sufficiently, causing the slightly molten puddle to flow into the unwelded base metal. See Figure 25-21. Very often if the puddle is allowed to become too large, this too will result in cold laps. For proper fusion, the arc should be kept at the leading edge of the puddle. When directed in such a manner, the molten puddle is prevented from flowing ahead of the welding arc. Also remember that the size of the puddle can be reduced by increasing the travel speed or reducing the wire speed feed.

Surface porosity. Generally, surface porosity is the direct result of atmospheric contamination. See Figure 25-22. It is caused by having the shielding gas set too low or too high. If it is too low the air in the arc area is not full displaced; if the gas flow is too much, an air turbulence is generated which prevents complete shielding. On occasion, porosity will occur if the welding is being done in a windy area. Without some protective wind shield the gas envelope may be blown away, exposing the molten puddle to the contaminating air.

Crater porosity or cracks. The chief cause of crater defects is removing the gun and shielding gas before the puddle has solidified. See Figure 25-23. Other possible causes of crater porosity or cracks are moisture in the gas, dirt, oil, rust or paint on the base metal, or excessive tip-to-work distance.

Insufficient penetration. Lack of penetration is due to a low heat input in the weld area or failing to keep the arc properly located on the leading edge of the puddle. See Figure 25-24. If the heat input is too slow, increase the wire feed speed to get a higher

TABLE 25-4. RECOMMENDED VARIABLE ADJUSTMENTS FOR GAS METAL-ARC WELDING.

Change Required \ Welding Variable	Arc Voltage	Welding Current (See footnote)	Travel Speed	Travel Angle	Electrode Extension (Stickout)	Wire Size	Gas Type
Deeper Penetration		Increase (1)		Drag Max. 25°	Decrease (2)	Smaller* (5)	CO_2 (4)
Shallower Penetration		Decrease (1)		Push (3)	Increase (2)	Larger* (5)	AR + CO_2
Bead Height and Bead Width — Larger Bead		Increase (1)	Decrease (2)		Increase* (3)		
Smaller Bead		Decrease (1)	Increase (2)		Decrease* (3)		
Higher Narrower Bead	Decrease (1)			Drag (2)	Increase (3)		
Flatter Wider Bead	Increase (1)			90° or Push (2)	Decrease (3)		
Faster Deposition Rate		Increase (1)			Increase* (2)	Smaller (3)	
Slower Deposition Rate		Decrease (1)			Decrease* (2)	Larger (3)	

Footnote: Same adjustment is required for wire feed speed.
Key: (1) First choice, (2) Second Choice, (3) Third choice, (4) Fourth Choice, (5) Fifth choice.
*When these variables are changed, the wire feed speed must be adjusted so that the welding current remains constant.

(Hobart Brothers Co.)

amperage.

Excessive penetration. Too much penetration or *burn-through* is caused by having excessive heat in the weld zone. See Figure 25-25. By reducing the wire speed feed, the amperage is lowered and there will be less heat. Excessive penetration can also be avoided by increasing the travel speed. If the root opening in the joint is too wide too much burn-through may result. Usually improper joint design can be remedied by increasing the stickout and by weaving the welding gun.

Whiskers. Whiskers are short lengths of electrode wire sticking through the weld joint. See Figure 25-26. They are caused by pushing the wire past the leading edge of the puddle. The small section of wire then protrudes inside the joint and becomes welded to the deposited metal. The best way to remedy this is to reduce travel speed, increase slightly the tip-to-work distance, or reduce the wire feed speed.

GAS METAL-ARC WELDING COMMON METALS

Gas metal-arc welding has become one of the most universally accepted processes for joining all types of metals. The ease with which sound welds can be produced by gas metal-arc welding has in many instances revolutionized welding practices in numerous industries. One of its particularly outstanding features is the ease with which production welding can be mechanized, thereby substantially reducing manufacturing costs.

Generally speaking, the same type of equipment and welding techniques apply to joining all metals.

Figure 25-20. Good electrical contact is necessary when starting the arc. (*Hobart Brothers Company*)

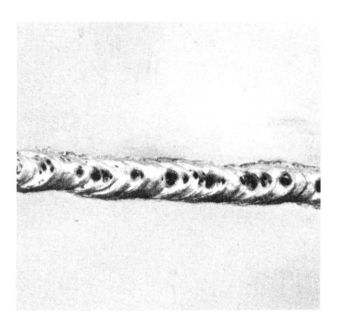

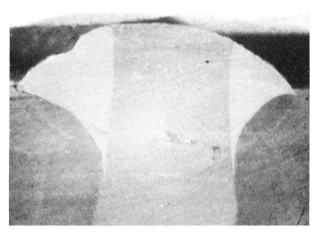

Figure 25-21. Cold lap occurs when the arc does not melt the base metal sufficiently.

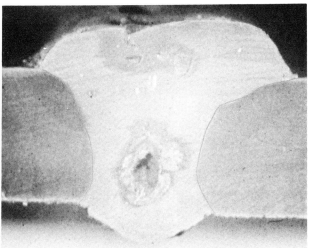

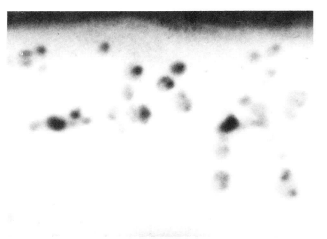

A few specific characteristics for welding several common metals are included in the paragraphs that follow.

Carbon Steels

Both the spray arc and short arc produce excellent welds in carbon steels. For spray-arc welding a mixture containing 5 percent oxygen with argon is generally recommended. The addition of oxygen provides a more stable arc, minimizes undercutting and permits greater speeds.

Considerable amount of steel welding is done with a mixture of argon and CO_2. A straight CO_2 gas is sometimes used, especially for high-speed produc-

Figure 25-22. Surface and subsurface porosity is caused by atmospheric contamination, the result of insufficient shielding gas.

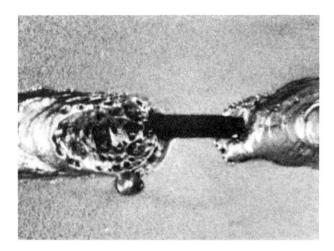

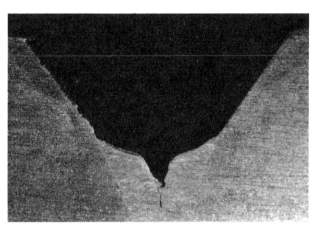

Figure 25-23. Crater porosity and cracks occur when the gun is removed before the weld puddle has solidified.

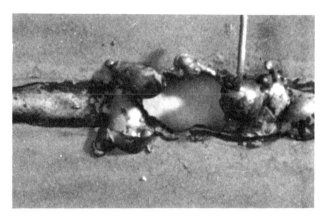

Figure 25-25. Excessive penetration and burn-through can be avoided by reducing the wire speed feed and increasing the travel speed. (*Hobart Brothers Company*)

Figure 25-24. Insufficient penetration can result from low heat input to the weld area or a failure to keep the arc located properly on the leading edge of the puddle. (*Hobart Brothers Company*)

tion welding. However, with CO_2 the arc is not a true spray arc.

For short-arc welding of carbon and low-alloy steels a 25 percent carbon dioxide and 75 percent argon mixture is preferred. The dioxide mixture improves arc stability and minimizes spatter.

Thin steel plates 0.035″ to ⅛″ in thickness may be butt-welded with square edges. Usually an opening of ¹⁄₁₆″ or less is recommended. For wider openings the short arc is better since relatively large gaps are more easily bridged without excessive penetration. Plates ³⁄₁₆″ and ¼″ may be square butt-welded with a ¹⁄₁₆″ to ³⁄₃₂″ root opening but usually two passes are necessary. For quality welds some beveling is desired. Plates ¼″ and greater require single or double-V grooves with 50 to 60 degree included angles. U-grooves having a root spacing of ¹⁄₃₂″ to ³⁄₃₂″ are necessary on plates thicker than one inch. See Figure 25-27.

Table 25-5 lists specific requirements for gas metal-arc welding carbon steels. On multipass welds the sequence of bead deposits is similar to shielded metal-arc welding. See Exercises 1-7.

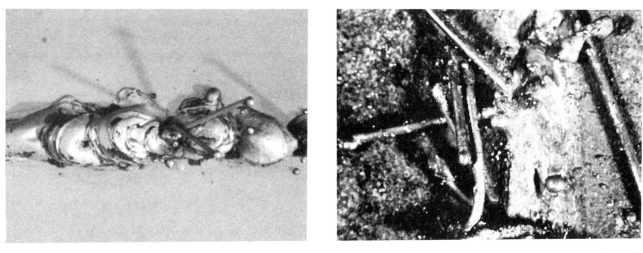

Figure 25-26. Whiskers (short lengths of electrode wire) are caused by pushing the wire past the leading edge of the puddle. (*Hobart Brothers Company*)

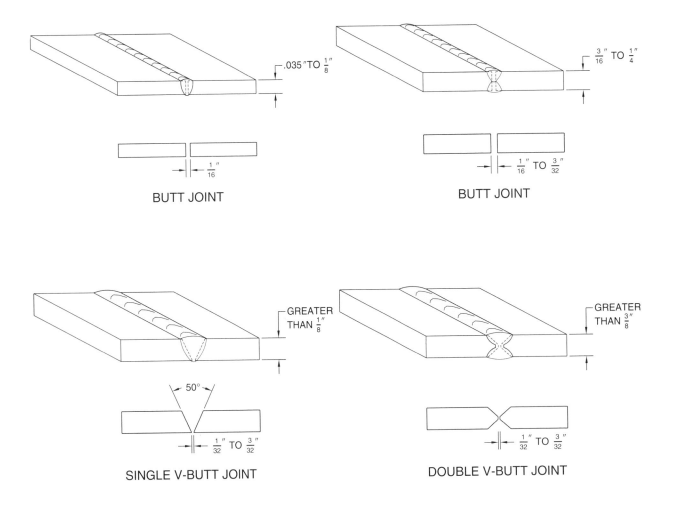

Figure 25-27. For gas metal-arc welding butt joints in carbon steels, proper edge preparation and root opening are necessary.

TABLE 25-5. GAS METAL-ARC WELDING—CARBON STEEL.

PLATE THICKNESS (inches)	JOINT AND EDGE PREPARATION	WIRE dia (inches)	GAS FLOW (cfh)	DCRP CURRENT (amps)	WIRE FEED (ipm)	
0.035				55	16*	117
.047				65	17*	140
.063	Non-positioned fillet or lap	.030	10-15	85	17*	170
.078				105	18*	225
.100				110	18*	225
1/8				130	19*	300
1/8	Butt (square edge)	1/16		280	—	165
3/16	Butt (square edge)	1/16		375	—	260
3/16	Fillet or lap	1/16		350	—	230
1/4	Double V butt (60° included angle, no nose)		40-50	375 (1st pass) 430 (2nd pass)	27	83 (1st) 95 (2nd)
5/16	Double V butt (60° included angle, no nose)			400 (1st pass) 420 (2nd pass)	28	87 (1st) 92 (2nd)
5/16	Non-positioned fillet			400		87
1/2	Double V butt (60° included angle, no nose)	3/32		400 (1st pass) 450 (2nd pass)		87 (1st) 100 (2nd)
1/2	Non-positioned fillet			450	28	100
3/4	Double V butt (90° included angle, no nose)			450 (all 4 passes)	29	100
3/4	Positioned fillet			475	30	110
1	Fillet			450 (all 4 passes)	28	100

Gas column: Mixture (75% A + 25% CO₂) / Mixture (95% argon + 5% O₂)

*short arc
Linde Co.

Running Beads on Mild Steel Using the GMAW (Gas Metal-Arc Welding) Process | Exercise 1

1 Use .035″ diameter wire, E-70S-3.

2 Wire stickout (electrode extension) should be ¼″ to ⅜″.

3 Set the machine for DCRP (direct current reverse polarity).

4 Amperage should be 100-120 amps, voltage 19-21 volts.

5 Carbon dioxide is used as shielding gas at 20 cfm.

6 Set the wire feed control so that the ammeter reads 100-120 amps. To obtain the correct reading have another person ob-

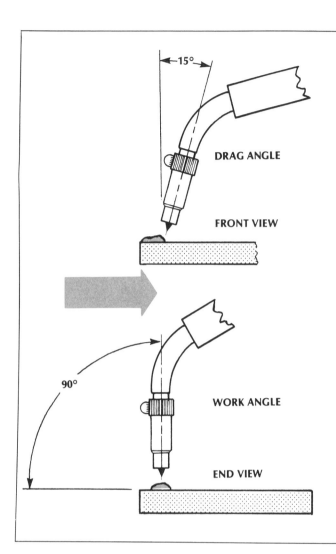

serve the amperage while the welding is being done.

7 Use the same procedure to set the voltage at the power source.

8 Adjust the voltage until wire is feeding properly and bead width is ⁵⁄₁₆″ wide and ⅛″ high.

9 Obtain a piece of mild steel ³⁄₁₆″ to ¼″ x 4″ x 6″.

10 Use a 90° work angle and a 10° to 15° drag angle. Drag angle refers to the travel angle when the gun is pointing backward while welding. If the gun is pointing forward, the travel angle is referred to as push angle. Drag angle results in better penetration than push angle does.

11 If the electrode extension is increased during welding, amperage is decreased. If the electrode extension is decreased, amperage is increased.

12 Run a series of straight consistent beads approximately ⅜″ apart.

Running a Padding Plate on Mild Steel Using the GMAW Process   Exercise 2

1 Refer to Exercise 1 for equipment set-up and adjustment.

2 Obtain a piece of ¼″ x 4″ x 6″ mild steel.

3 Using a 90° work angle and a 10° to 15° drag angle, deposit a bead ¼″ from the edge of the piece. The bead should be ⁵⁄₁₆″ wide and ⅛″ high.

4 The second bead should be deposited overlapping the first bead in half. Use an 80° to 90° work angle and a 10° to 15° drag angle.

5 Run consistent overlapping beads until the plate is covered.

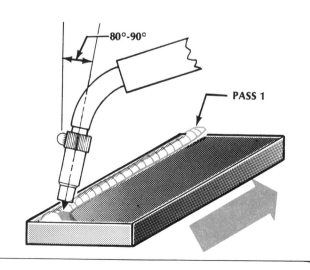

Welding a Butt Joint on Mild Steel in Flat Position Using the GMAW Process
Exercise 3

1 Refer to Exercise 1 for equipment set-up and adjustment.

2 Obtain two pieces of ³⁄₁₆″ to ¼″ x 1½″ x 6″ mild steel.

3 Tack weld the two pieces to form a butt joint with a ³⁄₃₂″ to ⅛″ root opening.

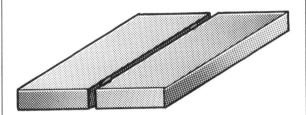

4 Use a 90° work angle and a 10° drag angle.

5 The speed of travel should allow complete penetration. A slight weaving motion may be used to control the puddle.

6 Maintain the electrode on the leading edge of the puddle to avoid whiskers caused by the wire penetrating through the weld puddle.

7 If there is too much penetration, lengthen the electrode extension, decrease the amperage, and/or adjust the voltage for a smooth running arc.

Welding a Lap Joint on Mild Steel in Flat Position Using the GMAW Process
Exercise 4

1 Refer to Exercise 1 for equipment set-up and adjustment.

2 Obtain two pieces of ³⁄₁₆″ to ¼″ x 1½″ x 6″ mild steel.

3 Tack weld the two pieces to form a lap joint. Rest the piece against a firebrick to obtain flat position.

4 Position the gun at a 45° work angle and a 10° to 15° drag angle. Use a slight weaving motion.

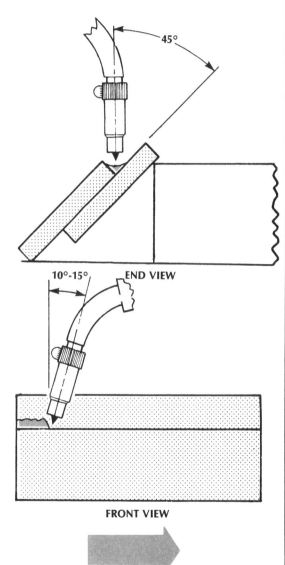

5 The bead face should be flat to slightly convex.

Welding a Multiple-Pass T-Joint on Mild Steel in Horizontal Position Using the GMAW Process

Exercise 5

1 Refer to Exercise 1 for equipment set-up and adjustment.

2 Obtain two pieces of ³⁄₁₆″ to ¼″ x 2″ x 6″ mild steel.

3 Tack weld the two pieces to form a T-joint.

4 Position the gun at a 45° work angle and a 10° to 15° drag angle. Deposit the first pass on both sides of the T-joint.

5 Position the gun at a 55° work angle and a 10° to 15° drag angle. Deposit the second pass overlapping half the width of the first pass on both sides of the T-joint.

6 Position the gun at a 35° work angle and a 10° to 15° drag angle. Deposit the third pass overlapping the first and second passes on both sides of the T-joint.

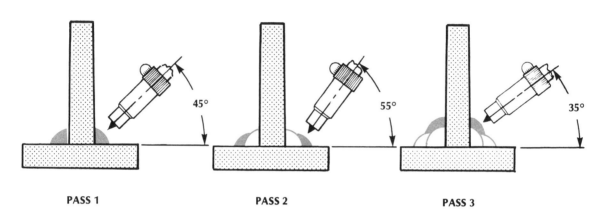

PASS 1   PASS 2   PASS 3

Welding a T-Joint on Mild Steel in Vertical Position Using the GMAW Process

Exercise 6

1 Refer to Exercise 1 for equipment set-up and adjustment.

2 Obtain two pieces of ³⁄₁₆″ to ¼″ x 2″ x 6″ mild steel.

3 Tack weld the two pieces to form a T-joint.

4 Position the gun with a work angle of 45° and a 10° to 20° drag angle.

5 Weld in a downward direction using a slight weaving motion. Pause at the toes of the weld to avoid undercutting.

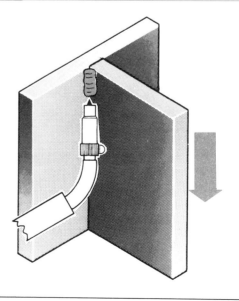

Welding a T-Joint on Mild Steel in Overhead Position Using the GMAW Process — Exercise 7

1 Refer to Exercise 1 for equipment set-up and adjustment.

2 Obtain two pieces of ³⁄₁₆″ to ¼″ x 2″ x 6″ mild steel.

3 Tack weld the two pieces to form a T-joint.

4 Position the gun with a work angle of 45° and a 5° to 10° drag angle. Deposit the first pass on both sides of the T-joint.

5 Position the gun with a work angle of 50° and a 5° to 10° drag angle. Deposit the second pass on both sides of the T-joint with a slight weaving motion.

6 Position the gun with a work angle of 40° and a 5° to 10° drag angle. Deposit the third pass on both sides of the T-joint with a slight weaving motion.

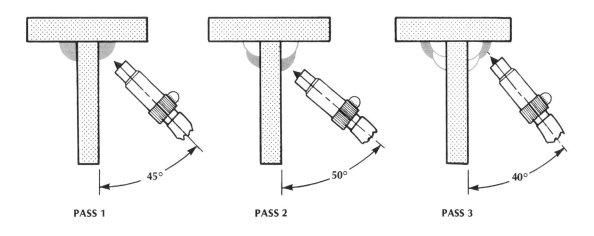

PASS 1 — 45°   PASS 2 — 50°   PASS 3 — 40°

Aluminum

The design of seams for aluminum is similar to that for steel. However, narrower joint spacing and lower welding currents are recommended due to the higher fluidity of the metal.

Argon gas is preferred for gas metal-arc welding plates up to 1″ in thickness since it provides better metal transfer and arc stability with less spatter. Sometimes in position welding of 1100 and 3003 aluminum the addition of a small amount of oxygen to the argon increases spray transfer and improves coalescence (flow of metals together).

When welding plates between 1″ and 2″ a mixture of 90 percent argon and 10 percent helium will often prove advantageous. This provides a higher heat input associated with helium and good cleaning action obtained with argon.

The *short-arc welding* of aluminum produces a colder arc than the spray type arc and thereby permits the weld puddle to solidify rapidly.

This action is especially advantageous in vertical, overhead, and horizontal welding, and in welding lighter materials. In vertical welding, a downhill technique is preferred.

Spray-arc welding aluminum is especially suitable for thick sections. With spray arc more heat is produced to melt the wire and base metal. As a rule vertical, horizontal and overhead welds are more difficult to make than with the short arc.

Practice welding—lap, T, and butt joints:

1. Tack two pieces of aluminum to form proper joint.

2. Set voltage, wire feed, and gas flow according to conditions in Tables 25-6 and 25-7.

3. Angle gun nozzle about 15° with a wire stickout of ½″ to ¾″.

4. Use a pushing motion.

Stainless Steel

Copper back-up strips should be used to weld

TABLE 25-6. GAS METAL-ARC WELDING—ALUMINUM (SPRAY-ARC).

PLATE THICKNESS (inches)	TYPE OF JOINT	WIRE dia (inches)	ARGON FLOW (cfh)	DCRP (amperes)	VOLTAGE (volts)	APPROXIMATE WIRE FEED (ipm)
0.040	Fillet or tight butt	0.030	30	40	15	240
.050	Fillet or tight butt	.030	15	50	15	290
.063	Fillet or tight butt	.030	15	60	15	340
.093	Fillet or tight butt	.030	15	90	15	410

Linde Co.

TABLE 25-7.   GAS METAL-ARC WELDING—ALUMINUM (SPRAY-ARC)

PLATE THICKNESS	PREPARATION	WIRE DIAMETER (inches)	ARGON FLOW (cfh)	DCRP (amperes)	VOLTAGE
0.250	Single V butt (60° included angle) sharp nose back-up strip used	3/64	35	180	24
	Square butt with back-up strip	3/64	40	250	26
	Square butt with no back-up strip	3/64	35	220	24
.375	Single V butt (60° included angle) shop nose, back-up strip used	1/16	40	280	27
	Double V butt (75° included angle, 1/16″ nose.) No back-up. Back chip after root pass	1/16	40	260	26
	Square butt with no back-up strip	1/16	50	270	26
.500	Single V butt (60° included angle) sharp nose. Back-up strip used	1/16	50	310	27
	Double V butt (75° included angle 1/16″ nose.) No back-up. Back chip after root pass	1/16	50	300	27

Linde Co.

stainless steel up to ¹⁄₁₆″ in thickness. Precaution must be taken to prevent air from reaching the underside of the weld while the puddle is solidifying since the oxygen and nitrogen will weaken the weld. To prevent air from contacting the underside of the weld an argon back-up gas is often used.

Spray-arc welding with ¹⁄₁₆″ diameter wire and high current produces good welds. DCRP with a 1 or 2 percent argon-oxygen mixture is recommended for most stainless steel welding.

The forehand or pushing technique is generally used for welding stainless steel. On plates ¼″ or more in thickness the gun should be moved back and forth with a slight side to side movement. Thin materials are best welded with just a slight back-and-forth motion along the joint. See Figure 25-28. As a rule the

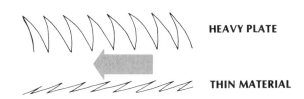

Figure 25-28. A slight weaving motion is used when gas metal-arc welding stainless steel plate 1/4″ or more in thickness.

short arc produces better welds on thin materials when overhead or vertical welding is required.

Tables 25-8 and 25-9 list specific requirements for gas metal-arc welding stainless steels.

TABLE 25-8. GENERAL WELDING CONDITIONS FOR SPRAY-ARC WELDING OF STAINLESS STEEL.

PLATE THICKNESS (inches)	JOINT AND EDGE PREPARATION	WIRE dia	GAS FLOW	CURRENT DCRP (amps)	WIRE FEED (ipm)	SPEED (ipm)	WELDING PASSES
0.125	Square butt with backing	1/16	35	200-250	110-150	20	1
.250	Single V butt 60° inc. angle no nose	1/16	35	250-300	150-200	15	2
.375	Single V butt 60° inc. angle 1/16″ nose	1/16	(O$_2$-1)	275-325	225-250	20	2
.500	Single V butt 60° inc. angle 1/16″ nose	3/32	(O$_2$-1)	300-350	75-85	5	3-4
.750	Single V butt 90° inc. angle 1/16″ nose	3/32	(O$_2$-1)	350-375	85-95	4	5-6
1.000	Single V butt 90° welded angle 1/16″ nose	3/32	(O$_2$-1)	350-375	85-95	2	7-8

Linde Co.

Copper

Gas metal-arc welding of copper is usually confined to the deoxidized types. Welding electrolytic copper is not advisable because such welds exhibit low strength.

Argon is preferred as the shielding gas for thin materials. For materials 1″ or more in thickness a mixture of 65 percent helium and 35 percent argon is recommended.

Steel back-up blocks are required for welding sheets ⅛″ or less in thickness. Although no preheat is necessary for materials of this thickness, some preheating (400°F or 204°C) is advisable on sections ⅜″ or more in thickness.

See Table 25-10 for special gas metal-arc welding conditions of copper.

TABLE 25-9. GENERAL WELDING CONDITIONS FOR SHORT ARC WELDING OF STAINLESS STEEL.

PLATE THICKNESS	JOINT AND EDGE PREPARATION	WIRE dia (inches)	GAS FLOW (cfh)	CURRENT DCRP (amps)	VOLTAGE	WIRE FEED (ipm)	WELDING SPEED (ipm)	PASSES
0.063	Non-positioned fillet or lap	0.030	15-20	85	15	184	18	1
.063	Butt (square edge)	.030	O_2-2	85	15	184	20	1
.078	Non-positioned fillet or lap	.030	O_2-2	90	15	192	14	1
.078	Butt (square edge)	.030	O_2-2	90	15	192	12	1
.093	Non-positioned fillet or lap	.030	O_2-2	105	17	232	15	1
.125	Non-positioned fillet or lap	.030	O_2-2	125	17	280	16	1

*Voltage values are for C-25 gas or O_2-2 gas. For 90% HE—10% C-25, voltage will be 6 to 7 volts higher
Linde Co.

TABLE 25-10. CONDITIONS FOR WELDING COPPER.

THICKNESS (inches)	CURRENT DCRP (amps)	VOLTS	TRAVEL (ipm)	WIRE dia (inches)	WIRE FEED (ipm)	JOINT DESIGN
1/8	310	27	30	1/16	200	Square butt, steel back-up strip required
1/4 (1)	460	26	20	3/32	135	Square butt
1/4 (2)	500				150	
3/8 (1)	500	27	14	3/32	150	Double bevel, 90° included angle,
3/8 (2)	550				170	3/16" nose
1/2 (1)	540	27	12	3/32	165	Double bevel, 90° included angle,
1/2 (2)	600		10		180	1/4" nose

Linde Co.

Points to Remember

1 Gas metal-arc welding (GMAW) is often referred to as MIG and by the manufacturer's trade name as Micro-wire, Air-comatic, Sigma, and Millermatic Welding.

2 Gas metal-arc welding is faster than stick electrode welding and is easier to learn.

3 Spray transfer type of welding is particularly adapted for welding heavy-gauge metals.

4 Short-circuiting transfer welding is best for welding light-gauge metals.

5 For optimum efficiency, DCRP current is required for gas metal-arc welding.

6 For gas metal-arc welding, a constant potential power supply with a nearly flat volt-ampere characteristic produces the best results.

7 The use of CO_2 as a shielding gas is most effective and least expensive when welding steel.

8 Argon or a mixture of argon and oxygen will produce the most effective results in welding aluminum and stainless steel.

9 The rate of gas flow for welding most metals is about 20 to 35 cu ft/hr. However, this rate may have to be varied somewhat, depending on the type, electrode size, and thickness of metal.

10 The effectiveness of the shielding gas is often governed by the distance of the

Points to Remember, cont.

gun from the workpiece. Generally the gas nozzle should not be spaced more than 2″ from the workpiece.

11 The use of correct diameter wire electrodes is necessary for good welds. Check recommendations for correct electrode diameters.

12 The correct current for welding must often be determined by trial. Check recommendations for starting current.

13 Be sure the wire feed is set for the amperage which is to be used for welding.

14 For most gas metal-arc welding applications, the wire stickout should be about ⅜″ to ¾″.

15 Keep the gun properly positioned to insure uniform weld with proper penetration.

16 Cold laps will occur if the arc does not melt the base metal sufficiently.

17 Check the weld for surface porosity. Surface porosity is usually caused by improper gas shielding.

18 Do not remove the gun from the weld area until the puddle has solidified; otherwise cracks may develop.

19 Remember, insufficient or excessive penetration is the result of failure to control heat input.

QUESTIONS FOR STUDY AND DISCUSSION

1 How does gas metal-arc welding differ from gas tungsten-arc welding?

2 What are some of the specific advantages of gas metal-arc welding?

3 What is the difference between spray and globular metal transfer?

4 Why is globular transfer ineffective for welding heavy-gauge metals?

5 What is meant by short-circuiting transfer? For what type of welding is this the most effective means?

6 Why is DCRP current best for doing gas metal-arc welding?

7 What results can be expected if DCSP current is used?

8 How does a constant potential power supply unit differ from the conventional constant current welding machine?

9 What is the advantage of using a constant potential power supply unit for gas metal-arc welding?

10 What is meant by slope control?

11 How is the electrode wire fed to the welding gun?

12 What are the elements that make up air?

13 Why is oxygen generally a harmful element in welding?

14 Why does nitrogen cause the most serious problems in welding?

15 When is argon or a mixture of argon and oxygen considered the ideal gas for shielding purposes?

16 When is CO_2 better for shielding purposes than an inert gas?

17 How is it possible to determine if the gas flow is proper for shielding?

18 What is likely to happen if the gas flow is allowed to drift from the weld area?

19 What factors must be taken into consideration in selecting the correct diameter (size) electrode?

20 What determines the rate at which the wire feed should be set?

21 Why is correct wire stickout important?

22 In what position should the gun be held for horizontal fillet welding?

23 In what position should the gun be held for flat fillet welding?

24 What determines whether a pulling or pushing technique should be used?

25 What is the probable cause for the formation of cold laps in a weld?

26 What should be done to avoid surface porosity in a weld?

27 How can crater porosity or cracks be avoided?

28 If weld penetration is insufficient, what should be done?

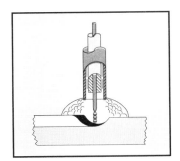

In addition to the gas metal-arc process described in Chapter 25, several other gas shielded-arc techniques have been developed to meet the ever increasing industrial demands for joining metals more effectively and at lower cost. Some of the more common ones are described in this chapter.

Buried-Arc CO$_2$ Welding Process

Buried-arc CO$_2$ welding is a high-energy, fast-weld method in which the end of the wire electrode is held either level or below the surface of the work with practically a zero arc length. See Figure 26-1. This process is designed for high-speed welding of mild steel. Although it can be employed for manual welding its greatest application is in mechanized welding. The process is widely used in many industries where the fabrication of parts requires deep penetration and fast deposition of weld metal without critical control of bead contour. In most instances welding wire diameters range from 0.045″ to ⅛″.

Standard gas metal-arc welding equipment is utilized for the buried-arc process. The shielding gas is pure carbon dioxide which provides additional economy over the more expensive argon. The metal transfer is globular, but since the wire is buried and a high-density current is used, a deep cavity is formed. This deep cavity traps the molten globules that normally would be ejected sideways through the arc. Thus, splatter which otherwise would be severe is minimized and does not affect the welding process.

Gas Metal-Arc Welding: Pulsed Arc

The pulsed-arc process is an extension of spray-transfer welding to a current level much below that required for continuous spray transfer. The pulsing current used may be considered as having its peak current in the spray-transfer current range and its minimum value in the globular transfer current range.

The need for current values less than the transition level becomes apparent when attempting to weld under heat transfer conditions which are inadequate for spray transfer. For example, when welding out-of-position, high current will result in a molten pool

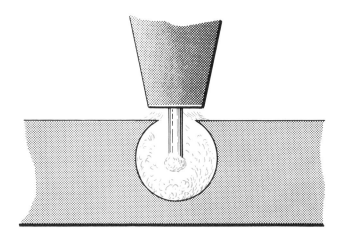

Figure 26-1. In the buried-arc CO2 welding process, the electrode is below the work surface.

which cannot be retained in position unless the material being welded has adequate thermal-conductivity (coupled with the joint type and plate thickness).

The same factors explain the burn-through obtained when a weld on thin material is attempted with too high a welding current. While smaller diameter electrodes have lower transition currents, their basic limitation cannot be avoided. The net result is that the spray transfer process is very applicable to flat position use but is rather more limited in its use for out-of-position and thin material welding.

Pulsed-spray transfer is achieved by pulsing the current back and forth between the spray transfer and globular transfer current ranges. See Figure 26-2. On the left, the current-time relationships for two power sources, A and B, are shown with A putting out a current in the globular transfer range and B putting out a current in the spray-transfer range. On the right, the two outputs are shown combined to produce a simple pulsed output by electrically switching back and forth between them.

Transfer is restricted to the spray mode. Globular transfer is suppressed by not allowing sufficient time for transfer by this mode to occur. Conversely, at the high-current level, spray transfer is insured by allowing more than sufficient time for transfer to occur.

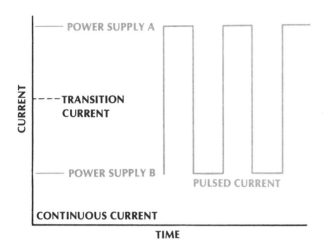

Figure 26-2. A switching system can convert two steady-state DC output currents in a simple pulsing-current output waveform. (*Airco*)

For the given electrode deposited by the pulsed-arc method, all the advantages of the spray-transfer process are available at average current levels from the minimum possible with continuous spray transfer down to values low in the globular transfer range.

Features of pulsed-arc welding. Pulsed-arc welding provides many features not previously available.

1. The heat input range bridges the gap between, and laps over into, the heat input ranges available from the spray and short-circuiting arc processes. Into its lower heat input range the pulsed-spray process brings the advantages of the continuous spray-transfer process. Also, due to lower heat input, the use of spray transfer is extended greatly into poor heat transfer areas, mainly related to welding out-of-position and on thinner materials.

2. The area of overlap with the spray-transfer process occurs because, having a higher transition current, a larger diameter electrode leaves the continuous spray and enters the pulsed-spray range at a higher current than a smaller electrode. Further, the use of a larger diameter electrode can be continued down to a current value considerably below the transition current associated with using a smaller diameter electrode.

The pulsed-arc process will not displace the short-circuiting arc process in those areas where the short-circuiting arc process is properly applicable and more economical.

3. The pulsed-arc method produces a higher ratio of heat input to metal deposition, permits the use of a completely inert gas shield where necessary, and is essentially free from spatter.

4. The pulsed-arc process is characterized by a uniformity of root penetration which is comparable to the gas tungsten-arc process; because of this feature, the process may permit deletion of weld backing in some cases.

Power source for pulsed-arc welding. The power source combines a standard, three phase, full-wave unit with a single phase, half-wave unit, both of the constant-potential type. See Figure 26-3. The three phase unit is termed the background unit and the single phase unit is termed the pulsing unit. These units are connected in parallel but commutate (to form a unidirectional current) in operation. The waveform of the pulsing current output determines the sequence of metal transfer. See Figure 26-4.

The units are made to switch back and forth in operation by means of the varying output voltage of the pulsing unit. The diode rectifiers in each unit alternately permit or block the passage of current depending upon whether there is a positive or negative voltage difference across their terminals. When

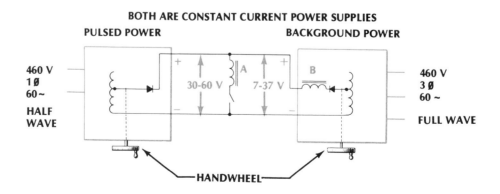

Figure 26-3. Diagram of the essential features of the pulsed-current power supply. (*Airco*)

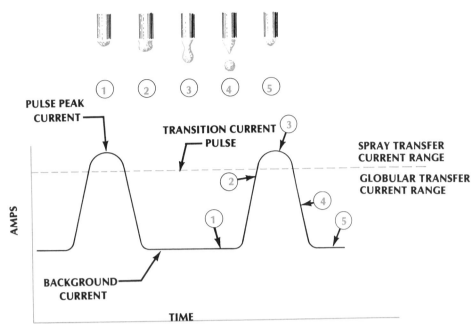

Figure 26-4. The output current waveform of the pulsed-current power supply determines the metal transfer sequence. (*Airco*)

the pulse is OFF or its voltage is less than the background voltage, the diode rectifiers of the background unit pass the full value of the instantaneous current. Conversely, when the pulse voltage exceeds the background voltage, blocking the background diode rectifiers, the pulse diode rectifiers pass the full value of the instantaneous current.

Two chokes are used. See Figure 26-3. The choke labeled A performs a commutation function. When the pulse voltage drops below the background voltage, it sustains the welding current momentarily, giving the background unit time to respond to the demand for current. Choke B, in series with the output of the background unit, filters the background current and prevents undesirable arc outages at low background current levels. See Figure 26-5.

Operation of the pulsed-arc power source is similar to that of conventional constant potential sources. The pulse peak voltage is determined by the electrode type and diameter. See Table 26-1. The wire feeder is set at the value which will produce the required current and is determined from Table 26-2 for the type and diameter of electrode to be used. The meters on the power supply read the average voltage and the average current which are the values familiar to gas shielded-arc welders.

Flux-Cored Arc Welding

Flux-cored arc welding (FCAW) is a gas metal-

Figure 26-5. The wire feeder unit can be mounted on top of the pulsed-arc power supply. (*Airco*)

TABLE 26-1. TYPICAL PULSED ARC POWER SUPPLY SETTINGS.

ELECTRODE TYPE	ELECTRODE DIA (inches)	PULSE PEAK RANGE (volts)	AVERAGE CURRENT RANGE (amps)	AVERAGE VOLTAGE (volts)
Mild and low-alloy steel	0.035	34-36	55-130	18-20
	0.045	37-39	90-180	19-23
	1/16	42-44	110-250	20-25
Stainless steel	0.035	33-35	55-130	18-20
	0.045	36-38	90-180	19-23
	1/16	41-43	110-250	20-25
Aluminum	1/16	34-36	80-250	20-30

TABLE 26-2. AVERAGE CURRENT VS. ELECTRODE FEED SPEED FOR TYPICAL DIAMETERS OF STEEL AND ALUMINUM ELECTRODES WHEN USING PULSED ARC UNIT.

ELECTRODE FEED (ipm)	CORRESPONDING AVERAGE CURRENT (amps) MILD STEEL AND STAINLESS STEEL*			ALUMINUM†
	0.035" dia	0.045" dia	1/16" dia	1/16" dia
70	—	70	115	70
90	—	90	175	90
105	50	105	215	105
125	60	125	—	125
135	70	135	—	135
155	80	155	—	155
185	90	185	—	185
220	110	220	—	220
235	120			
255	130			
275	140			
300	150			
325	160			
345	170			
365	180			
380	190			
425	210			
500	—			

*Argon + 290 O_2
†Argon

arc welding process that uses a continuous flux-cored wire as the electrode. The electrode varies in size from .045" to 5/32". The equipment necessary for the flux-cored arc welding process is the same as that for gas metal-arc welding with the exception of greater current needs and a larger gun. See Figure 26-6. The wire can be used with or without CO_2 as a shielding gas depending on the type of electrode wire used. See Table 26-3. The flux ingredients in the wire include ionizers to stabilize the arc, deoxidizers to purge the deposit of gas and slag, and other metals to produce high strength, ductility and toughness in weld deposits. The flux generates a gas shield, which is augmented

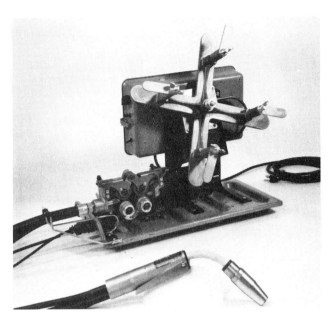

Figure 26-6. The equipment used for flux-cored arc welding is similar to the equipment used for gas metal-arc welding. (*Airco*)

by the regular CO_2 shield, and a slag blanket that retards the cooling rate and protects the weld deposit as it solidifies.

Tubular wire is designed for high current densities and deposition rates which when combined with high duty cycles result in sharply increased production speeds. It is especially intended for application in large fillet single and multiple-pass welds in either a horizontal or flat position using DCRP current. Because of its deep penetrating qualities into the weld root, tubular wire fillet welds of smaller leg size will have the same strength as stick fillet welds of larger size. For instance, double-welded butt joints up to ½″ thick can be welded without edge preparation.

The actual operation of flux-cored wire welding is similar to other gas metal-arc welding processes. See Table 26-4. See Exercises 1 to 3.

TABLE 26-3. CHARACTERISTICS OF CARBON STEEL FLUX CORED ELECTRODES

AWS Classification	Welding Current	Shielding	Single or Multiple Pass
EXXT-1	DCEP	CO_2	Multiple
EXXT-2	DCEP	CO_2	Single
EXXT-3	DCEP	None	Single
EXXT-4	DCEP	None	Multiple
EXXT-5	DCEP	CO_2	Multiple
EXXT-6	DCEP	None	Multiple
EXXT-7	DCEN	None	Multiple
EXXT-8	DCEN	None	Multiple
EXXT-10	DCEN	None	Single
EXXT-11	DCEN	None	Multiple
EXXT-G	a	a	Multiple
EXXT-GS	a	a	Single

a. As agreed between purchaser and supplier
(Hobart Brothers)

TABLE 26-4. FLUX-CORED ARC WELDING CONDITIONS—DCRP.

	MATERIAL THICKNESS (inches)[1]	CURRENT DCRP (amps)	ARC VOLTAGE	WIRE FEED ipm	SHIELDING GAS FLOW CFH[2]	TRAVEL SPEED ipm	NO. OF PASSES	WIRE STICK-OUT (inches)
flux-cored	1/8	300-350	24-26	100-120	35-40	25-30	1	3/4 to 1 1/2
arc welding	3/16	350-400	24-28	120-150	35-40	25-35	1	3/4 to 1 1/2
of steel	1/4	350-400	24-28	120-150	35-40	20-30	1	3/4 to 1 1/2
using 3/32"	3/8	475-500	28-30	180-210	35-40	15-20	1	3/4 to 1 1/2
diameter electrode	1/2	400-450	25-28	150-170	35-40	18-20	2-3	3/4 to 1 1/2
wire size	5/8	400-450	25-28	150-170	35-40	14-18	2-3	3/4 to 1 1/2
flat and horizontal positions	3/4	400-450	25-28	150-170	35-40	14-18	5-6	3/4 to 1 1/2

ELECTRODE SIZE	FLAT POSITION[3]		HORIZONTAL POSITION[3]		VERTICAL POSITION[3]	
(inches)	(ampere)[4]	(voltage)[5]	(ampere)[4]	(voltage)[5]	(ampere)[4]	(voltage)[5]
0.045	150-225	22-27	150-225	22-26	125-200	22-25
1/16	175-300	24-29	175-275	25-28	150-200	24-27
5/64	200-400	25-30	200-375	26-30	175-225	25-29
3/32	300-500	25-32	300-450	25-30	—	—
7/64	400-525	26-33	—	—	—	—
1/8	450-650	28-34	—	—	—	—

[1]For groove and fillet welds. Material thickness also indicates fillet size. Use V groove for 1/4 inch and thicker. Double V for 1/2 inch and thicker.
[2]Welding grade CO_2.
[3]Applies to groove, bead or fillet welds in position shown.
[4]Ampere range can be expanded. Higher currents can be used, especially with automatic travel.
[5]Voltage range can be expanded. It will increase when larger stick-out tip to work distance is employed.

(Hobart Brothers)

Running Beads on Mild Steel Using the FCAW (Flux-Cored Arc Welding) Process

Exercise 1

1 Use a ³⁄₃₂″ diameter electrode E70T-1. Electrode extension (stickout) should be 1″ to 1½″.

2 Set the machine for DCRP (direct current reverse polarity).

3 Amperage should be 390 to 410 amps, voltage 26-28 volts.

4 Carbon dioxide is used as a shielding gas at 40 cfm.

5 Set the wire feed control so that the ammeter reads 390-410 amps. To obtain the correct reading have another person observe the amperage while welding is being done.

6 Use the same procedure to set the voltage to 26-28 volts at the power source.

7 Obtain a piece of ½″ to 1″ x 4″ x 6″ mild steel plate.

8 Position the welding gun for a 90° work angle and a 20° to 30° drag angle.

9 The finished bead should be approximately ¾″ wide and ⅛″ to ¼″ high.

Exercise 1, cont.

10 Run a series of straight consistent beads on the plate approximately ⅜" apart.

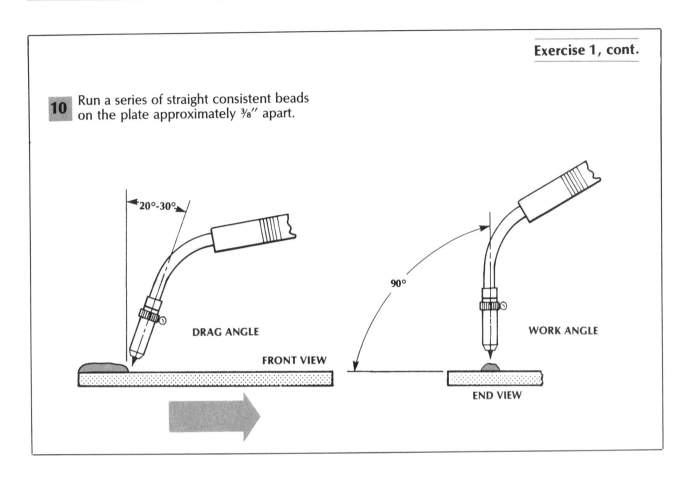

20°-30°

DRAG ANGLE

FRONT VIEW

90°

WORK ANGLE

END VIEW

Welding a Lap Joint on Mild Steel Using the FCAW Process

Exercise 2

1 Refer to Exercise 1 for equipment set-up and adjustment.

2 Obtain two pieces of ¾" to 1" x 2" x 6" mild steel.

3 Position the pieces and tack weld to form a lap joint. Rest the piece against a firebrick to obtain flat position.

4 Deposit the first pass with the gun at a 45° work angle and a 20° drag angle. Weld the root pass on both sides of the lap joint.

5 Deposit the second pass using a weaving motion on both sides of the lap joint. Pause at the toes of the weld to prevent undercutting.

6 Deposit the third pass using the same procedure as for the second pass on both sides of the lap joint.

45°

END VIEW

Exercise 2, cont.

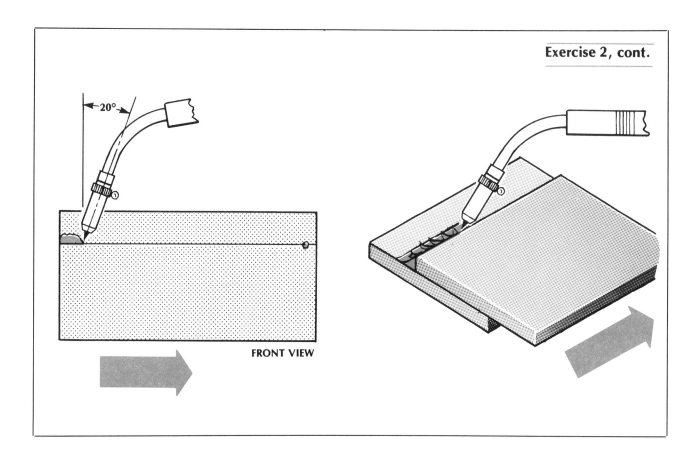

FRONT VIEW

Welding a Multiple-Pass T-Joint in Horizontal Position Using the FCAW Process

Exercise 3

1 Refer to Exercise 1 for equipment set-up and adjustment.

2 Obtain two pieces of ¾" to 1" x 2" x 6" mild steel.

3 Position the pieces and tack weld to form a T-joint.

4 Deposit the first pass with the gun at a 45° work angle and a 20° drag angle. Weld the root pass on both sides of the T-joint.

5 Position the electrode one electrode diameter below the bottom toe of the root pass and deposit the second pass using a 50° to 60° work angle and a 20° drag angle on both sides of the T-joint.

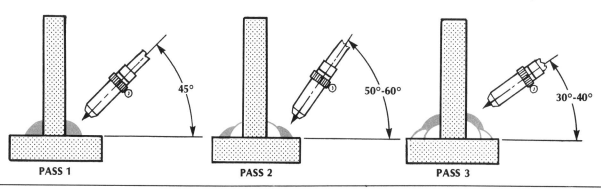

PASS 1   PASS 2   PASS 3

welding multiple-pass welds to avoid buildup, resulting in lower ductility.

Figure 26-7. The FCAW-S process does not require shielding gas equipment. (*The Lincoln Electric Company*)

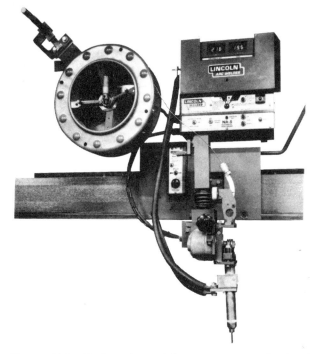

Figure 26-8. During the welding process, mechanized FCAW-S equipment is monitored and adjusted by the welder to maintain weld quality. (*The Lincoln Electric Company*)

Exercise 3, cont.

6 Deposit the third pass using a 30° to 40° work angle and a 20° drag angle on both sides of the T-joint.

Self-Shielded Flux-Cored Arc Welding

The self-shielded flux-cored arc welding process (FCAW-S) is a flux-cored arc welding process variation in which shielding gas is provided exclusively by the flux within the electrode. A shielding gas supply system is not required in FCAW-S, and guns do not have a shielding gas nozzle. This allows greater access to the weld joint. Higher amperages than gas-shielded flux-cored arc welding (FCAW-G) processes are used. However, higher amperages and shielding gas created by burning flux can produce excessive smoke in the weld area. Specially designed guns can be used to evacuate smoke from the weld area for maximum visibility.

Equipment. Equipment for the FCAW-S process consists of a DC power source, a wire feeder, cables, and gun. The equipment can be designed for semiautomatic or mechanized operation. See Figures 26-7 and 26-8. With semiautomatic equipment, the welder moves the welding gun along the weld joint. With mechanized equipment, the welder makes adjustments as required while observing the welding operation.

The trigger on the gun is pulled to start current flow and wire feed. Proper electrode extension is required throughout the weld to maintain the required amperage. To stop welding, the trigger is released to stop current flow and wire feed. The gun has a metal shield to provide protection from molten weld metal. See Figure 26-9.

Filler Wire. The tubular mild steel filler wire serves as a filler rod and contains a core of ingredients for producing a gaseous shield, deoxidizing agents, and slag. Some core ingredients melt before the electrode from the intense heat of the arc. Metallic salts in the core vaporize to provide a gaseous shield around the weld area. This prevents contamination of the weld before solidification and protection from slag. Other additives in the filler wire core provide a deoxidizing function. See Figure 26-10. Because of the high amount of deoxidizing additives in FCAW-S filler wire, special care is required when

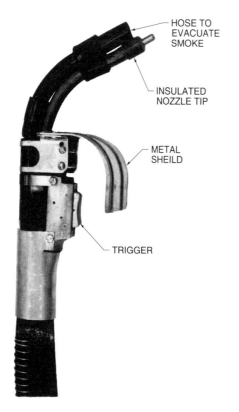

Figure 26-9. A hose on the FCAW-S gun is used to evacuate smoke from the weld area. (*The Lincoln Electric Company*)

electric arc is submerged or hidden beneath a granular material. The electric arc provides the necessary heat to melt and fuse the metal. The granular material, called flux, completely surrounds the electric arc thus shielding the arc and the metal from the atmosphere. A metallic wire is fed into the welding zone underneath the flux.

The welding process can be either semiautomatic or fully automatic. In the semiautomatic, a special welding gun is used. See Figure 26-11. Any regular gas metal-arc welding DC power source can be adapted for submerged-arc welding.

The difference between submerged-arc welding and other forms of gas metal-arc welding is that no inert shielding gas is required. The gun is pointed over the weld area and the gun trigger depressed. As soon as the trigger is pulled, the wire is energized and the arc is started. At the same time the flux begins to flow. The actual welding operation is now carried out in the same manner as in the procedure for gas metal-arc welding.

As the wire is fed into the weld zone the gun deposits the granulated flux over the weld puddle and completely shields the welding action. The arc is not visible since it is buried in the flux, thus there is no flash or spatter. That portion of the granular flux immediately around the arc fuses and covers the molten metal, but after it has solidified it can be tapped

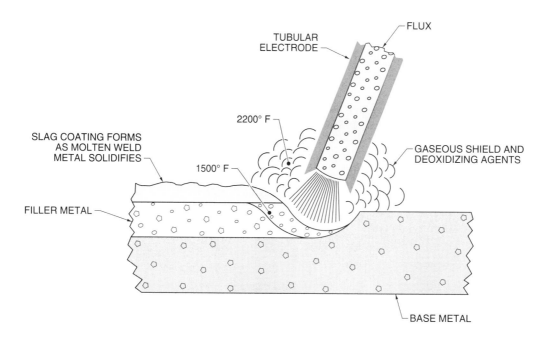

Figure 26-10. In FCAW-S, a tubular electrode contains flux ingredients used to produce a gaseous shield, deoxidizing agents, and slag. (*The Lincoln Electric Company*)

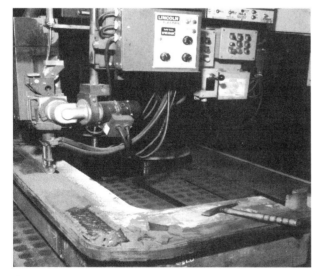

Figure 26-11. The gun used for submerged-arc welding deposits granular flux while feeding the electrode. (*The Lincoln Electric Company*).

Figure 26-12. Automatic submerged-arc welding equipment deposits weld metal at a controlled speed. (*The Lincoln Electric Company*)

off easily with a chipping hammer.

With the fully automatic process, the welding unit is arranged to move over the weld area at a controlled speed. On some machines, the welding head moves and the work remains stationary. In others the head is stationary and the work moves. See Figures 26-12 and 26-13.

Submerged-arc welding can be used for metals from 1/16" thick. It is usually used for welding thicker metals and where deep penetration is required. For example, it is possible to weld 3" plate in a single pass. However, caution is necessary as impurities in the weld collect toward the center of the weld, developing a weak area. Little edge penetration is necessary on material under 1/2" in thickness. Generally back-up support is essential for welding heavy plate.

Welding positions are limited because of the large amount of fluid molten metal.

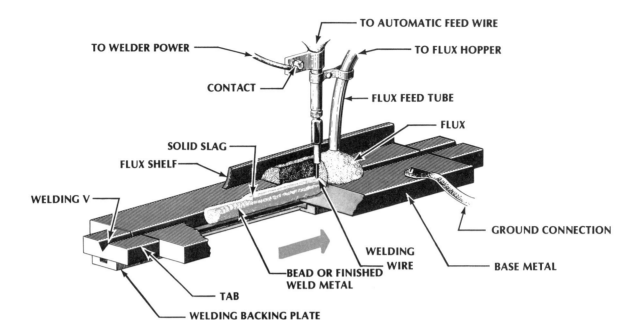

Figure 26-13. This cutaway view of a submerged-arc welding machine shows how the granulated flux shields the welding action and covers the molten metal. (*Linde Company*)

Points to Remember

1 Buried-arc CO_2 is a high energy, fast welding process used for fabricating parts that require fast deposition and deep penetration.

2 Keep the wire completely buried in the weld cavity when using the buried-arc CO_2 welding process.

3 The pulsed-arc process permits welding thin materials and making out-of-position welds with less danger of burn-through.

4 Operation of a pulsed-arc power source is the same as the conventional constant potential welding machine.

5 Flux-cored wire is used principally in combination with CO_2 and is intended for high current densities.

6 Use a DCRP current with flux-cored wire.

7 Vapor-shielded arc welding uses a vapor instead of a gas for shielding purposes. The vapor is generated by a continuous tubular electrode containing vapor-producing materials, such as salts and oxides.

8 In submerged-arc welding, the electric arc is completely hidden beneath a flux.

9 Submerged-arc welding has its greatest application in welding thick metals where deep penetration is required.

10 No inert shielding gas is required for submerged-arc welding since the flux completely surrounds the electric arc.

QUESTIONS FOR STUDY AND DISCUSSION

1 How does the buried-arc CO_2 process differ from gas metal-arc welding?

2 Where does the buried-arc CO_2 process have its greatest application?

3 What is pulsed-arc welding?

4 Pulsed-arc welding is particularly adapted for what kinds of operations?

5 What is tubular wire welding?

6 In combination with CO_2, flux-cored wire is designed for what types of welding?

7 What equipment is required for the FCAW-S process?

8 How does flux protect molten weld metal from contaminants?

9 What is submerged-arc welding and by what material is it shielded?

10 What are some of the special advantages of submerged-arc welding?

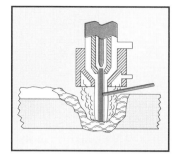

According to the American Welding Society, *brazing* is defined as a group of welding processes wherein coalescence (forming together in one mass) is produced by heating the metal to suitable temperatures above 800°F (427°C) and by using a nonferrous filler metal having a melting point below that of the base metals. The filler metal is distributed between the closely fitted surfaces of the joint by capillary attraction (power of a heated surface to draw and spread molten metal).

Most commercial metals can be brazed, such as copper and copper alloys, stainless steels, magnesium base metals, aluminum, low-carbon and low-alloy steels, cast iron, titanium, zirconium, and beryllium. Although most brazed joints have a relatively high tensile strength they do not possess the full strength properties of other conventional welding techniques.

The outstanding characteristic of brazing is that the mechanical properties of the base metal are not impaired since lower bonding temperatures are used than in fusion-welding. Brazing also has wide application in joining dissimilar metals. The one notable exception is that copper and copper alloys cannot be brazed directly to aluminum.

BRAZING

To achieve sound brazed joints the following are the preliminary requirements which must be recognized and followed a step at a time in order:
1. Joint design
2. Clean surfaces
3. Correct fluxes
4. Correct filler metals
5. Proper heating equipment

Joint Design. The design of the joint used is based on the adhesive qualities of the filler metal used. The two basic joints used for brazing are the butt and lap (T and corner are considered as butt welds). The lap joint probably is most common because it offers greatest strength. For maximum efficiency the overlap should equal or exceed three times the thickness of the thinnest member. The main drawback of the lap joint is that metal thickness at the joint is increased.

The butt joint does not provide the same degree of strength as the lap joint since its cross-sectional area is equal only to the cross-sectional area of the thinnest member. Although higher strengths can be achieved by scarfing the edges, this involves greater care in preparing the joint and keeping the pieces in alignment. The strength of a butt joint can also be improved by using a sleeve. See Figure 27-1.

Another factor associated with joint design is joint clearance. If the surfaces are too tight together the plastic flow of the filler metal is hindered. Too great a distance will prevent the full effects of any capillary action, thus leaving voids and poor distribution of filler metal. Adequate joint clearance should fall in the range between 0.001″ and 0.10″.

Surface preparation. Clean, oxide-free surfaces are absolutely necessary to make sound brazed joints. Uniform capillary action is possible only when surfaces are completely free of foreign substances such as dirt, oil, grease, and oxide. Dirt, oil, and grease can be removed by immersing the parts in some commercial cleaning solvent or salt bath, by pickling in acid (sulfuric, nitric, or hydrochloric) or using some type of vapor-degreasing unit. Surface oxides can be eliminated by sanding, grinding, filing, machining, blasting, or wire brushing. The method used will depend on the contaminants, the joint design, and type of metal to be brazed.

During any cleaning process care must be taken to avoid getting the faying (prepared adjoining) surfaces too smooth. Surfaces that are too smooth will prevent the filler metal from effectively wetting the joining areas. Such surfaces can be roughened slightly by rubbing with 30 or 40 grit (coarse) emery cloth. Brazing should be undertaken as soon as the metal is cleaned to prevent contamination from atmospheric exposure or handling.

Fluxes. The purpose of a flux is to prevent or inhibit the formation of oxide during the brazing process. It is not intended to remove oxides already formed, or dirt, grease, and oil. Metal surfaces are easily contaminated in the atmosphere after they are cleaned. Some metals are more susceptible than others to attack. Moreover, any chemical reaction

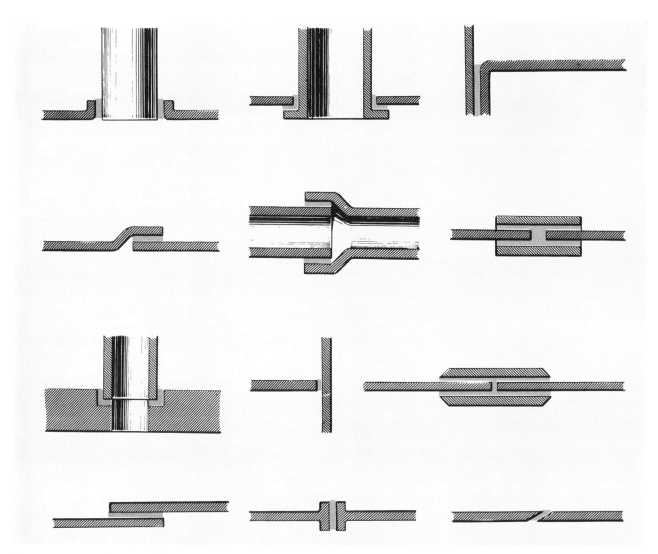

Figure 27-1 Brazed joints rely on joint surface area and the capillary action necessary for the adhesive qualities of the filler metal to provide strength.

resulting from air exposure is accelerated as the temperature is raised during the brazing process. Hence a flux is needed to dissolve and remove oxides which may form during brazing.

An important requirement of a flux is its ability to readily dissolve and promote the fluidity of the filler metal. Equally important is its surface tension since this affects the wetability of the base metal and its flow in the joint. Finally a flux must last long enough to counteract any reactive effects developed during the brazing operation.

Some brazing rods are coated with a flux. Other fluxes are available in powder, paste, or liquid form. The main ingredients of fluxes are boric acid, borates, fluorides, fluoroborates, chlorides, and wetting agents. Fluxes must be selected to suit a particular metal. See Table 27-1 for matching of flux and metal.

A paste flux is the most common for brazing. It can be applied with good adherence to a joint before brazing. See Figure 27-2. Powder flux is sprinkled on the joint or applied to the heated end of the filler rod by dipping it into the flux container. A liquid flux is used mostly in torch brazing where the fuel gas is passed through the liquid flux, carried along with it, and deposited wherever the flame is applied.

Flux removal. Once a brazing operation is completed, all the flux residue must be removed; otherwise corrosion will set in. This residue can usually be recognized by its glass-like surface appearance.

Flux can be removed by washing the part in hot water. In some instances the joint can be immersed in cold water before it has completely cooled from the brazing temperature. The thermal shock of the cold water will usually crack off the residue. For heavy

TABLE 27-1. FLUXES FOR BRAZING.

AWS BRAZING FLUX (Type No.)	RECOMMENDED BASE METALS	RECOMMENDED FILLER METALS	RECOMMENDED USEFUL RANGE (°F)	INGREDIENTS	FORMS SUPPLIED
1	All brazeable aluminum alloys	BAlSi	700-1190	Chlorides Fluorides	Powder
2	All brazeable magnesium alloys	BMg	900-1200	Chlorides Fluorides	Powder
3A	All except those listed under 1, 2 and 4	BCuP, BAg	1050-1600	Boric Acid Borates Fluorides Fluoroborates Wetting Agent	Powder Paste Liquid
3B	All except those listed under 1, 2 and 4	BCu, BCuP, BAg, BAu, RBCuZn, BNi	1350-2100	Boric Acid Borates Fluorides Fluoroborates Wetting Agent	Powder Paste Liquid
4	Aluminum bronze, aluminum brass and iron or nickel base alloys containing minor amounts of Al and/or Ti	BAg (all) BCuP (Copper base alloys only)	1050-1600	Chlorides Fluorides Borates Wetting Agent	Powder Paste
5	All except those listed under 1, 2 and 4	Same as 3B (excluding BAg-1 through -7)	1400-2200	Borax Boric Acid Borates Wetting Agent	Powder Paste Liquid

AWS

residue a chemical dip is sometimes used. Wire or fiber brushing, steam jet, or blast cleaning are also effective means of removing heavy residues or when large objects are involved. On some soft metals such as aluminum, mechanical removal of residues must be followed by fluid cleaning because small flux particles may have become embedded in the surfaces.

Controlled atmosphere. Where mass production is involved and particularly when high quality joints are required the application of fluxes is a time consuming task. Consequently controlled atmospheres are often used to prevent the formation of oxides during brazing. In a controlled atmosphere a gas is continuously supplied to a furnace and circulated within it at slightly higher than atmospheric pressures. Gas may consist of high-purity hydrogen, carbon dioxide, carbon monoxide, nitrogen, argon, ammonia, or some form of combusted fuel gas.

Filler metals. The American Welding Society specifies that a brazing filler metal should have the following characteristics:

1. Ability to wet and make a strong bond on the base metal.

2. Suitable melting temperature to permit adequate distribution by capillary attraction.

3. Sufficient homogeneity and stability to minimize separation by liquation (separation of the solid and liquid portion) and not be excessively volatile.

4. Capable of producing a brazed joint to meet service requirements such as strength and corrosion resistance.

Brazing filler metals are available in wire, rod, strip, and powder forms. These various types are designed to braze different metals.

Filler metals may be designated by commercial names or AWS classification symbols. The AWS classification consists of the letter B, which identifies it as a brazing filler metal, followed by the chemical

Figure 27-2. Paste flux is applied with a brush to uncoated brazing rods.

symbols of the metallic elements included in the filler metal. If dash digits are shown at the end of the chemical symbols they simply specify some characteristic of the filler metal. Table 27-2 shows the basic AWS designated filler metals for brazing different metals.

Filler metal application. Brazing filler metals are applied either manually after the work is heated or are preplaced in a suitable position before the work is heated. Rod and wire are generally used for manual face-feeding. Preplaced filler metals are in the form of rings, washers, formed wire, shims, and powder. These are located in strategic places near the joint to assure a uniform flow of filler metal into the joining surfaces. Although preplaced filler metals can be used in manual brazing, their greatest application is in furnace, induction, or dip brazing. See Figure 27-3.

Manual heating methods. The required heat for brazing purposes may be applied in various ways. For most manual brazing operations a gas torch is considered the most practical. See Figure 27-4. The gas mixture may be oxyacetylene, air-gas, gas-oxygen, oxy-hydrogen, or Mapp-oxygen. To a large extent the type of gas mixture used depends on the thermal conductivity, type, and thickness of the metal

to be brazed.

Oxyacetylene or Mapp-oxygen is generally more versatile because of its wide range of heat control. With this gas mixture a slightly reducing flame is required. A single or multiflame tip may be used. In either case, only the outer envelope of the flame and not the inner core should be applied to the work.

The air-gas torch provides the lowest heat and has greater applications in brazing thin sections. The air-gas mixture may consist of air at atmospheric pressure and city gas or air and acetylene.

The gas-oxygen torch uses oxygen with city gas, bottled gas, propane, or butane. This mixture produces a high flame temperature and is effective where greater brazing heat is required.

The oxy-hydrogen torch is particularly adaptable for brazing aluminum and other nonferrous metals due to its low heat. The low temperature prevents overheating the metal and the hydrogen provides additional cleaning action and shielding during the brazing process.

Filler metal heat. A brazing filler metal must be completely molten before it flows into a joint. The melting temperature of filler metals will vary, depending on the type of filler used which in turn is governed by the kind of base metal to be brazed. In any event, the melting temperature *(liquidus)* of the filler metal must be lower than the *solidus* of the base metal. Solidus temperature is the highest temperature that the base metal can reach and still remain in a solid state. See Table 27-1. The lowest

TABLE 27-2 BRAZING FILLER METALS.

AWS CLASSIFICATION OF BRAZING FILLER METALS	TYPES OF METALS TO BE BRAZED
BAlSi (aluminum-silicon)	Aluminum, aluminum alloys
BCuP (copper-phosphorus)	Copper, copper alloys
BAg (silver)	Ferrous and nonferrous metals except aluminum and magnesium
BAu (precious metals)	Iron, nickel and cobalt base metals
BCu (copper)	Ferrous and nonferrous metals
BCuZn (copper-zinc)	Ferrous and nonferrous metals
BNi (nickel)	Stainless steels, carbon steels, low-alloy steels, copper
BMg (magnesium)	Magnesium, magnesium alloys

AWS

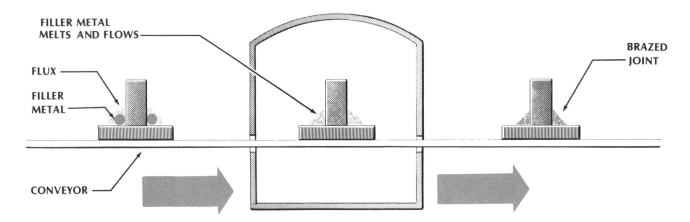

Figure 27-3. Production brazing is frequently done in a furnace.

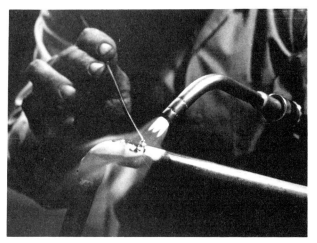

Figure 27-4. A multiflame gas torch is one of the possible sources of heat that can be used for brazing. (*Airco*)

effective brazing temperatures are preferred to minimize the effects of heat on the base metal such as grain growth, warpage, and hardness reduction.

Production heating methods. Although torch brazing can be mechanized for production purposes, high production rates usually are better accomplished by means of furnace, induction, resistance, or dipping techniques. With these methods accurate heat control can be achieved, thereby insuring high quality brazed joints.

Furnace heating involves positioning parts to be brazed on trays and then placed in a gas, electric, or oil-fired furnace. See Figure 27-3.

Induction brazing consists of placing the work near an induction coil. As the current flows through this coil, resistance of the object to the flow of current

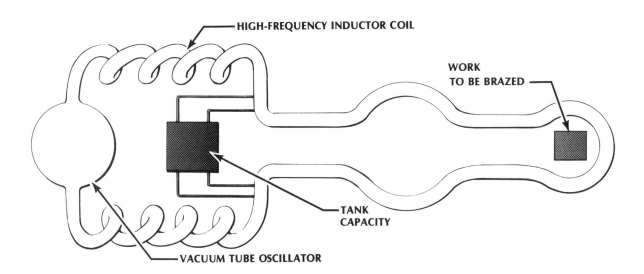

Figure 27-5. In induction brazing, current flows through an induction heating coil. Resistance of the work to the flow of current creates the necessary heat.

causes instant heat to occur. See Figure 27-5 for example.

Resistance brazing is very much like spot welding where heat is generated by the passage of a low-voltage current through carbon electrodes that are clamped around the work. See Figure 27-6.

Dip-brazing consists of immersing parts in a bath of molten brazing metal. The brazing material is contained in a crucible which is externally heated. This method is limited to brazing small assemblies such as wire connections or metal strips when they can be easily held in fixtures. Another dip-brazing method involves the placement of parts in a bath of molten salt. The salt bath is heated either by passing an electrical current through the bath or heating the outside of the container. See Figure 27-7.

Manual Brazing Procedure

The following are the basic steps which generally apply for brazing most metals:

1. Determine the appropriate type joint most suitable for the work to be brazed.

2. Remove all dirt, grease, oil, and oxides from the surfaces to be brazed.

3. Select the correct flux and apply it to both the work and filler metal by brushing, dipping, sprinkling, or spraying.

4. Assemble the pieces and keep them in alignment by means of clamps, fixtures, or jigs. See Figure 27-8. Avoid too much pressure since enough clearance between mating surfaces must exist to allow a free flow of filler metal.

5. Preheat the entire work by playing the torch over the surfaces to bring them up to a uniform brazing temprature.

6. As soon as the flux is completely fluid, touch the filler metal to the joint. Keep applying filler metal until it flows completely through the joint. Do not apply the inner cone of the flame directly to the filler metal or work and be sure the flame is slightly reducing.

7. Clean the brazed work to remove all the flux residue or debris.

BRAZE WELDING

Braze welding, or bronze welding as it is sometimes called, is a little different from regular brazing. Whereas conventional brazing processes involve the joining of two surfaces by a thin bond of brazing material, braze welding is carried out much as in fusion welding except that the base metal is not melted and the filler metal is not distributed by cap-

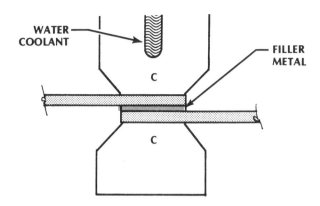

Figure 27-6. In resistance brazing, current passes through carbon electrodes clamped around the work. (*AWS*)

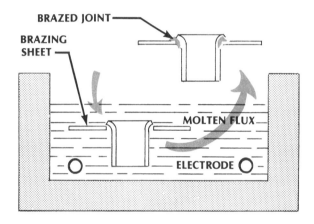

Figure 27-7. In dip brazing, parts are immersed in molten brazing metal inside an externally heated crucible. (*AWS*)

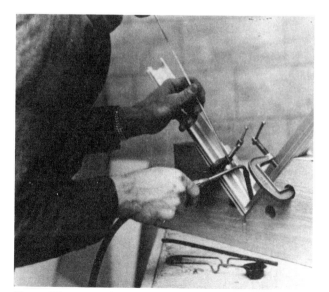

Figure 27-8. Use clamps to keep pieces properly aligned for brazing.

illary action. The base metal is simply brought up to what is known as a *tinning temperature* (dull red color) and a bead deposited over the seam with a brazing filler rod. Although the base metal is not melted, the unique characteristics of the bond formed by the brazing rod are such that the results are often comparable to those secured through fusion welding.

The brazing operation is generally performed with an oxyacetylene torch, although other gases may be used to obtain the necessary source of heat.

Braze welding is particularly adaptable for joining or repairing such metals as cast iron, malleable iron, copper, brass, and various dissimilar metals (cast iron and steel, etc.). See Figure 27-9. *The one precaution that must be considered in braze welding is not to weld a metal that will be subjected to a high temperature later.* The filler metal loses its strength when exposed to high temperatures. Also, braze welding should not be used on steel parts that must withstand unusually high stresses.

Advantages. Since the metal does not have to be heated to a molten condition in braze welding, there is less possibility of destroying the main characteristics of the base metal. Thus in repairing malleable castings the danger of affecting its ductility is minimized.

Equally important is the elimination of stored-up stresses which are often present in fusion welding. This is especially critical in repairing castings. The low degree of heat in braze welding reduces to a minimum these expansion and contraction forces.

With braze welding there is less need for extensive preheating. On thick sections where some preheating may be desirable the temperature is brought up only to a black heat.

Filler rod. The main elements of a brazing filler rod used in braze welding are copper and zinc, which produce high tensile strength and ductility. In addition, the rod contains small quantities of tin, iron, manganese, aluminum, lead, nickel, chromium, and silicon. These elements help to deoxidize the weld metal, decrease the tendency to fume, and increase the free-flowing action of the molten metal.

Flux. A clean metal surface is essential for braze welding. If the filler metal is to provide a strong bond, it must flow smoothly and evenly over the entire weld area. Adhesion of the molten filler metal to the base metal will take place only if the surface is chemically clean. The flow action of the braze over a surface is much like the flow of water over a piece of glass. If the glass is perfectly clean, the water will spread out into a thin, even film, whereas if the glass is dirty the water tends to gather into tiny drops and

Figure 27-9. Braze welding can be used to build up a missing gear tooth. (*Linde Company*)

roll off.

Even after a surface has been thoroughly cleaned by mechanical means, certain oxides may still be present on the metal surfaces. These oxides can only be removed by means of the correct flux. Prefluxed brazing rods eliminate the need to apply flux while brazing.

The flux may also be applied by dipping the heated rod into the powdered flux. The flux adheres to the surface of the rod and thus can be transferred to the weld. Another method is to dissolve the flux in boiling water and brush it on the rod before welding is started.

Procedure:

1. Clean the surfaces thoroughly with a stiff wire brush. Remove all scale, dirt, or grease; otherwise the braze will not stick. If a surface has oil or grease on it, remove these substances by heating the area to a bright red color and thus burn them off.

2. On thick sections, especially in repairing castings, bevel the edges to form a 90° V-groove. This can be done by chipping, machining, filing, or grinding.

3. Arrange the work in flat position.

4. Adjust to a neutral flame. Then gently heat the surfaces of the weld area.

5. Heat the brazing rod and dip it in the flux. (This step is not necessary if the rods have been prefluxed.) In heating the rod, do not apply the inner cone of the flame directly to the rod.

6. Concentrate the flame on the starting end until the metal begins to turn red. Melt a little brazing rod onto the surface and allow it to spread along the

entire seam. The flow of this thin film of filler metal is known as the *tinning* operation. Unless the surfaces are tinned properly the brazing procedure to follow cannot be carried out successfully. You will find that if the base metal is too hot, the filler metal will tend to bubble or run around like drops of water on a warm stove. If the filler metal forms into balls which tend to roll off just as water would if placed on a greasy surface, then the base metal is not hot enough. When the metal is at the proper temperature the filler metal spreads out evenly over the base metal.

7. Once the base metal is tinned sufficiently, start depositing the proper size beads over the joint. Use a slight circular torch motion and run the beads as in regular fusion welding with a filler rod. Keep dipping the rod in the flux as the weld progresses forward. See Figure 27-10. Be sure that the base metal is never permitted to get too hot. See Figure 27-11.

8. If the pieces to be welded are grooved, use several passes to fill the V. On the first pass make certain that the tinning action takes place along the entire bottom surface of the V and about half way up on each side. The number of passes to be made will depend on the depth of the V. When depositing several layers of beads be sure that each layer is fused into the previous one.

9. When making a braze weld with the work in a vertical position, first build up a slight shelf at the bottom. The shelf then acts as a support for further filler metal. As the weld is carried upward, swing the flame from side to side to maintain uniform tinning and to produce even beads. See Figure 27-12.

SOLDERING

Soldering is a form of brazing in which nonferrous filler metals having melting temperatures below 800°F (427°C) are used. The filler metal is called solder and is distributed between surfaces by capillary action.

There are two classifications of soldering: soft soldering and hard soldering. Soft soldering has greater applications in sheetmetal and plumbing industries. It uses filler metals composed of tin and lead and produces joints with relatively low tensile strength. Hard soldering is actually a regular brazing process very much like those previously described where filler metals use melting temperatures above 800°F (427°C). This soldering process is often referred to as silver soldering since the main element of the filler metal is silver. Silver soldering is used more in the electrical field, jewelry making, arts and crafts, and where higher strength joints are required than are obtained with soft soldering. The actual silver sol-

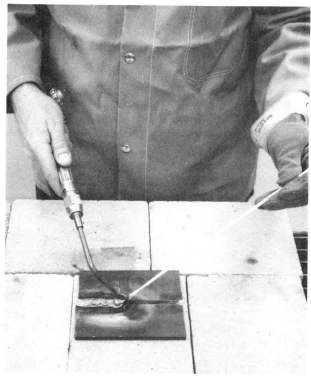

Figure 27-10. Use a slight circular motion with the torch when brazing.

Figure 27-11. If not overheated, the brazed bead will have uniform and consistent ripples. (*Linde Company*)

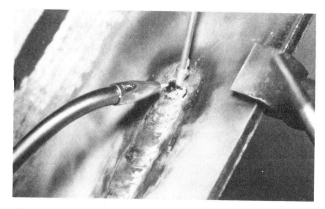

Figure 27-12. When making a brazed weld in a vertical or semivertical position, first build up a shelf at the bottom. (*Linde Company*)

dering is carried out like any ordinary brazing process.

Soldering Conditions:

1. Parts to be soldered must fit perfectly so the solder can travel by capillary action between the two surfaces. Solder will cease to flow where there is a gap between the two pieces.

2. Parts to be soldered must be absolutely clean because the solder will not stick to a dirty, oily or oxide-coated surface. Dirt and grease can be removed with a cleaning solvent. Steel wool or some form of abrasive cloth is used to eliminate the oxide. Application of a flux completes the cleaning process and keeps the metal free from oxide during the heating and soldering operation.

3. Parts must be held together during soldering so there is no movement. Any movement during the heating will cause the pieces to be misaligned and the slightest disturbance of the solder will cause it to solidify without forming a bond. The result is a weak joint.

4. Parts to be soldered must have a suitable joint design to withstand the necessary load imposed on it. A lap joint is the most satisfactory for most purposes. If greater strength is needed, some type of mechanical joint should be made before soldering. Typical joints for soldering are shown in Figure 27-13.

5. Parts must be washed in hot water after soldering is completed to stop the corrosive action.

Soft solders. Tin-lead alloy solders have a melting range from about 370°F (188°C), for a mixture of 70% tin and 30% lead, to about 590°F (310°C) for a 5% tin and 95% lead mixture. The most common general-purpose solder is known as half-and-half or 50-50 solder. It contains 50% lead and 50% tin and melts at approximately 471°F (244°C).

Alloys with a low tin content have higher melting points and do not flow as readily as the high tin alloys. Solders with a high amount of tin have better wetting properties and produce less cracking.

Special solders are also available for specific purposes. Thus a tin-antimony solder is designed to solder food-handling vessels where lead contamination must be avoided. Tin-zinc solders are intended primarily for joining aluminum. Lead-silver solders are used where strength at elevated temperatures is required.

Solders are available in bar, cake, solid wire, flux-core wire, ribbon and paste forms. Flux-core wire solder has an acid or rosin flux in the center of the wire. With these solders no additional flux is needed.

Most metals such as steel, galvanized sheet steel, tin plate, stainless steel, copper, brass, and bronze can be joined with a soft solder.

Fluxes. Just as in brazing, a flux is required for soldering. The flux prevents the formation of oxides during the soldering operation and increases the wetting action so the solder can flow more freely.

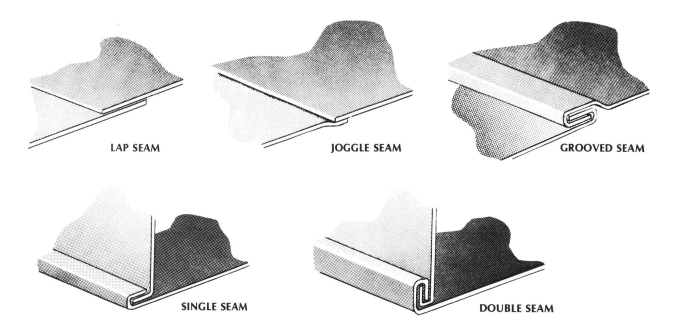

LAP SEAM   JOGGLE SEAM   GROOVED SEAM

SINGLE SEAM   DOUBLE SEAM

Figure 27-13. The type of soldered joint design is determined by the strength requirements of the joint.

There are many excellent commercial fluxes available in paste, liquid, powder, and cake form. Some are general-purpose fluxes usable on most metals. Others are special fluxes such as those for soldering aluminum.

All fluxes are classified as corrosive or noncorrosive. Rosin is the most common noncorrosive flux. Zinc chloride is the most frequently used corrosive flux. Although the corrosive types are most effective, they must be washed away from the metal after soldering. They should never be used for electrical or electronics work.

Zinc chloride is prepared by adding small pieces of zinc to muriatic (commercial hydrochloric) acid until the zinc will no longer dissolve. The cut or killed acid is then diluted with an equal quantity of water.

WARNING: In diluting the acid, the acid must always be added to the water. ALWAYS ADD ACID! Pouring water into the acid may result in a violent and dangerous action.

WARNING: It is also important to remember that when zinc is dissolved in muriatic acid, injurious chlorine fumes are given off! Therefore the preparation must always be carried out near an open window or under a ventilation hood. Uncut or raw acid (straight or diluted) is preferred for galvanized iron but cut acid may be used and is safer to handle.

Heating devices. In any soldering operation both pieces of metal to be joined must be hot enough to melt the solder. A strong bond is achieved only if the molten solder spreads evenly over the surface. A number of devices are available for heating purposes. The type used depends on the size and configuration of the assembly.

Soldering coppers. A soldering copper consists of a piece of copper fastened to an iron rod with a wooden handle. These coppers vary in size with heads forged in several shapes. Generally a lightweight copper is used for soldering light-gauge metal and a heavyweight copper for soldering heavy-gauge metal. A lightweight copper on heavy metal does not hold enough heat to heat the metal or allow the solder to flow smoothly. Soldering coppers are heated in a furnace or with a blowtorch.

Tinning a soldering copper. The point of a soldering copper must be covered with a thin coat of solder to work properly. Overheating or failure to keep the copper clean causes the point to become covered with oxide. The process of replacing this coat of solder is called *tinning*. To tin a copper:

1. File each side of the point until all oxide and pits are removed.

2. Heat the soldering copper until it is hot enough to melt solder.

3. Rub the point on a block of sal-ammoniac and apply a little solder as you continue to rub. Sal-ammoniac, which is ammonium chloride, helps to clean the point of the copper. An alternate way is to dip the point in a liquid or paste flux and then apply the solder.

4. Remove the excess solder by wiping the point with a clean cloth.

Electric soldering irons and pencils. These devices are often more convenient than soldering coppers because they maintain a uniform heat. See Figure 27-14. They vary in size from 25 watts to 550 watts. Lightweight, low-voltage irons with replaceable heating elements and tips are called soldering pencils and are preferred for electric and electronic work. Electric soldering guns produce instant heat at the tip of a long small point when the trigger is pulled. On some guns the trigger also turns on a light which focuses at the point. For these reasons the soldering gun is very popular for electronics soldering work.

Flame-burning devices. Some soldering operations are impossible or very difficult to perform with a soldering copper or iron. For such tasks a flame is used as the source of heat. See Figure 27-15. The flame can be produced with a gas torch. The gases used depend on the nature of the job. The most efficient, safe, and versatile gas torch is one that uses a variety of gases such as acetylene, Mapp, natural gas, propane, and compressed air. These torches are equipped with changeable tips which can produce a wide range of flame sizes. The gas-air torch has two needle valves; one valve controls the pressure of gas and the other valve the compressed air. See Figure 27-16. To light this torch the gas-needle valve is opened slightly and ignited with a spark lighter. Then the air-valve is turned on and adjusted until a blue flame results. The length of the flame is controlled by the amount of gas and air allowed to flow to the tip.

Bottled-gas torches are also used for soldering especially when a stationary torch is not available. The bottled-gas torch must be operated with care; therefore the manufacturer's instructions should always be followed carefully. See Figure 27-17.

Soldering Techniques

The two manually performed soldering operations are known as seam-soldering and sweat-soldering. See Figure 27-18.

Seam-soldering. Seam-soldering involves running a layer of solder along the outside edge of the joint. To solder a seam directly, place the fluxed pieces together and tack the seam in several places.

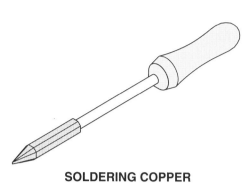

SOLDERING COPPER

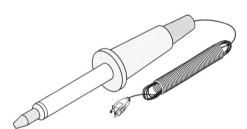

SOLDERING IRON

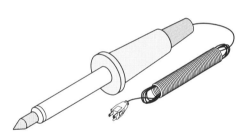

SOLDERING PENCIL

SOLDERING GUN

Figure 27-14. A number of devices are available to provide the necessary heat for soldering.

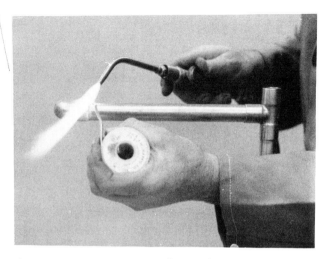

Figure 27-15. A gas-air torch can be used for sweat-soldering copper pipe.

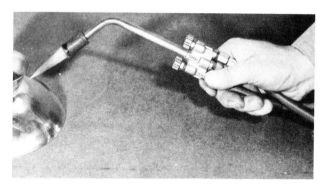

Figure 27-16. The gas-air torch has two needle valves—one for controlling gas pressure, the other for the compressed air.

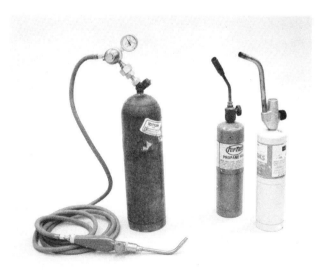

Figure 27-17. Bottled-gas torches are used for soldering when a stationary torch is not available.

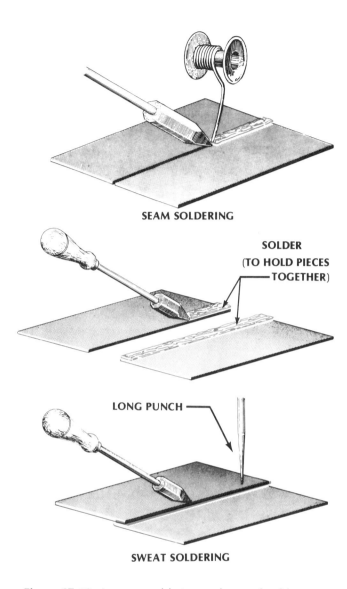

SEAM SOLDERING

SOLDER
(TO HOLD PIECES
TOGETHER)

LONG PUNCH

SWEAT SOLDERING

Figure 27-18. In seam-soldering, a layer of solder runs along the outside edge of the joint. In sweat-soldering, two pieces are joined without any solder being visible.

This is done by holding the copper on the metal until the flux begins to sizzle. Then apply a small amount of solder directly in front of the point. Do not apply the solder to the point. The hot metal should melt the solder. Now start at one end of the seam. Heat the metal and apply solder as needed in front of the point. If necessary press each newly soldered portion together.

Sweat-soldering. Sweat-soldering is a process whereby two surfaces are soldered together without allowing the solder to be seen. To perform such an operation proceed as follows:

1. Coat the pieces to be soldered with flux after all dirt, oil, grease, and oxide have been removed.

2. Apply a uniform coating of solder to each of the surfaces to be joined.

3. Place the surfaces together with the soldered sides in contact.

4. Place the flat side of a heated copper on one end of the seam. To avoid smearing the exposed surfaces of the metal with solder, remove any excess solder on the copper by quickly wiping the point with a damp cloth before placing it on the joint.

5. As the solder between the two surfaces begins to melt and flow out from the edges, press down on the metal with a punch. Draw the copper slowly along the seam and follow with the punch. Do not move the copper any farther than the solder melts.

Points to Remember

Brazing:

1 Use joints designed for brazing.

2 Be sure surfaces to be brazed are completely free of oil, grease, dirt, and oxide.

3 Always use a flux that is prescribed for the metal to be brazed.

4 Remove all flux residue after the brazing operation is completed.

5 Use an appropriate filler metal for the work to be brazed.

6 When using a gas torch, heat the surfaces with the outer envelope of the flame and not the inner cone.

7 Use the lowest effective heat for brazing.

Braze welding:

1 Do not braze weld a metal that will be subjected to high temperatures or high stresses.

2 If some preheating is necessary, bring the temperature up only to black heat.

3 Use only manufacturers recommended rods for braze welding operations.

4 Use a special brazing flux for all braze welding jobs.

5 Clean surfaces thoroughly before applying the filler metal.

Points to Remember, cont.

6 Arrange the work in flat position.

7 Use a neutral flame unless otherwise specified.

8 Be sure the surfaces are properly tinned before depositing beads.

9 Do not melt the surfaces to be welded; heat them only to a dull red.

10 Use a circular torch motion.

Soldering:

1 Always use the recommended flux when soldering.

2 Never prepare zinc chloride in a confined space without ventilation.

3 Make sure parts to be soldered are clean and their surfaces fit closely together.

4 During the soldering process, do not allow the parts to move while the solder is in a liquid state.

5 Be sure the soldering heat is adequate for the soldering job to be done, including the types of metal and the fluxes.

6 Wash the soldered work in hot water to stop later corrosion action.

QUESTIONS FOR STUDY AND DISCUSSION

1 How does brazing differ from fusion welding?

2 What specific advantage does brazing have over regular fusion welding?

3 What is the main limiting factor of any brazing process?

4 Why is a lap joint better than a butt joint for brazing purposes?

5 Why is joint clearance an important factor in brazing?

6 What procedure should be used in cleaning surfaces to be brazed?

7 Why is a flux needed for brazing?

8 Why should all flux residue be removed after brazing is completed?

9 When is a controlled atmosphere used in place of a flux?

10 What do the AWS classification symbols for brazing filler metal represent?

11 How should the torch flame be applied to the work to carry out a brazing operation?

12 What is meant by *liquidus* and *solidus* temperatures?

13 What brazing techniques are used for high-rate production work?

14 What is the difference between braze welding and brazing?

15 What are some of the advantages of braze welding?

16 When should braze welding not be used?

17 What kind of rod is needed for braze welding?

18 What is the function of the flux in braze welding?

19 How should the flux be applied?

20 Why should the piece be in flat position when braze welding?

21 What kind of flame is recommended for braze welding?

22 What is meant by tinning?

23 How can you tell when the surface is hot enough for braze welding?

24 How does soldering differ from brazing?

25 How does the tin content of solder affect its flowing properties?

26 How is a zinc chloride flux prepared? With what material?

27 During the soldering process, why should parts be held firmly in place?

28 Why is a lightweight copper not suitable for soldering heavy-gauge metal?

29 What is meant by tinning a copper?

30 What are some of the basic requirements which contribute to effective soldering?

31 How does seam-soldering differ from sweat-soldering?

32 What type of heating devices can be used for soldering purposes?

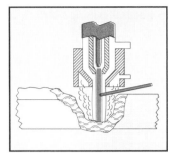

Surfacing is a process of applying a hard, wear-resistant layer of metal to surfaces or edges of worn-out parts. It is considered one of the most economical methods of conserving and extending the life of machines, tools, and construction equipment. Thus, the process may involve building up worn shafts, gears, or cutting edges of tools. See Figures 28-1 and 28-2.

icals, and oxidation or scaling at elevated temperatures.

An understanding of what caused the wear is important, since any surfacing application requires different techniques.

Properties of Parts to be Surfaced

An additional requirement of any hardfacing operation is a knowledge of the composition of the

Figure 28-1. The edges of these plowshares have been repaired with abrasive-resistant hardfacing material. (*The Lincoln Electric Company*)

The two main types of surfacing are known as *hardfacing* and *metallizing.* Hardfacing, or hardsurfacing as it is sometimes called, is a fusion technique whereby a hard, tough overlay of metal is actually fused with the worn unit. Metallizing is a spray coating procedure where finely divided particles of metal are deposited on worn away surfaces.

Types of Wear

Parts in service are subjected to three main types of wear: impact, abrasion, and corrosion.

Impact refers to crushing forces which cause parts to chip or crack.

Abrasion is associated with grinding, rubbing or gouging actions.

Corrosion involves the destruction of a surface because of atmospheric contamination (rusting), chem-

Figure 28-2. Surfacing with hardfacing material is an economical method for building up worn parts that are subject to great wear. (*The Lincoln Electric Company* and *The Wall Colmonoy Corporation*)

component to be serviced.

Metals of these parts can be grouped into two categories. In one group are the metals whose physical properties are not changed significantly or are subject to cracking when heated and cooled in a hardfacing operation. These metals include the low-range carbon and medium-carbon steels, the low-alloy steels, and the stainless steels. The second group includes metal parts made of steels whose physical characteristics are changed with the application of hardfacing materials. These metals usually have been hardened by some heat-treating process and any subsequent exposure to heat may jeopardize this hardness or produce cracks. Metals in such a group include the higher range medium-carbon steels, high-carbon steels, cast irons, and other alloy steels.

Metals in the first group can be hardfaced without any particular precautions since no harmful cracking will affect the hardness of adjacent welds. With metals in the second group special care must be taken to minimize the sudden shock of localized heat. This is done by reducing the hardness by annealing or through gradual and uniform preheating and postweld heat treatment. Preheating from 300° to 500°F (149° to 260°C) will usually prevent weld hardening in medium and high-carbon steels. High-carbon alloy steels and wear resistance alloy steels require preheating to the same temperature. After the surfacing operation is completed, postweld heat treat to a temperature of 800° to 1300°F (about 425° to 700°C) and allow to cool slowly.

HARDFACING

There are many different types of hardfacing materials. Most of them have a base of iron, nickel, copper, or cobalt. Auxiliary elements may be carbon, chromium, molybdenum, tungsten, silicon, manganese, nitrogen, vanadium, and/or titanium.

The alloying elements form hard carbides which contribute to the crystalline properties of the hardfacing metals. Thus a high percentage of tungsten or chromium with a high carbon content will form high carbide crystals that are harder than quartz. Materials having a high chromium content provide excellent resistance to oxidation and scaling. Nickel, cobalt, and chromium are particularly effective for corrosion resistance when added as hardfacing.

Hardfacing metals are either martensitic, pearlitic, or austenitic. Martensite is the hardest and strongest. Pearlite is moderately tough and hard. Austenite is soft, tough, and resists impact.

To measure the hardness of a particular metal, the Rockwell hardness tester may be used. The machine measures hardness by the amount of penetration a small diamond under pressure will make. The popular C scale is often used which is 150 kg. load applied to the test specimens. See Chapter 34.

Electrodes for hardfacing. Electrodes for hardfacing are divided into three main groups: severe abrasion resistant, moderate abrasion and impact resistant, and severe impact and moderately severe abrasion resistant.

Hardfacing metals are available as rods for oxyacetylene welding, electrodes for shielded metal-arc welding, and hard wires for automatic welding. Some hardfacing materials come in tubular rods which contain a mixture of powder metal, powder ferroalloys and fluxing ingredients. The same material is available in powder form for hardfacing with the carbon arc.

1. Severe abrasion-resistant. Electrodes in this group are of the tungsten carbide and chromium carbide types. These electrodes deposit a very hard abrasive-resistant material. They are not suitable for impact wear, since the deposit chips and cracks when subjected to shock.

Chromium and tungsten electrodes come either in coated tubular form or as regular coated cast alloy. The tube rods contain a mixture of powder metal, powder ferroalloys, and fluxing ingredients. The tubes are coated for arc stabilization and arc shielding. Both types are used with the shielded metal-arc process. The same material is available in powder form. The powder alloy is used when hardsurfacing is performed with the plasma arc.

Tungsten electrodes have tiny crystals of tungsten carbide embedded in the steel alloy. When applied on a surface, the steel wears away, leaving toothlike particles of tungsten carbide exposed. Since tungsten carbide is very hard, the exposed particles make the edge of the part self-sharpening. This property is particularly desirable for earth digging equipment, scraping tools, plowshares, rotary digger blades, cultivator sweeps, and other similar machinery. See Figure 28-3.

Still another type of tungsten carbide electrode deposits fine particles of tungsten carbide that are so close that they form a smooth cutting edge. These electrodes are useful in repairing steel cutting edges such as lathe tool bits.

Chromium carbide electrodes are slightly less hard and less abrasion-resistant than the tungsten carbide type but are tougher. Most of them are not affected by heat treatment and are too hard to be machined. In addition to being hard, chromium carbide electrodes produce surfaces that provide better protection against oxidative corrosion.

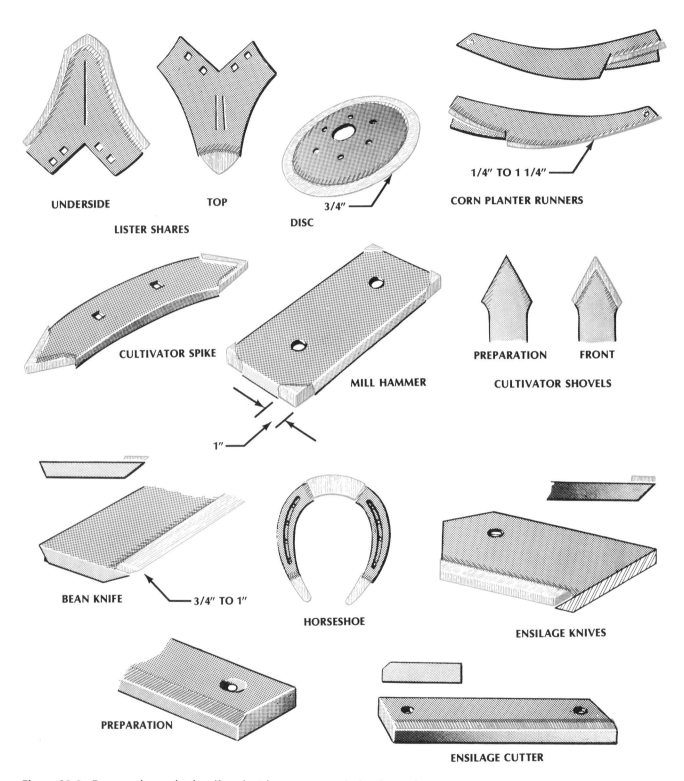

Figure 28-3. Farm tools can be hardfaced with tungsten carbide electrodes to increase their useful life. (*The Lincoln Electric Company*)

2. Moderate abrasion and impact resistant. Electrodes in this group are of the high-carbon type and leave a very hard and tough deposit. They are excellent for repairing surfaces which must withstand both abrasion and impact forces, such as chisels, hammers, sprockets, gears, tractor lugs, bucket teeth on loaders, scraper blades, etc. See Figure 28-4. These electrodes are good for general purpose surfacing and cost considerably less than the tungsten carbide electrode.

Deposits from high-carbon electrodes can be heat treated to produce even harder surfaces, or they can be annealed to soften them for machining. The hardness of the deposit depends on the rate of cooling. The faster the part is cooled, the harder will be the deposit.

In this group are also the electrodes having a high percentage of manganese as an alloying element. Manganese electrodes are tougher but not abrasion-resistant like high-carbon types.

3. Severe impact and moderately severe abrasion-resistant. Deposits of these electrodes are tough but not as hard. They are highly resistant to impact and produce good resistance to abrasion. They are often referred to as self-hardening because the deposited surface hardens as it is pounded. While the outside surface is hard, the material underneath remains soft. This prevents cracking, even though there may be some deformation. These electrodes are especially adaptable for hardsurfacing rock crusher parts, chain hooks, scraper blades, pins, and links. See Figure 28-5.

Stainless steel electrodes are often used for hardfacing parts that must resist impact forces without cracking. These electrodes offer the least resistance to abrasion in the "as deposited" condition; however, they will work-harden (cold working). In addition to their toughness they are very corrosion-resistant. Stainless steel electrodes are often used as base layers for other hardfacing electrodes.

Hardfacing with Shielded Metal-Arc

Hardfacing with the shielded metal-arc is probably used more extensively because of its high deposition rate. It also has wide application where large areas have to be surfaced or for heavy parts that normally would require excessive time to heat with the oxyacetylene flame. This method of hardfacing is especially suitable for depositing overlays on manganese steel and other steel alloys where heat build-up must be restricted.

Either AC or DC current produces satisfactory hardfacing welds. The electrodes may be of the coated solid wire type or hollow tube containing alloy powder and flux. Most of them are referred to by a manufacturer's trade name.

Procedure:

1. Clean the surface thoroughly of rust, scale, and all other foreign matter.

Figure 28-4. Tools that must withstand both impact and abrasion forces are surfaced with high-carbon electrodes.

2. Use only enough amperage to provide sufficient heat to maintain the arc. This is very important to prevent dilution of the deposit by the base metal.

3. Arrange the work so it is in a flat position. Most hardfacing electrodes are designed to be run in the flat position only. See Figure 28-6.

4. Maintain a medium long arc and do not allow the coating of the electrode to touch the base metal. In making the deposit, use either a straight or weaving bead. A weaving bead is preferred when only a thin deposit is required. Do not extend the width of the weave over ¾".

5. Remove all the slag from the surface before depositing additional layers.

6. Manipulate the electrode carefully to secure

Figure 28-5. Hardfacing is commonly used on heavy equipment to retain the ductility of the base metal while providing a surface resistant to wear from abrasion. *(The Lincoln Electric Company)*

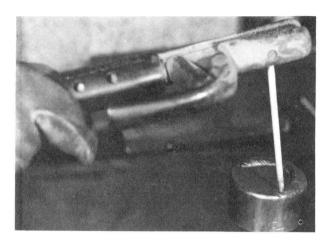

Figure 28-6. Hardfacing is usually done in the flat position.

adequate penetration into the adjoining beads. This can be done by holding the electrode a moment over the deposited bead to allow the heat to build up in the adjoining beads. Such a procedure will also minimize undercutting.

A whipping action is often used when surfacing an area along a thin edge. The arc is held over the heavy portion and then whipped out to the thin edge. In this manner a shallow deposit is made before the heat builds up enough in the base metal to burn through.

Hardfacing with Oxyacetylene

The oxyacetylene hardfacing technique is very useful in depositing overlays on small parts such as engine valves, plowshares, tools, and other similar items. With the oxyacetylene flame, tiny areas can be surfaced and thin layers applied smoothly. Preheating and slow cooling are readily controlled minimizing cracking even with brittle wear-resistant surfacing materials. The principal limitation of the oxyacetylene technique is its low deposition speed and need for heat control.

Metals for hardfacing with oxyacetylene generally consist of low melting high-carbon filler rods. As a rule, a slightly reducing flame is recommended as this will add carbon to the deposit.

The hardfacing operation is started by preheating the surface to produce a sweating condition. During the preheating cycle the tip of the hardfacing rod is held on the fringe of the flame. The rod is then moved into the center of the flame and melted. The actual deposition of the filler rod is carried out with a regular forehand welding technique using a slightly weaving motion.

Hardfacing with Gas Shielded Arc

Both the gas tungsten-arc and gas metal-arc processes are ideal for hardfacing. In many cases, the gas shielded-arc processes are considered superior because of the ease with which an overlay can be made. The surfacing materials are readily deposited to form smooth, uniform, porosity-free surfaces.

Hardfacing with gas tungsten arc is somewhat slower than with gas metal arc, but the overlays are of slightly higher quality. Gas tungsten arc is particularly effective in applying cobalt-base alloys. The process ordinarily requires very little preheating. Since the heat build-up is minimal there is less distortion and very little of the base metal is affected by the heat of the process.

The gas metal-arc process with its continuous wire is faster than gas tungsten arc and produces excellent overlays. With both gas tungsten arc and gas metal arc the shielding gas provides an added feature when aluminum and bronze surfacing materials are used. The shielding gas prevents oxidation and loss of alloying ingredients. A variety of special wires are available for practically every conceivable hardfacing operation.

Care must be taken in using gas tungsten arc and gas metal arc for surfacing to avoid dilution of the deposited weld metal. Helium or a mixture of helium-argon generally produces a higher arc voltage than pure argon and therefore increases the tendency for greater dilution; hence, argon or a mixture of argon and oxygen is recommended for hardfacing with the gas shielded-arc processes.

Hardfacing with Submerged Arc

The submerged-arc process is used when hardfacing parts where heavy deposits are required and extensive areas are to be surfaced. Since the submerged arc utilizes high welding current its deposition rate is high and its deposits are of high quality. Smooth overlays can be made with little or no welding experience required of the operator.

The filler metal may be either solid or tubular and is especially suitable for surfacing that requires high compression strength. However, the relatively deep penetration of the submerged arc plus its protective flux covering usually develops more intensive heat in the welded area. Consequently, greater precautions must be taken to provide suitable preheat and postweld heat treatment for stress relief.

Very often the full strength of the hardfacing metal is attained only by depositing two or more layers. The initial layer frequently becomes diluted when fused into the base metal and therefore an additional layer is necessary to secure the required results.

Hardfacing with Plasma Arc

Plasma-arc surfacing is a mechanized tungsten-arc process that uses a metal powder as surfacing material. The metal powder is carried from a hopper to the electrode holder in an argon gas stream. See Figure 28-7. From the torch the powder moves into the arc stream where it is melted and then fused to the base metal. The surfacing is an actual welding process and not a metal spray process. A wide variety of cobalt, nickel, and iron-base surfacing powders are available from manufacturers of welding supplies. These powders are fused materials and consequently are homogeneous in composition. They are classified as high-alloy materials having varying degrees of impact resistance qualities, abrasion resistance qualities, and corrosion resistance qualities. The type of application should be determined before selecting the powder to be used.

The power source consists of a conventional DC power supply unit with straight polarity. A second DC unit is connected between the tungsten electrode and the arc-constricting orifice to support a nontransferred arc. The second power supply supplements the heat of the transferred arc and serves as a pilot arc to start the transferred arc. Argon gas is used to form the plasma as well as the shield.

METALLIZING

Metallizing, sometimes referred to as metal spraying, is a process of depositing fine semimolten metal particles or metal powder onto the surface of a metal to form an adherent coating. The powder or metal particles are shot through an intense heat and are spread on the surface where they form thin layers of metal. The powder or wire rod is fed into an oxygen-fuel gas flame and the small semimolten droplets are driven onto the surface by a stream of high pressure air. The minute particles strike the surface at estimated speeds of 250 to 500 feet per second, depending on the gun design. Cohesion is achieved by the mechanical interlocking and fusion of the tiny metallic particles and the bonding of the thin oxide film which forms on the particles while in motion.

Metallizing is an important process in the machine field where worn surfaces need to be restored to their original sizes but where low tensile strength and porosity are not objectionable. It is a very functional

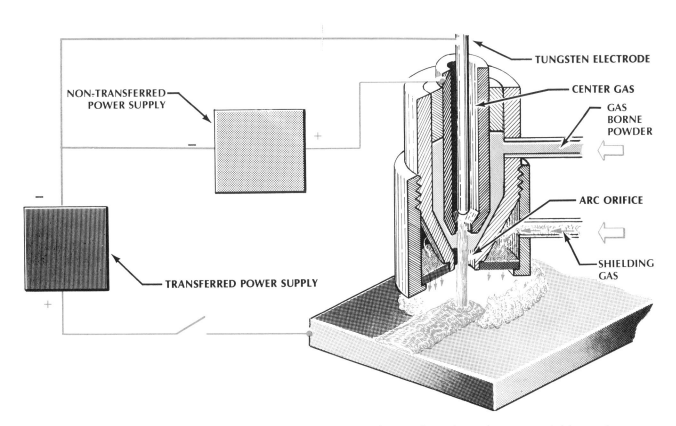

Figure 28-7. In plasma-arc surfacing, an argon gas stream carries the metal powder surfacing material from a hopper to the electrode holder. The powder then moves into the arc stream where it is melted and fused to the base metal.

process for jobs where welding or brazing heat is impractical or for applying deposits of dissimilar metals which otherwise are not possible. The process is unique since there is no limit to the size of the object or structure which can be coated. The added feature of metallizing is that no preheating or postweld heat treatment is required. There is little or no distortion resulting from the spray. See Figure 28-8.

Metallizing Process

The success of any spraying process depends on having a clean surface that has been properly roughed. All traces of oil, dirt, scale, rust, etc., must be removed. The roughing of the surface provides mechanical anchorages for the sprayed metal particles. Grit blasting is probably the most common method used for surface roughing. Blasting abrasives may be steel grit, hard sand, aluminum oxide, or silicon carbide. Sometimes after a surface is prepared for spraying, a thin layer of molybdenum is sprayed on. This produces a fusion bond which gives greater adherence to the subsequent spraying coats. Another method employed in roughing a surface is to run a chasing tool over the area. The chasing tool

produces ragged threads.

The sprayed metal coatings are somewhat porous but in machine elements this is an advantage. The porous coating absorbs oil which provides more complete lubrication. Porosity becomes more critical in applications that are subject to severe attacks by acids and other corrosive materials.

To some extent porosity can be controlled by the adjustment of gas and air as well as the distance of the gun from the workpiece. However, too much effort to reduce porosity will usually result in hard, brittle and highly oxidized coatings which will fail in service.

Oxidation normally occurs in the melting flame and during the flight of the metal particles to the surface. As a rule, little oxidation will take place as the metal is melted unless the gas-fuel mixture is oxidizing. The greatest causes of oxidation are overheating of coating, excessive use of oxygen, and spraying too great a distance from the workpiece.

Wire spray guns. Metallizing is done with a special spray gun which weighs from 3 to 6 pounds and handles 20 gauge to ³⁄₁₆″ dia. wire. Usually these guns will spray 4 to 12 pounds of metal per hour. Larger type guns are often mounted on a fixture and designed for spraying large machine components.

Figure 28-8. Metallizing involves spraying tiny particles of metal onto the surface of the part to be built up.

The gun consists of two major parts: the power unit and the gas head. See Figure 28-9. The power unit feeds the wire into the nozzle of the gun. The gas head controls the flow of oxygen, fuel gas, and the compressed air. The nozzle has a center orifice through which the wire is fed. Around this orifice are a number of gas jets which provide the flame and high velocity air stream. As the wire comes through the orifice, it is melted and atomized by the flame. The fine molten particles are picked up by the air stream and projected against the work. The most common gas for the oxy-fuel flame is acetylene which produces a temperature exceeding 5600°F (3094°C), although hydrogen or propane is sometimes used for metals that melt at a lower temperature.

Spraying operation. Any metal spraying operation should be carried out in a well-ventilated area. *WARNING: Adequate ventilation is necessary to remove dust particles and fumes which are extremely hazardous to health. If positive ventilation is not possible, the operator should wear an effective respirator.*

The wire speed, amount of spray, gas, and oxygen pressure must be regulated according to the recommendations established for the equipment to be used and the type of metallizing to be done. Air pressure is normally set for 60 psi. The use of a flowmeter will insure more accurate control of the gases. A slight increase in air pressure provides a finer coating and, similarly, a decrease in air pressure produces a coarser coating.

The tip of the melting wire should project beyond the end of the air cap. This length depends to a large extent on the material being used. A recommended practice is to speed up the wire until chunks of it are being ejected. Then the wire feed is reduced until the ejection of chunks discontinues.

Each coating should be kept as light as possible, somewhere around 0.003" to 0.005" in thickness. Too heavy a coat will produce an irregular and stratified surface. The actual movement of the gun is very similar to paint spraying. The nozzle should be kept approximately 4" to 10" away from the surface and moved with a uniform motion. If the gun is held too close to the work minute cracks will form in the coating. Too great a distance will produce a soft spongy deposit with low physical properties. The rate of gun travel is also important. When the travel is too rapid the coating develops a high oxide content.

In spraying a flat surface the gun is moved back and forth to allow a full uniform deposit. Spraying should begin beyond the edge of the area to be covered and continued beyond the end of the area.

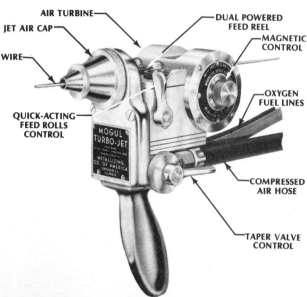

Figure 28-9. Metallizing guns may be mounted or hand held. (*Metallizing Company of America*)

After the first layer the work or gun is often rotated 90° and this technique repeated for each subsequent coating until the required thickness is built up. On cylindrical pieces the work is generally fastened in a lathe with the gun mounted on the traveling carriage.

Electric-Arc Metallizing

The electro-spray unit deposits particles that are hotter and more fluid than those resulting from oxyacetylene metallizing equipment. See Figure 28-10. Heat to melt the wire is generated by an electric arc instead of oxyacetylene. The arc, which reaches a temperature of approximately 7000°F (3870°C), produces greater bond strength since the highly heated particles are able to create better fusion. Also the hotter particles will form coatings which will have a lower oxide content.

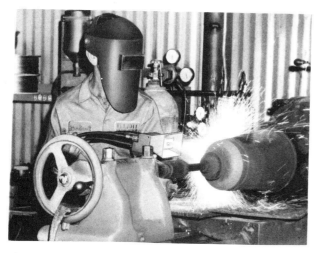

Figure 28-10. Electric-arc metallizing equipment produces coatings with greater bond strength and lower oxide content than coatings created with oxyacetylene metallizing guns. (*The Wall Colmonoy Corporation*)

Oxyacetylene Metal-Spraying Torch

Another metal-spraying unit is the oxyacetylene torch. See Figure 28-11. A hopper mounted on top of the torch body feeds powdered metal alloy into the gas stream. The flow of the powdered alloy is controlled by the operator. Alloy particles become molten as they are sprayed through the flame and onto the workpiece. The same flame is used to preheat the work, to spray the part, and to fuse the deposited powder.

Figure 28-11. An oxyacetylene metal-spraying torch has a hopper on top that feeds powdered metal alloy into the gas stream. The metal particles, melted as they are sprayed through the flame, are fused to the workpiece. (*The Wall Colmonoy Corporation*)

Points to Remember

1 Determine what type of wear the part must withstand: impact or abrasion or both.

2 Use a tungsten carbide or chromium carbide surfacing electrode for parts that must withstand abrasion.

3 Use high-carbon or manganese electrodes for parts which will be subjected to moderate abrasion and impact.

4 Use manufacturers recommended electrodes for hardfacing parts which must withstand severe impact loads.

5 Clean the surface thoroughly before hardfacing.

6 Place the work in a flat position.

7 Maintain a medium long arc length.

8 Use a minimum amount of heat when using the shielded metal-arc process.

9 Do not allow the coating of the electrode to contact the base metal.

10 With the shielded metal-arc process, be sure to remove all slag before depositing additional layers.

11 If the shielded metal-arc process is employed to hardface thin-edge material, use a weaving or whipping action.

QUESTIONS FOR STUDY AND DISCUSSION

1 What is hardfacing?

2 Of what value is hardfacing?

3 What types of wear do parts encounter in service?

4 Why must the correct type of electrode be used for hardfacing?

5 Why are tungsten carbide or chromium carbide electrodes used for hardfacing parts which must withstand heavy abrasion forces?

6 What are the advantages of chromium carbide electrodes over tungsten carbide?

7 For what type of surfacing are high-carbon electrodes used?

8 When are stainless steel electrodes used for hardfacing?

9 What are some factors in choosing the method of hardfacing?

10 Why avoid excessive heat when hardfacing?

11 Why should hardfacing be done in a flat position?

12 How should the shielded metal-arc be manipulated in hardfacing large objects where a high deposit rate is required?

13 When is a whipping action used in hardfacing with a shielded metal-arc welding process?

14 What is the difference between hardfacing and metallizing?

15 When is the submerged-arc process used for hardfacing?

16 How does hardfacing with the plasma arc differ from that produced by the regular shielded metal-arc process?

17 When is it more advantageous to use oxyacetylene for hardfacing purposes?

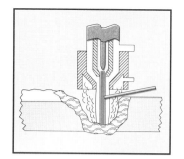

Special Welding Processes

Pipe of all types and sizes is used a great deal in transporting oil, gas, and water. It is used extensively for piping systems in buildings, refineries, and industrial plants. Furthermore, pipe has gained acceptance in construction, and often takes the place of beams, channels, angles, and other standard shapes. See Figure 29-1.

Welding is the easiest and simplest method of joining sections of pipe together since it eliminates complicated threaded joint designs, permits free flow of liquids, and reduces installation costs.

Welding is also considered a practical and effec-

Figure 29-1. Welding is used extensively in joining pipe. (*Hobart Brothers Company*)

tive cost-cutting technique in joining noncritical low-pressure piping systems for refrigeration, air-conditioning or heating applications.

PIPE WELDING PROCESSES

Although some pipe sizes are occasionally welded with oxyacetylene, most pipe welding is done with the shielded metal-arc process. However, a considerable amount of pipe is currently being welded with gas shielded arc, either manually, semiautomatic or fully automatic.

Pipe welding is recognized as a specialty in itself. See Figure 29-2. Although many of the skills and practices are similar to other types of welding, pipe welders usually must develop certain techniques that are characteristic of pipe welding alone. Furthermore, since public health, environmental restrictions and safety are involved, especially in welding cross-country transmission pipelines and high-pressure lines that are to convey steam, oil, air, and corrosive materials, pipe welders always have to pass certain tests to be certified.

It is impossible to list here any one set of specific certification standards because they will vary for different welding jobs and often are supplemented or modified by local specifications. Accordingly, anyone interested in qualifying as a pipe welder should be aware of certification requirements as they apply

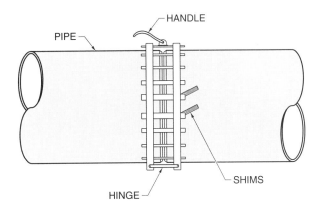

Figure 29-3. Pipe clamps hold short sections of pipe in alignment while tack welds are made. After the short sections are welded the long pipe is placed in line.

for the pipe welding job involved. Some general certification requirements for several classifications of welding are included in Chapter 36.

Roll and Position Welding

Pipe welding in the field is done in several ways. In one method two or more sections are lined up and tack welded. Special pipe clamps, as shown in Figure 29-3, are used to hold the pipe in alignment until they are tacked. The weld is then completed in the flat position while helpers rotate the pipe. After the short sections are joined, the long pipe is placed in line with the connecting pipe and the weld made with the entire length in a stationary position. This operation is called *roll welding*. See Figure 29-4.

The *stove pipe* or *position* method consists of lining up each section, length by length, and welding each joint while the pipe remains stationary. Since the pipe is not revolved, the welding has to be done in various positions—flat, horizontal, vertical, and overhead.

Pipe weld joint positions are identified as test positions. Because pipe welds are usually groove welds, they are identified by the letter G. Test position 1G is roll welding with the axis of pipe horizontal, the welding done in flat position with the pipe rotating under the arc. See Figure 29-5. Test position 2G is identified as horizontal welding with the axis of the pipe in vertical position and the axis of the weld in horizontal position.

There is no 3G or 4G test position in pipe welding.

Figure 29-2. Pipe welding is recognized as a trade in itself. (*The Lincoln Electric Company*)

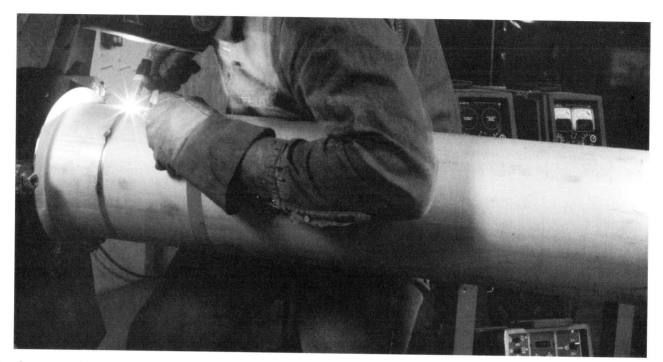

Figure 29-4. In roll welding, pipe is rotated so that all welds are made in the flat position. (*Miller Electric Manufacturing Company*)

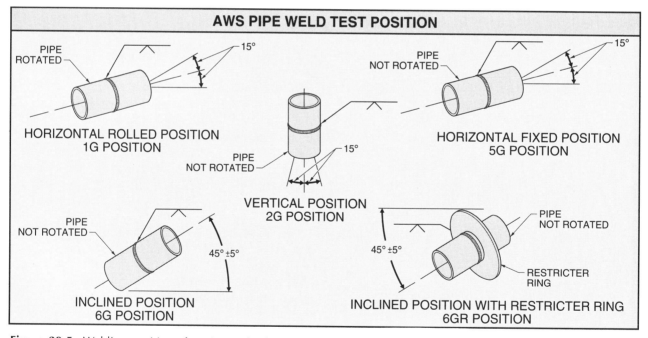

Figure 29-5. Welding positions for pipe and tubing are identified as test position 1G, test position 2G, test position 5G, test position 6G, and test position 6GR.

Test position 5G is identified as horizontal fixed position. The axis of the pipe is horizontal, but the pipe is not turned or rolled during the welding operation. Test position 6G for pipe has the axis of the pipe at 45° and the pipe is not to be turned while welding.

Restricted accessibility is often added (6GR) by placing a ring near the weld. The axis of the pipe may vary 15° (plus or minus) for the 1G, 2G, and 5G test positions, but only 5° (plus or minus) for the 6G position.

Figure 29-6. Gas tungsten-arc welding is often used to join small diameter pipe. (*Hobart Brothers Company*)

As mentioned most pipe welding is done either with the shielded metal-arc or gas metal-arc processes. Gas tungsten arc is occasionally used in shop welding of small diameter pipes. See Figure 29-6. However, gas tungsten arc is also used in certain pipe jobs to lay the root bead in large diameter pipes.

The advantage of gas metal arc over stick welding as described in Chapter 25 is that no slag inclusions will occur in the weld. Furthermore, because of the excellent gas protection shield over the weld area, there is less danger of atmospheric contamination, thereby producing sound welds. Since no slag has to be removed in gas metal-arc welding, less welding time is required.

Insofar as welding techniques and procedures are concerned there is no significant difference between the shielded metal-arc and gas metal-arc processes. Therefore, the general description of pipe welding techniques which follows applies to both stick and gas metal-arc welding.

Pipe Joint Preparation

For welding most pipes a single-V joint is used. The beveling is usually done with a special oxyace-

tylene beveling machine. See Figure 29-7.

Pipes having wall thicknesses of 1/8" to 5/16" are classified as thin-wall pipe, and pipes with wall thick-

Figure 29-7. An oxyacetylene pipe beveling machine is used to prepare the single-V joint for welding. (*DND Corporation*)

nesses over $\frac{5}{16}''$ are thick-wall pipe regardless of diameter. The included angle, root face, and root opening will vary for both thin-wall and thick-wall pipe.

Small diameter pipes with wall thicknesses of less than $\frac{1}{8}''$ are normally welded without any edge preparation. The ends are simply butted together with a small separation to ensure complete fusion. This classification of pipe is frequently welded by the gas metal-arc or gas tungsten-arc methods. See Figure 29-8.

Occasionally, liners or backing rings are fitted into the pipe before welding. These rings assist the welder in securing penetration without burning through the surface as well as in preventing spatter and slag from entering the pipe at the joint. Backing rings also are useful in keeping pipes in alignment and stopping metal icicles from forming on the inside of the joint. See Figure 29-9.

Consumable insert rings are sometimes used to insure accurate root opening dimensions before welding. See Figure 29-10. When the joint is welded these rings are consumed into the weld and become

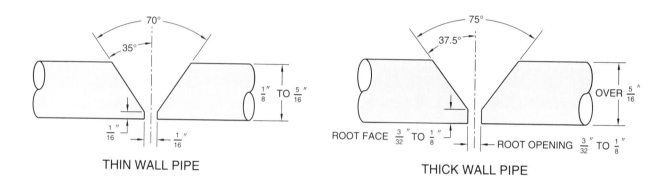

THIN WALL PIPE   THICK WALL PIPE

NOMINAL INSIDE DIAMETER SIZE (IN.)	OUTSIDE DIAMETER (BW GUAGE)	INSIDE DIAMETER (BW GUAGE)			NOMINAL WALL THICKNESS		
		STANDARD	EXTRA-STRONG	DOUBLE EXTRA-STRONG	SCHEDULE 40	SCHEDULE 60	SCHEDULE 80
$\frac{1}{8}$	0.405	0.269	0.215		0.068	0.095	
$\frac{1}{4}$	0.540	0.364	0.302		0.088	0.119	
$\frac{3}{8}$	0.675	0.493	0.423		0.091	0.126	
$\frac{1}{2}$	0.840	0.622	0.546	0.252	0.109	0.147	0.294
$\frac{3}{4}$	1.050	0.824	0.742	0.434	0.113	0.154	0.308
1	1.315	1.049	0.957	0.599	0.133	0.179	0.358
$1\frac{1}{4}$	1.660	1.380	1.278	0.896	0.140	0.191	0.382
$1\frac{1}{2}$	1.900	1.610	1.500	1.100	0.145	0.200	0.400
2	2.375	2.067	1.939	1.503	0.154	0.218	0.436
$2\frac{1}{2}$	2.875	2.469	2.323	1.771	0.203	0.276	0.552
3	3.500	3.068	2.900	2.300	0.216	0.300	0.600
$3\frac{1}{2}$	4.000	3.548	3.364	2.728	0.226	0.318	
4	4.500	4.026	3.826	3.152	0.237	0.337	0.674
5	5.563	5.047	4.813	4.063	0.258	0.375	0.750
6	6.625	6.065	5.761	4.897	0.280	0.432	0.864
8	8.625	7.981	7.625	6.875	0.322	0.500	0.875
10	10.750	10.020	9.750	8.750	0.365	0.500	
12	12.750	12.000	11.750	10.750	0.406	0.500	

Figure 29-8. The included angle, root face, and root opening vary according to wall thickness of the pipe.

a part of the completed weld. Different shapes and compositions of inserts are used as required by specific jobs.

Tack welding. Before welding, pipes are properly aligned and then tack welded. Special line-up clamps are used to insure correct alignment. See Figure 29-11.

To maintain the required spacing between pipe sections a spacing tool is necessary. A wire of correct diameter placed between the pipes will provide the proper spacing.

For most pipe welding, four tack welds are made. These tack welds are evenly spaced around the pipe and approximately ¾" long. Tack welds should penetrate to the root of the groove since they become part of the root bead.

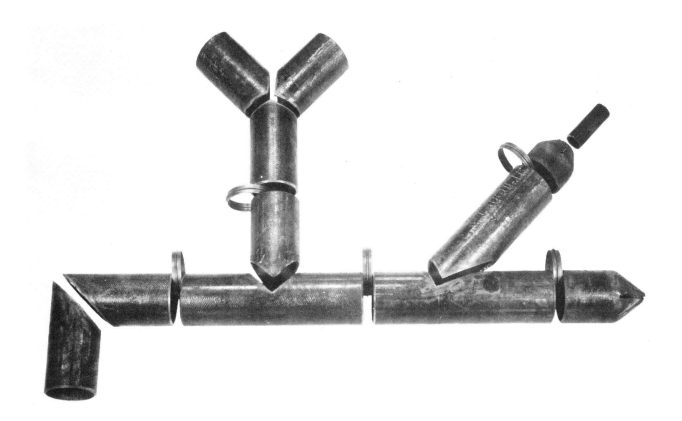

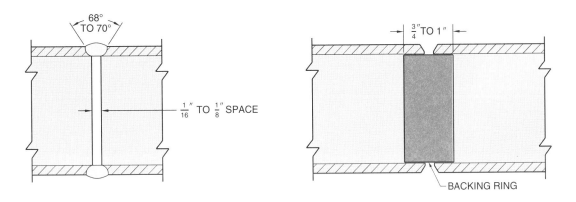

Figure 29-9. Backing rings, fitted in the pipe before welding, keep the sections of pipe in alignment and prevent excessive penetration. (*Hobart Brothers Company*)

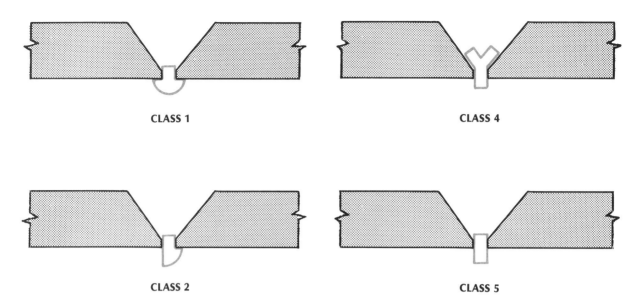

CLASS 1   **CLASS 4**

CLASS 2   **CLASS 5**

Figure 29-10. Consumable insert rings are consumed into the weld and are used for maintaining accurate root opening dimensions.

Figure 29-11. Special line-up clamps hold pipe sections in position while tack welds are made. *(Left: G.A.L. Gage Company; Right: Mathey/Leland)*

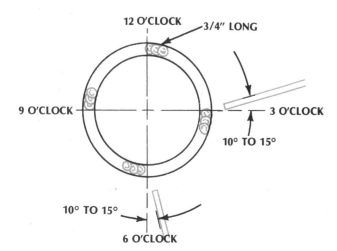

Figure 29-12. Four tack welds are made, evenly spaced at 90° intervals around the pipe.

To make a tack weld, the electrode is inclined 10 to 15 degrees as shown in Figure 29-12. The arc is struck in the joint slightly ahead of where the weld is to be made. The arc is then quickly lengthened so as to stabilize it and give it time to form the protective gas shield. Now the rod is pushed into the joint with a light pressure and a sliding motion started in the groove. If the electrode has a tendency to stick, it should be wiggled slightly but kept buried in the groove. When the tack weld is completed the electrode is pulled away. This procedure produces a strong and fully penetrating tack weld.

Electrodes

Most shielded metal-arc pipe welding is done with E-6010 or E-6011 electrodes except where higher strength welds are required. When higher strengths are needed, especially on low-alloy steel pipe, electrodes in the E-70xx series are used.

The recommended electrode sizes for various pipe weld passes are as follows:[1]

PASS	ELECTRODE	*CURRENT
Stringer bead	5/32	140-165 amps†
Hot pass	5/32	170-200 amps
Fillers	3/16	170-200 amps
Cap pass	3/16	160-180 amps

*The ideal current should be selected from within the range shown. In general, the best quality will be obtained by operating at the lower end of the range.
†Weld stringer bead at 24-26 arc volts and 10-16 ipm arc speed.

Welding Thin-Wall Pipe (Downhill Technique)

Most thin-wall pipe (1/8″–5/16″) is welded by using the downhill technique. It is usually preferred for welding cross-country pipelines because it is faster.

The weld is started at the top or 12 o'clock position and carried downward to the bottom or 6 o'clock point of the pipe. After the 6 o'clock position is reached the same procedure is followed on the opposite side. See Figure 29-13.

One of the problems in downhill welding is controlling the heat input. This is particularly true in welding small diameter pipe where the heat does not dissipate fast enough and excessive heat builds up in the weld zone. Generally heat input can be regulated by using a smaller diameter electrode and reducing the current setting.

Another problem in downhill welding is proper control of the puddle. The molten metal tends to flow downward in the same direction the arc is moving. If this is not controlled penetration cannot be achieved and slag becomes entrapped in the molten metal, thereby producing slag inclusions in the weld. Slag inclusion, of course, is no problem in gas metal-arc welding. Control of metal flow is accomplished by keeping the arc ahead of the puddle. This can be done by using a fast travel speed and a high-current setting.

Welding procedure. After the pipes are securely tacked, a root bead is made completely around the joint. The electrode is held in approximately the same position as in making tack welds. The arc is struck slightly ahead of the weld to preheat the area where the weld bead is to be started. After the arc has stabilized, the electrode is lowered into the root opening and dragged along the edges of the groove. If the electrode has a tendency to stick and fails to glide smoothly because of the built-up heat, a slight side-to-side oscillating motion will usually correct the problem.

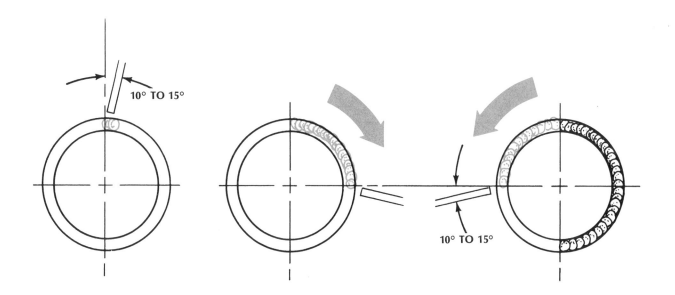

Figure 29-13. In downhill welding, the weld is started at the 12 o'clock position and carried downward to the 6 o'clock position. The same procedure is followed on the opposite side.

1. The Lincoln Electric Company

A properly deposited root bead should penetrate to the root and leave a solid bead below the surface with a slight crown not to exceed ⅟₁₆″. See Figure 29-14. Some undercutting may occur on the faces of the groove but this is not objectionable since this defect will be eliminated by successive passes.

There will be times when the root opening will vary due to poor fit-up. If the root opening is narrow the speed of travel and electrode angle should be reduced. Where a widened root opening exists the travel should be increased.

The success of a pipe weld depends on the correct penetration of the root bead because it forms the base upon which the successive layers are made. Without full penetration the final weld joint will not be sound.

Starting and stopping. Since there is a certain amount of starting and stopping a weld due to changing electrodes or weld position, careful attention must be given to tying the ends of the weld together. A keyhole is made by the penetration of the root pass and must be completely filled when restarting the root pass. See Figure 29-15. To restart a weld the arc is struck about ½″ back of the bead and then moved forward with a long arc. As soon as the arc is stabilized the electrode is momentarily buried in the crater of the last beads to generate a pool of molten metal. The electrode is then raised slightly and the weld continued.

When a weld approaches the end and must be tied into the other deposited bead, the electrode is moved up the sloping sides of the previous bead and after the molten puddle blends smoothly between the two beads the direction of travel is briefly reversed. The arc is then withdrawn quickly by flicking the electrode downward and away from the center.

Successive passes. Upon completion of the root bead, additional layers of weld are deposited. The number of passes will depend on the thickness of the pipe. Usually the subsequent layers will consist of one *hot pass,* one or more *filler passes* and a final *cover* or *cap pass.* The specific function of the *hot pass* is to burn out the remaining particles of slag that may exist in the groove and to achieve a more complete fusion of the base metal and the root bead. This pass usually consists of only a light bead deposit and is made with a whipping motion as shown in Figure 29-16. The electrode is moved down and up a distance of approximately 1½ electrode diameters, pausing momentarily on top of the up motion. The whipping action permits better control of the weld puddle and forces the metal to flow into the undercuts. The hot pass is made with the same di-

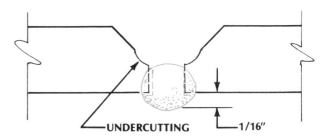

Figure 29-14. A properly deposited root bead should penetrate to the root and leave a solid bead below the surface with a slight crown not to exceed 1/16″.

Figure 29-15. The keyhole left from a stopped root pass must be filled in completely when restarting the arc.

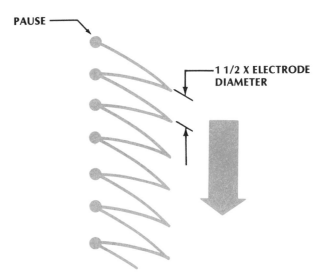

Figure 29-16. A hot pass is made using a whipping motion of the electrode.

ameter electrode used for the root bead but with slightly higher current.

The intermediate or filler passes are deposited with larger diameter electrodes and are intended to fill the weld joint. After each pass is completed the slag must be entirely removed if stick welding is used. Each layer should start and end at a different point to insure a uniform strong weld bond. Thus if the first pass is made at the 12 o'clock position the next pass should begin about one inch below this spot.

Filler passes are normally made with a slant motion. See Figure 29-17A. The side-to-side weave should travel downward a distance of about one electrode diameter per stroke. The electrode should pause at the end of each stroke to insure good fusion at each edge of the weld. As the electrode reaches the bottom of the weld or in the 6 o'clock position a semicircle or horseshoe weave is often used. See Figure 29-17B. This motion permits better control of the puddle since it reduces the fluidity of the molten metal.

The final cover or cap pass is intended to provide maximum reinforcement to the weld joint and at the same time give the weld a neat appearance. The cover pass should have a slight crown extending about ¹⁄₁₆″ above the surface of the pipe. Either a slant or semicircular motion can be used. However, the weave must be wide enough to cover the entire weld joint.

Welding Heavy-Wall Pipe (Uphill Technique)

Uphill welding is generally intended to join heavy-wall pipe. The welding progresses upward on one side of the pipe and then upward on the opposite side. See Figure 29-18. As in downhill welding, a root

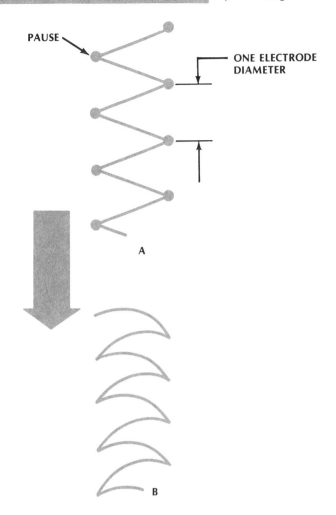

Figure 29-17. A: Filler passes are usually made with a side-to-side weaving motion. B: As the heat increases near the bottom of the weld, a horseshoe weave is often used to control the puddle.

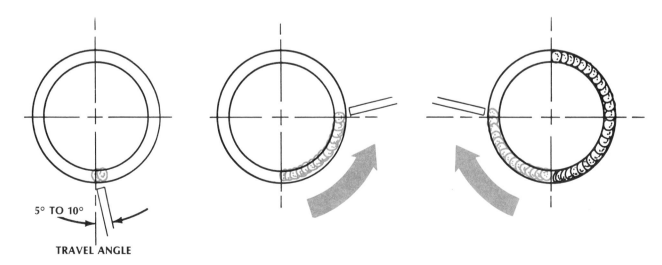

Figure 29-18. In vertical up pipe welding, the welding progresses up on one side of the pipe and then up on the opposite side. The electrode is advanced at a travel angle of 5° to 10°.

bead is deposited first after tack welding. The weld is started just back of the bottom position or what is commonly known as the 6:30 position. Actually the arc is struck ahead of this spot and a long arc maintained for a short period to preheat the surface; then it is brought back and the weld begun. While the root bead is being deposited, no electrode weaving motion is necessary. The electrode is simply advanced at a travel angle of 5° to 10° with a slow and uniform movement along the joint. As the electrode approaches the upper level, the molten metal begins to flow downward at a faster rate. When this happens a slight whipping action is desirable to achieve better puddle control.

After the root bead is completed, one or more filler layers are deposited followed by the final cover pass. Both fill and cover passes are made with a slant weave as described in downhill welding.

AUTOMATIC GAS METAL-ARC PIPE WELDING

A considerable amount of large diameter (24″ and over) pipe is welded with automatic gas metal-arc welding equipment, Figure 29-19. Automatic gas metal-arc welding machines not only speed up the welding process but also produce welds without any danger of slag inclusions which constantly must be

Figure 29-19. Large diameter pipe is often joined with automatic gas metal-arc welding equipment. (*Crutcher Resources Corporation*)

guarded against in regular stick welding.

Unlike the conventional pipe welding procedure where the root bead is deposited externally, in automatic welding the root bead is applied inside the pipe. A special bevel is made on the pipe for this purpose. See Figure 29-20. Usually four welding heads mounted on an internal line-up clamp are used to make the internal root bead in a single pass.

The internal welding unit is self-propelled through the pipe and held in place by clamp shoes extending from the pipe. See Figure 29-21. The welding heads are positioned precisely over the joints by means of special aligner blocks. Once the unit is correctly stationed, the next section of pipe is slipped over the reach rod of the unit. The joint is now properly spaced and another set of clamp shoes is actuated to hold the joint in place for welding. Then a button on the control box mounted on the handle of the reach rod is pushed and the welding commences.

Each welding head welds a 90° arc. All welds are made downhill with two heads moving clockwise and the other two counterclockwise. Shielding gas for internal welding consists of 75 percent argon and 25 percent CO_2.

External welding. The external welding process includes the hot pass, fill pass, and cap pass. These are made with special welding units traveling externally around the pipe on prepositioned circumferential pipe bands. Two welding machines, sometimes referred to as bugs, move simultaneously on the pipe. One starts at 12 o'clock and travels downward to 6 o'clock. The other welding head starts at the 3 or 9 o'clock position and stops at 6 o'clock. The bug is then moved to the 12 o'clock position to complete the pass at the horizontal. See Figure 29-22. All external welds are made with 100 percent CO_2 because of its higher rate of deposition and because it produces better penetrating qualities.

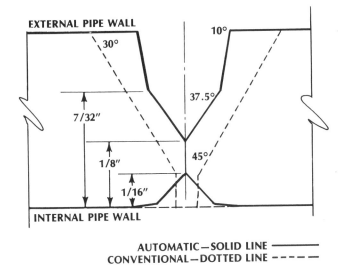

AUTOMATIC—SOLID LINE ————
CONVENTIONAL—DOTTED LINE — — — —

Figure 29-20. The bevel for automatic versus conventional pipe welding is different to allow for differences in penetration. (*Crutcher Resources Corporation*)

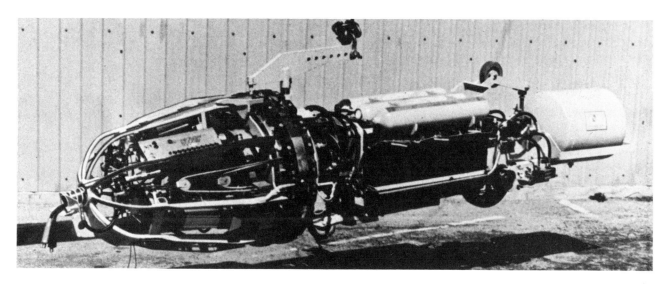

Figure 29-21. An internal welding unit is used to make the initial root bead while positioned inside the pipe. (*Crutcher Resources Corporation*)

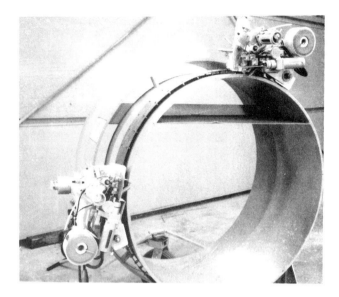

Figure 29-22. External welding bugs are used to make the hot fill and cover passes while positioned outside the pipe. (*Crutcher Resources Corporation*)

Points to Remember

1 Most pipe welding jobs require that you be certified.

2 Usually small diameter pipe with wall thickness of less than ⅛″ are not beveled.

3 Tack the pipes before welding them.

4 Use a downhill technique to weld thin-wall pipe.

5 Be sure the root bead completely penetrates into the root of the joint.

6 Make certain that the ends of the weld are always tied together.

7 Use sufficient passes to fill the weld joint.

8 Finish a pipe weld with a final cap pass.

9 Use an uphill technique on heavy-wall pipe.

10 In uphill welding, start the weld just back of the 6:30 position.

QUESTIONS FOR STUDY AND DISCUSSION

1 Why do pipe welders usually have to meet stiff certification requirements?

2 How is thin-wall pipe distinguished from thick-wall pipe?

3 What is the function of a backing ring?

4 As a rule, how many tack welds are made on pipe?

5 When making a tack weld and the electrode sticks in the groove, what should be done?

6 Why is a proper root opening very important in pipe welding?

7 What is meant when the weld should be started at the 6 o'clock position?

8 What is the difference between *uphill* and *downhill* welding?

9 What are some of the problems which may be encountered in downhill welding?

10 Downhill welding is used for welding what kind of pipe?

11 What is a root bead?

12 What is meant by tying the ends of a weld together?

13 When is the hot pass used and what is its function?

14 How many filler passes are deposited in a pipe weld?

15 What is the function of the cap pass?

16 Why is a whipping action of the electrode sometimes used in making a root bead?

17 Why should each layer start and stop at difference points?

18 What rod motions are used in making fill passes?

19 At what angle should the electrode be held for downhill welding?

20 Why is gas metal arc being used more and more in pipe welding?

21 The external welding process includes what passes?

22 Discuss the considerations in starting and stopping a pipe weld.

23 What is the starting position for thin-wall pipe welded by the downhill technique?

24 Which electrodes are used for most shielded metal-arc pipe welding?

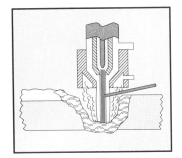

CUTTING OPERATIONS

Special Welding Processes

The most popular methods of cutting metal today are oxyacetylene flame cutting, plasma arc, and air arc. The type of process used depends largely on the kind of metal to be severed or the economy of the operation. The cutting may be done manually or with mechanized equipment. In manual cutting, the operator manipulates a torch over the area to be cut. In machine cutting, the torch is guided entirely by automatic controls.

Flame Cutting

Cutting metal with the use of a flame is widely used in many industrial fields. The cutting is done by means of a simple, hand cutting torch or by a more complicated, automatically controlled cutting machine as shown in Figure 30-1.

Several gases may be used in combination with oxygen to flame cut. These include Mapp, natural, propane, and acetylene. In this chapter the oxyacetylene process will be discussed. Severing metal by this process is made possible when ferrous metals are subjected to rapid oxidation. When a piece of wrought iron or steel is left exposed to various atmospheric conditions, a reaction known as rusting begins to take place. Rust is simply the result of the oxygen in the air uniting with the metal, causing it gradually to change its nature and sometimes to wear away. Naturally this action is very slow. But if the metal is heated and permitted to cool, heavy rust scales form on the surfaces, showing that the iron oxidizes much faster when subjected to heat. If a piece of steel were to be heated red hot and dropped in a vessel containing oxygen, a burning action would immediately take place, reducing the metal to an iron oxide commonly known as slag.

In order to make possible the rapid cutting of metal, it is necessary to use a cutting torch that will heat the iron or steel to a certain temperature and then

Figure 30-1. A flame-cutting machine is guided entirely by automatic controls. (*Linde Company*)

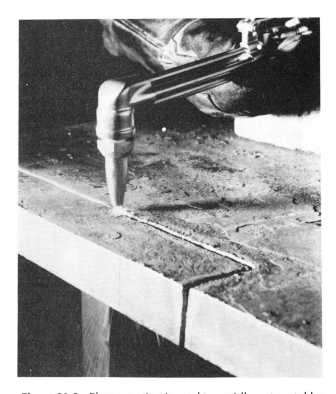

Figure 30-2. Flame cutting is used to rapidly cut metal by subjecting preheated metal to rapid oxidation. (*Linde Company*)

throw a blast of oxygen on the heated section. See Figure 30-2.

The cutting torch. The cutting torch varies from the regular welding torch in that it has an additional lever for the control of the oxygen used to burn the metal. It is possible to convert the welding torch into a cutting torch by replacing the mixing head with a cutting attachment.

The cutting torch has conventional oxygen and acetylene needle valves. See Figure 30-3. These are used to control the passage of oxygen and acetylene when heating the metal. Many cutting torches have two oxygen needle valves for securing a finer adjustment of the neutral flame. The cutting tip is made with an orifice in the center surrounded by several smaller ones. The center opening permits the flow of the cutting oxygen and the smaller holes are for

the heating flame. See Figure 30-4.

A number of different tip sizes are provided for cutting metals of varying thicknesses. In addition, special tips are made for other purposes, such as for cleaning metal, cutting rusty, scaly or painted surfaces, rivet washing, etc.

Determining pressure. The pressure of oxygen and acetylene needed will depend upon the size tip used, which in turn is governed by the thickness of the metal to be cut. Table 30-1 points out the approximate pressure for various tip sizes for a particular type of cutting torch.

Always consult the manufacturer's recommendations for your particular torch as to the proper oxygen and acetylene pressure. The given oxygen pressure cannot always be strictly followed because, for example, steels may have an exceptionally heavy coat-

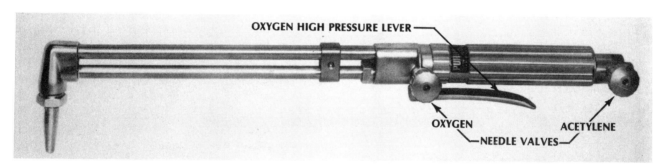

Figure 30-3. This flame-cutting torch has one oxygen needle valve (left) and one acetylene needle valve (right). (*Linde Company*)

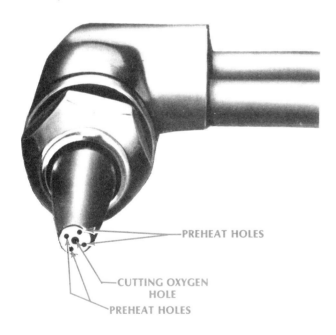

Figure 30-4. The parts of the flame-cutting torch tip include the cutting oxygen hole and the preheat holes.

ing of rust or scale and may require a somewhat greater oxygen pressure to burn entirely through the metal.

TABLE 30-1. CUTTING PRESSURE.

TIP NO.	THICKNESS (inches)	ACETYLENE PRESSURE (pounds)	OXYGEN PRESSURE (pounds)
0	1/4	3	30
1	3/8	3	30
1	1/2	3	40
2	3/4	3	40
2	1	3	50
3	1 1/2	3	45
4	2	3	50
5	3	4	45
5	4	4	60
6	5	5	50
6	6	5	55
7	8	6	60
7	10	6	70

Cutting Steel Using the Flame Cutting Torch   Exercise 1

1 Turn on the acetylene needle valve and light the gas with a spark lighter as if for welding.

2 Turn on the oxygen valve and adjust it for a neutral flame. This flame is the one used to bring the metal to a kindling temperature which, for example in the case of plain carbon steel, is 1400°F to 1600°F (760° to 871°C).

3 Observe the nature of the cutting flame by pressing down the oxygen control lever. When the oxygen pressure lever is turned on it may be necessary to make an additional adjustment to keep the preheating cones burning with a neutral flame.

4 Rule a chalk line about ¾" from one edge of the plate.

5 Place the plate so this line clears the edge of the welding bench.

6 With the torch adjusted to a neutral flame, grasp the torch handle with the right hand in such a position as to permit instant access to the oxygen control lever. The valve is usually operated either with the thumb or forefinger.

7 Making a clean, straight cut depends on how steady you hold the torch. Naturally, when the tip wavers from side to side, a wide kerf (cutting slit) will result, which means a rough cut, slower speed, and greater oxygen consumption. To help keep the torch steady, hold the elbow or forearm on some convenient support.

8 Start the cut at the edge of the plate. Hold the torch with the tip vertical to the surface of the metal, with the inner cone of the heating flame approximately ⅟₁₆" above the chalk line. Keep the torch in this position until a spot in the metal has been heated to a bright red heat.

9 Gradually press down the oxygen pressure lever and move the torch forward slowly along the chalk line. The movement of the torch should be just rapid enough to insure a fast but continuous cut. A shower of sparks will be seen to fall from the underside, indicating that the penetration is complete and the cut is proceeding correctly.

10 If the cut does not seem to go through the metal, close the oxygen pressure lever and reheat the metal until it is a bright red again. If the edges of the cut appear to melt and have a very ragged appearance, the metal is not burning through and the torch is being moved too slowly.

11 When an exceptionally straight cut is desired, clamp a bar across the plate alongside the cutting line to act as a guide for the torch to follow.

12 At the start, the pieces may stick together even when the cut has penetrated through. This is due to the slag, produced by the cutting, flowing across the metal pieces. However, this is not serious, as the slag is quite brittle and a slight blow with a hammer will separate the two sections.

Exercise 1, cont.

13 It may be necessary sometimes to start the cut in from the edge of the plate. In such a case, hold the preheating flame a little longer on the metal; then raise the cutting nozzle about ½″ and pull the oxygen lever. When a hole is cut through, lower the torch to its normal position and proceed with the cut in the usual manner.

Cutting round-bar steel.　To sever round stock, start the cut about 90° from the top edge. See Figure 30-5.

Keep the torch in a straight-up (perpendicular) position and gradually lift it to follow the circular outline of the bar. Maintain this position of the torch while ascending as well as descending on the opposite side.

Beveling.　To make a bevel cut on a steel plate, incline the head of the torch to the desired angle instead of holding it vertically. An even bevel may be made by resting the edge of the torch tip on the work as a support. See Figure 30-6. The torch can also be guided by clamping a piece of angle iron across the plate.

Piercing holes.　Hold the torch over the spot where the hole is to be cut until the flame has heated a small, round spot. Gradually press down the oxygen lever and at the same time raise the tip slightly. In this manner a small, round hole can be pierced quickly through the metal. See Figure 30-7.

When larger holes and circular shapes are required, trace the shapes with a piece of chalk. If the

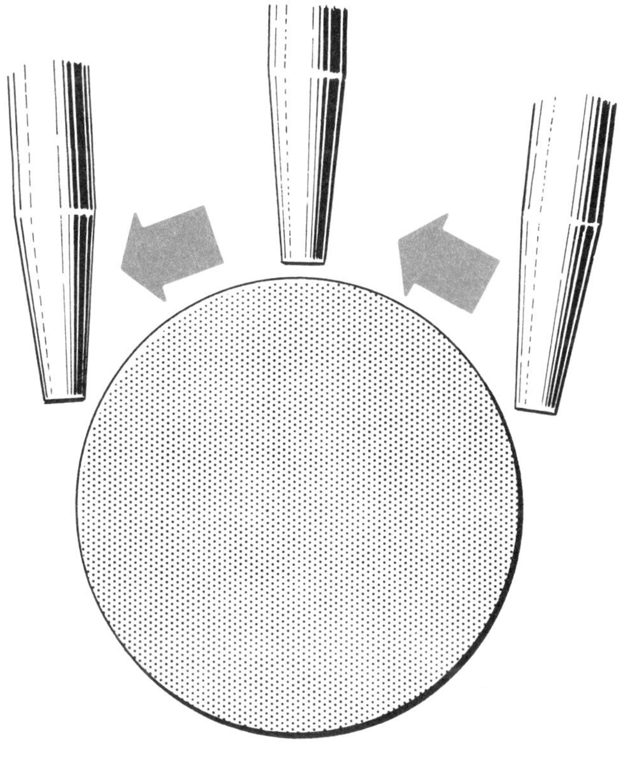

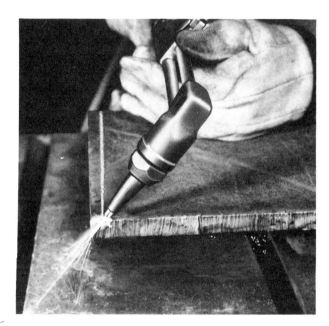

Figure 30-6. To make a bevel, position the cutting torch at an angle.

Figure 30-7. A cutting torch can be used to pierce a hole through metal.

Figure 30-5. When cutting round stock, start 90° from the top edge. Then follow around the bar as the arrow indicates.

holes are located away from the edge of the plate, first pierce a small hole, and then start the cut from this point, gradually working to the chalk line and continuing around the outline. See Figure 30-8.

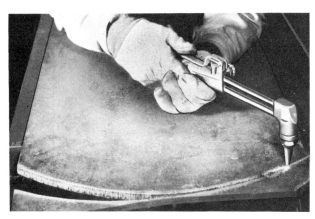

Figure 30-8. The cutting torch must be held steady when cutting small circles and large curves. (*Linde Company*)

Cutting Cast Iron with Oxyacetylene[1]

When cutting cast iron the chemical composition must be considered. Since cast iron has such a wide range of use, a wide difference in quality and chemical composition can be expected. The better grades of castings are more easily cut, and those of random grades of scrap, such as counterweights, gratebars, floorplates, etc., present greater difficulty, requiring more gas, a wider kerf, and a corresponding slower rate of cutting speed.

Preparation. There is a rule for heavy steel cutting which certainly applies in every way to cast-iron cutting, and that is,

DO NOT START THE CUT UNTIL YOU ARE CERTAIN THAT YOU CAN COMPLETE IT.

If the cut is stopped on a heavy section, it is ex-

tremely difficult to start again. More heat, as well as sparks and slag, are generated when cutting cast iron. Proper protection for the body, face, and limbs is recommended when cutting. Welding gloves are essential, and a fire brick or suitable torch rest is desirable.

Cutting Cast Iron Using the Flame Cutting Torch	Exercise 2

1 For cutting the better grades of cast or gray iron, set the regulator to deliver the proper pressure as indicated. The regulator adjustment is made with the high-pressure cutting valve open to compensate for the customary drop in pressure when the gas is released.

Exercise 2   Step 1.
PRESSURE TABLE FOR CUTTING CAST IRON.

TIP SIZE	THICKNESS (inches)	OXYGEN PRESSURE (pounds)	ACETYLENE PRESSURE (pounds)
	1/2	40	
	3/4	45	
L-3	1	50	7-8
	1 1/2	60	
	2	70	
	3	80	
	4	90	
L-4	6	110	8-10
	8	120	
	10	150	
	12	170	

2 Light the torch and adjust the preheating flame so that it will show an excess of acetylene. It is important that this is done with the high-pressure valve wide open to avoid any change in the character of the flame during the actual cutting operation.

The excess acetylene, as indicated by the length of the white cone, must be varied to best suit the grade and thickness of the material cut. Experience is the best guide on this point. However, it will generally vary from little or no ex-

1. The Modern Engineering Co.

Exercise 2, cont.

cess of acetylene for the extremely light sections to an excess indicated by a one to two inch white cone for the heavier sections cut.

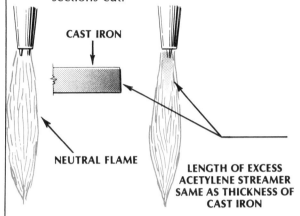

CAST IRON

NEUTRAL FLAME

LENGTH OF EXCESS ACETYLENE STREAMER SAME AS THICKNESS OF CAST IRON

3 Bring the tip of the torch to the top or starting point. Hold the torch on an angle of approximately 40° to 50° and heat a spot about ½″ in diameter to a molten condition.

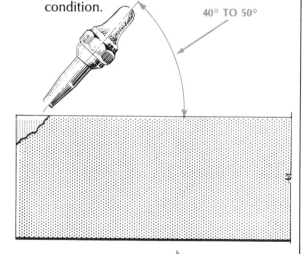

40° TO 50°

4 With the end of the shorter, preheating cone about ³⁄₁₆″ from the metal, start to move the torch with a swinging motion and open the high-pressure cutting valve.

5 Gradually bring the torch along the line of the cut, continuing the swinging motion. As the cut progresses, gradually straighten the torch to an angle of 65° to 70°, which will help to facilitate the penetration.

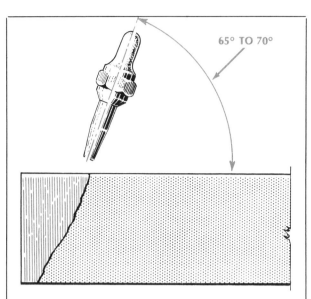

65° TO 70°

6 The same swinging movement is continued throughout the entire length of the cut.

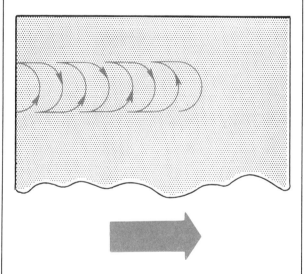

7 On the heavier sections, sufficient heat is usually necessary to allow the cut to proceed without interruption. On the lighter sections, it is easy to lose the cut as the surface of the metal cools too rapidly and only a slight groove is made with the flame.

Start over again by heating a small circle as previously described. Gradually raise the torch and incline it as to cut away the lower portion of the section. Proceed as before with the exposed side of the cutting groove appearing bright. Continue the cut until it is completed.

Plasma-Arc Cutting[1]

Plasma-arc cutting is regarded as one of the best processes for high-speed cutting of nonferrous metals and stainless steels. It cuts carbon steel up to ten times faster than any oxyfuel, with equal quality and with greater economy. See Figure 30-9.

Plasma is often considered the fourth state of matter. The other three are gas, liquid, and solid. Plasma results when a gas is heated to a high temperature, and changes into positive ions, neutral atoms and negative electrons. When matter passes from one state to another, latent heat is generated. Latent heat is required to change water into steam, and similarly, the plasma torch supplies energy to a gas to change it into plasma. When the plasma changes back to a gas, the heat is released.

Figure 30-9. The plasma-arc cutting process can cut carbon steel ten times faster than an oxyacetylene cutting torch can.

In a plasma-arc torch, the tip of the electrode is located within the nozzle. The nozzle has a relatively small opening (orifice) which constricts the arc. The high-pressure gas must flow through the arc where it is heated to the plasma temperature range. Since the gas cannot expand, due to the construction of the nozzle, it is forced through the opening, and emerges in the form of a supersonic jet, hotter than any flame. This heat melts any known metal and its velocity blasts the molten metal through the kerf. See Figure 30-10.

1. Linde Company

Because maximum transfer of heat to work is essential in cutting, plasma-arc torches use a transferred arc (the workpiece itself becomes an electrode in the electrical circuit). The work is thus subjected to both plasma heat and arc heat. Direct current, straight polarity is used. Precise control of the plasma jet is feasible by controlling the variables—current, voltage, type of gas, gas velocity, and gas flow (cfh).

The power supply for cutting is a special rectifier type with an open-circuit rating of 400 volts. A control unit automatically controls the sequence of operations—pilot arc, gas flow, and carriage travel.

A water pressure input of 60 to 80 psi for gas cutting, and 100 psi for air cutting is necessary to keep the torch cool.

When cutting aluminum and stainless steel, best results are obtained with an argon-hydrogen, or nitrogen-hydrogen gas mixture. Carbon steels require an oxidizing gas. Air has proven to be the most efficient gas; however, oxygen can also be used.

The first steps in making a plasma-arc cut are—adjust the power supply and the gas flow to the appropriate settings. See Table 30-2. Then, when the operator pushes the START button on the remote-control panel, the control unit performs all ON-OFF and sequencing functions. The cooling water must also be turned on, or the waterflow interlock will block the starting circuit.

To make a mechanized cut, the operator locates the center of the torch about ¼″ above the surface of the plate to be cut and pushes the start button. Current flows from the high-frequency generator to establish the pilot arc between the electrode (workpiece) and the cathode in the nozzle. Gas starts to flow, and welding current flows from the power supply.

The pilot arc sets up an ionized path for the cutting arc. As soon as the cutting arc is established, the high-frequency current is shut off, and the carriage starts to move. See Figure 30-11.

When the cutting operation is completed, the arc goes out because it has no ground, and the control stops the carriage, opens the main contactor, and shuts off the gas flow.

Air Carbon-Arc Cutting

Air carbon-arc cutting is a process wherein the severing of metals is accomplished by melting with the heat of an arc between a carbon electrode and the base metal. Power is supplied either with a regular AC or DC welding machine. However, the power requirements for a given diameter carbon electrode are higher than those for a comparable diameter

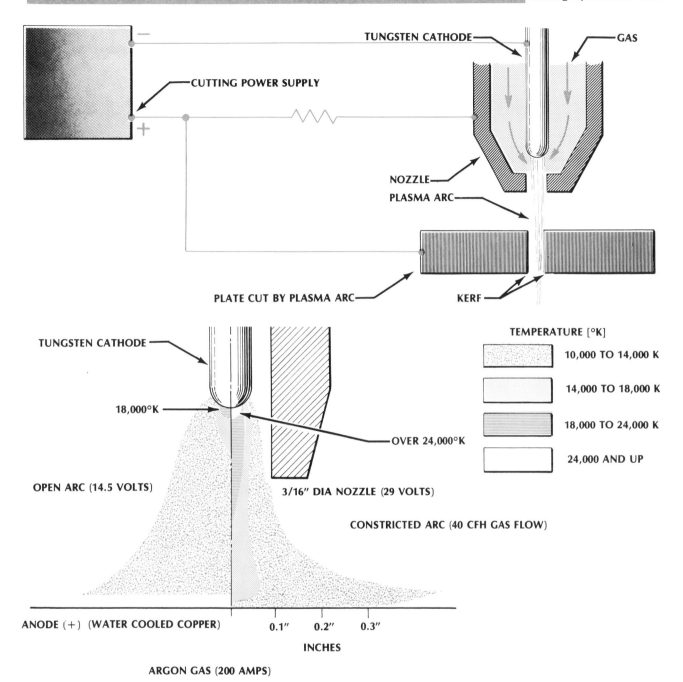

CUTTING POWER SUPPLY

TUNGSTEN CATHODE

GAS

NOZZLE

PLASMA ARC

PLATE CUT BY PLASMA ARC

KERF

TUNGSTEN CATHODE

18,000°K

OVER 24,000°K

OPEN ARC (14.5 VOLTS)

3/16″ DIA NOZZLE (29 VOLTS)

CONSTRICTED ARC (40 CFH GAS FLOW)

ANODE (+) (WATER COOLED COPPER)

0.1″   0.2″   0.3″

INCHES

ARGON GAS (200 AMPS)

TEMPERATURE [°K]	
	10,000 TO 14,000 K
	14,000 TO 18,000 K
	18,000 TO 24,000 K
	24,000 AND UP

Figure 30-10. In a plasma-arc torch, gas emerges from the nozzle in the form of a supersonic jet that can melt any known metal. (*Linde Company*)

shielded metal-arc welding. See Table 30-3.

The carbon-graphite electrode is held in a special holder as shown in Figure 30-12. Plain or copper-clad carbon-graphite rods are used. Plain electrodes are less expensive, but copper-clad electrodes last longer, carry higher currents and produce more uniform cuts.

As the metal melts, a jet of compressed air is directed at the point of arcing to blow the molten metal away. The compressed air line is fastened directly to the torch. The jet air stream is controlled by simply depressing a push-button on the holder. The jet orifices must be positioned under the electrode when using the torch. See Figure 30-13.

Air is supplied by an ordinary compressor. In general pressure will range from 40 to 80 psi. See Table 30-4.

Gouging. Grip the electrode so a maximum of

TABLE 30-2. TYPICAL PLASMA-ARC CUTTING CONDITIONS.

	THICKNESS (inches)	SPEED (lpm)	ORIFICE TYPE	INSERT (dia)	POWER (kw)	GAS FLOW (cfh)
STAINLESS STEEL	1/2	25	4 x 8	1/8	45	130 N_2
	1/2	70	4 x 8	1/8	60	130 N_2
	1 1/2	25	5 x 10	5/32	85	10 H_2
						175 N_2
	2 1/2	18	8 x 16	1/4	150	15 H_2
						175 N_2
	4	8	8 x 16	1/4	160	15 H_2
ALUMINUM	1/2	25	4 x 8	1/8	50	100*
	1/2	200	4 x 8	1/8	55	100*
	1 1/2	30	5 x 10	3/32	75	100*
	2 1/2	20	5 x 10	5/32	80	150*
	4	12	6 x 12	3/16	90	200*
CARBON STEEL	1/4	200	4 x 12M†	1/8	55	250
	1	50	5 x 14M†	5/32	70	300
	1 1/2	35	6 x 16M†	3/16	100	350
	2	25	6 x 16M†	3/16	100	350

* 65% Argon, 35% Hydrogen Mixture
† Multiport Orifice

Figure 30-11. A semiautomatic plasma-arc cutting unit. (*Linde Company*)

6″ extends from the electrode holder to the work. For aluminum alloys this distance should be reduced to 4″.

Hold the torch so the electrode slopes back from the direction of travel. The air blast should be behind the electrode. Maintain a short arc and move fast enough to keep up with metal removal. The arc must provide sufficient clearance so the air blast can sweep beneath the electrode and remove all molten metal. The depth and contour of the groove is controlled by the electrode angle and travel speed. For a narrow deep groove, a steep electrode angle and slow speed is used. A flat electrode angle and fast speed produce a wide shallow groove. The width of the groove is governed by the diameter of the electrode. In all instances, proper speed will produce a smooth, hissing sound.

For gouging in the vertical position, hold the electrode as in Figure 30-14 and move downward. This permits gravity to assist in removing the molten metal.

Gouging in the horizontal position can be done by moving the electrode either right or left. When traveling to the right, hold the electrode as shown in Figure 30-14. In gouging to the left, reverse the position of the electrode so the air jet is under the electrode.

Cutting. The technique for cutting is the same as gouging except that the electrode is held at a steeper angle and is directed at a point that permits the tip of the electrode to pierce the metal being cut.

For cutting thick nonferrous metals, hold the electrode in a vertical position with a push angle of 45° and, with the air jet above it, move the arc up and down through the metal with a sawing motion. Push angle refers to the angle the electrode makes behind a line perpendicular to the cut axis at the point of cutting.

Washing. Washing is a process of removing metal from large areas, such as removal of surfacing and of riser pads on castings. In using the air carbon-arc for this purpose, weave the electrode from side to

TABLE 30-3. SUGGESTED CURRENT RANGES FOR THE COMMONLY USED ELECTRODE TYPES AND SIZES.

ELECTRODE DIA (inches)	(mm)	DC ELECTRODE DCRP min/A	max/A	AC ELECTRODE AC min/A	max/A	AC ELECTRODE DCRP min/A	max/A
5/32	4.0	90	150	—	—	—	—
3/16	4.8	150	200	150	200	150	180
1/4	6.4	200	400	200	300	200	250
5/16	7.9	250	450	—	—	—	—
3/8	9.5	350	600	300	500	300	400
1/2	12.7	600	1000	400	600	400	500
5/8	15.9	800	1200	—	—	—	—
3/4	19.1	1200	1600	—	—	—	—
1	25.4	1800	2200	—	—	—	—

AWS

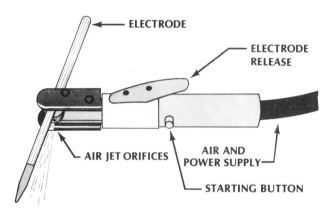

Figure 30-12. In air carbon-arc cutting, the carbon-graphite electrode is held in a special electrode holder. (*Arcair*)

Figure 30-13. The air jet orifices must be positioned under the electrode when using the air carbon-arc process. (*Arcair*)

TABLE 30-4. COMPRESSED AIR PRESSURE AND CONSUMPTION RATE FOR VARIOUS ELECTRODE SIZES.

ELECTRODE DIA (inches)	(mm)	AIR PRESSURE (psi)	(Pa)
Under 1/4	6.4	40	280
Under 1/4	6.4	80	550
Under 3/8	9.5	80	550
Under 3/4	19.1	80	550
Under 5/8	15.9	80	550
Under 5/8	15.9	80	550
Under 5/8	15.9	80	550

AWS

side in a forward direction to the depth desired. A push angle of 55° is recommended with the air stream behind the electrode. The steadiness of the operator determines the smoothness of the surface produced. See Figure 30-15.

Beveling. For beveling purposes hold the electrode at approximately a 45° angle, with the air blast between the electrode and the metal surface. Draw the electrode smoothly along the edge being beveled. See Figure 30-16.

Safety Precautions

In any cutting operation, a large amount of metal always falls on the floor.

WARNING: Be sure there are no combustible materials nearby.

Turn the cuffs of your trousers down over your shoes to prevent the possibility of molten metal getting inside the cuffs or shoes

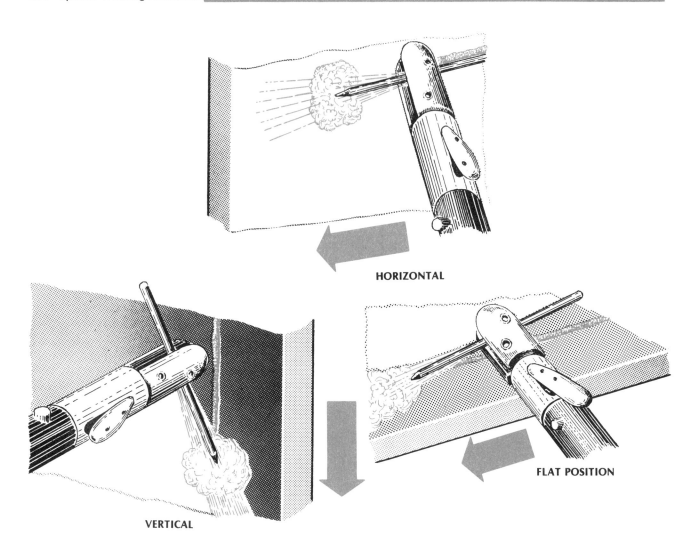

HORIZONTAL

VERTICAL

FLAT POSITION

Figure 30-14. In gouging, the torch is held so that the electrode slopes back from the direction of travel. (*Arcair*)

Figure 30-15. Weave the electrode from side to side when washing with the air carbon-arc.

Figure 30-16. When beveling a plate, hold the electrode at a 45° angle.

When an excessive amount of cutting is to be done, it is a good idea to sprinkle sand over the concrete floor. This prevents the molten metal from heating the concrete so that it cracks and causes particles to fly upward. Another method is to cut over a work bench tray partially filled with sand. If the bench lacks a tray, a sand-filled pan can be placed on the floor.

Points to Remember

1 In flame cutting, be sure to use the correct oxygen and acetylene pressure.

2 Keep the preheating cones burning with a neutral flame.

3 Hold the torch with the inner cone of the heating flame about $\frac{1}{16}$" above the metal until a spot is heated to a bright red heat.

4 Move the cutting torch just fast enough to make a fast but continuous cut.

5 If the cut does not go through the metal, start the cutting process over again, beginning at the original site.

6 In cutting cast iron, adjust the preheating flame so it is slightly carburizing.

7 For high-speed cutting of nonferrous metals, plasma-arc cutting is considered to be the most effective.

8 In plasma-arc cutting, use direct current with straight polarity.

9 Use plain or copper-clad carbon-graphite rods when cutting metals with the air carbon-arc process.

10 When using any cutting process, be sure there are no combustible materials which might be set on fire.

QUESTIONS FOR STUDY AND DISCUSSION

1 What causes metal to rust?

2 What principle makes possible the cutting of metal by means of oxygen and acetylene?

3 Describe the process for lighting and adjusting the flame for a cutting torch.

4 What is the function of the high-pressure oxygen valve?

5 How does the cutting tip differ from the welding tip?

6 What governs the pressure of oxygen and acetylene that must be used for cutting?

7 As a general rule, where should the cut be started? Why?

8 What aids may be used to facilitate an even cut?

9 How can it be determined that the cut is penetrating through the metal?

10 What happens when the cutting torch is moved too slowly?

11 What is the position of the torch when cutting round material?

12 How is it possible to make a bevel cut with a cutting torch?

13 Describe the operation followed in piercing small holes with a cutting torch.

14 What type of flame is used for cutting cast iron assuming a good grade of iron?

15 How is the torch held when cutting cast iron?

16 What torch motion is used for cutting cast iron?

17 What is meant by plasma-arc cutting?

18 What are some of the precautions that should be observed before engaging in any cutting operation?

19 What types of metals can be cut by plasma-arc cutting?

20 What type of electrode is used in the air carbon-arc cutting process?

21 What causes the removal of molten metal when air carbon-arc cutting?

22 What does the term *washing* mean when air carbon-arc cutting?

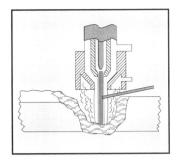

PRODUCTION WELDING

Special Welding Processes

31

Production welding refers to welding techniques used in the fabrication of goods on a mass production basis. Industries involved in manufacturing such products must rely on welding processes where hand manipulation is kept to a minimum and the joining of metal is performed rapidly and automatically. See Figure 31-1.

Since production techniques depend on the nature of the goods made, the kind of welding equipment used will vary from one industry to another. Very often special welding machines are designed for a particular industry. Thus an aircraft company

Figure 31-1. A machine-operated welding machine joins metal rapidly and automatically. (*Berkeley-Davis, Inc.*)

may need a spot-welding machine designed to join certain types of aluminum structures. An automotive manufacturer may require a resistance-type seam welder especially made to handle a body structure. Another concern may have to use a stud welding gun to fasten studs on some metal component. It is the purpose of this chapter to describe briefly some of the more common production welding techniques used in industry.

Resistance Welding

Of the many welding techniques applicable to production processes, resistance welding dominates the field. The fundamental principles upon which all resistance welding is based are these: (1) heat is generated by the resistance of the parts to be joined to the passage of a heavy electrical current, (2) this heat at the juncture of the two parts changes the metal to a plastic state, and (3) when combined with the correct amount of pressure, fusion takes place.

There is a close similarity in the construction of all resistance welding machines whether they are of simple design or are very complex and costly. The main difference is in the type of jaws or electrodes which hold the object to be welded. A standard resistance welder has four principal elements:

1. *The frame* is the main body of the machine which differs in size and shape for both stationary and portable types.

2. *The electrical circuit* consists of a step-down transformer which reduces the voltage and proportionally increases the amperage to provide the necessary heat at the point of the welding.

3. *The electrodes* include the mechanism for making and holding contact at the weld area.

4. *The timing controls* represent the switches which regulate the volume of current, length of current time, and the contact period.

The principal forms of resistance welding are classified as spot welding, seam welding, projection welding, flash welding, and butt welding.

Spot welding. This type of welding is probably the most commonly used type of resistance welding. The material to be joined is placed between two electrodes, pressure is applied, and a charge of electricity is sent from one electrode through the material to the other electrode.

There are three stages in making a spot weld. First the electrodes are brought together against the metal and pressure applied before the current is turned on. Next the current is turned on momentarily. This is followed by the third, or hold time, in which the current is turned off but the pressure continued. The

hold time forges the metal while it is cooling.

Regular spot welding usually leaves slight depressions on the metal which are often undesirable on the "show side" of the finished product. These depressions are minimized by the use of larger-sized electrode tips on the show side.

Spot welders are made for both direct and alternating current. The amount of current used is very important. Too little current produces a light tack giving insufficient strength; too much causes burned welds.

To dissipate the heat and cool the weld as quickly as possible, the electrodes are water-cooled.

The two basic types of spot-welding machines are single spot and multiple spot. The single spot, Figure 31-2, has two long horizontal horns, each holding a single electrode, with the upper arm providing the moving action.

Multiple-spot welders have a series of hydraulic or air-operated welding guns mounted in a framework or header but using a common or bar mandrel for the lower electrode. The guns are connected by flexible bands to individual transformers or to a common buss bar attached to the transformer. See Figure 31-3. Two or four guns are often attached to a transformer.

Spot welders are utilized extensively for welding steel, and when equipped with an electronic timer, can be used for other commercial metals such as aluminum, copper, and stainless steel. They are also very effective for welding galvanized metal.

Although many spot welders are of the stationary design, there is an increased demand for the portable type. The portable, or spot-welding gun as it is often called, consists of a welding head connected by flexible cables which run to the transformer. The jaws are operated either manually, pneumatically, or hydraulically. With this apparatus many spot welds may be made on irregular shaped objects as shown in Figure 31-4.

The self-contained portable spot welder contains a built-in timer, electrode contactors, and transformer which requires only a 115 volt power connection. See Figure 31-5. It is especially suitable for sheet metal and auto body welding.

On spot welders, the electrodes which conduct the current and apply the pressure are made of low-resistance copper alloy and are usually hollow to facilitate water cooling. These electrodes must be kept clean and shaped correctly to produce good results. Thus if a ¼" dia. electrode face is allowed to increase to ⅜" by wear or mushrooming, the contact area is doubled with a corresponding decrease in current density. Unless this is compensated for by

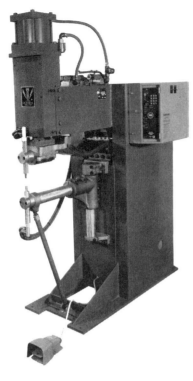

**VERTICAL PRESS-TYPE
SPOTWELDER**

**ROCKER-ARM
SPOTWELDER**

Figure 31-2. The single spot type of a spot-welding machine has two long horizontal arms, each holding a single electrode. The upper arm provides the moving action. *(LORS Machinery, Inc.)*

Figure 31-3. A multiple-spot welding machine has a series of hydraulic or air-operated welding guns mounted in a framework. *(Taylor Winfield Corporation)*

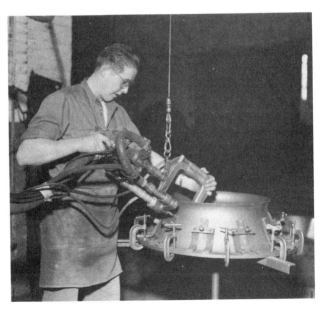

Figure 31-4. The spot-welding gun makes it possible to apply spot welds to irregular shaped objects such as this intake housing of a commercial fan. *(Clarage Fan Company)*

Figure 31-5. The portable spot welder contains a built-in timer, electrode contactors, and a transformer. (*Miller Electric Manufacturing Company*)

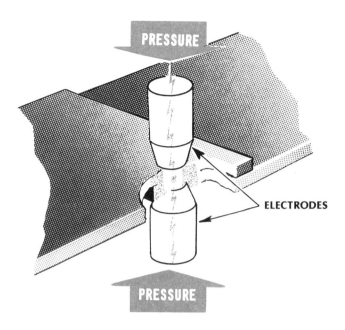

Figure 31-6. In pulsation spot welding, the current flow is interrupted by precise electronic control. (*LTV Steel*)

an increase in current setting, the result will be weak welds. Additional factors which cause poor welds are: misalignment of electrodes, improper electrode pressure, and convex or concave electrode surfaces.

Pulsation welding.   This is a form of spot welding. In regular spot welding, interruption of the flow of welding current is controlled manually; with pulsation welding the current is regulated to go on and off a given number of times during the process of making one weld. This method permits spot welding thicker material and increases the life of the electrode. The interrupted current helps to keep the electrodes cooler, thereby minimizing electrode distortion and reducing the tendency of the weld to spark. See Figure 31-6.

Seam welding.   Seam welding is like spot welding except that the spots overlap each other, making a continuous weld seam. In this process the metal pieces pass between roller type electrodes. See Figure 31-7. As the electrodes revolve, the current is automatically turned on and off at intervals corresponding to the speed at which the parts are set to move. With proper control, it is possible to obtain airtight seams suitable for containers such as barrels, water heaters, and fuel tanks. See Figure 31-8. When spots are not overlapped long enough to produce a continuous weld, the process is sometimes referred to as *roller spot welding* because of the intermittent current. See Figure 31-9.

Because of the short current cycle, seam welding has several advantages. The rollers may be cooled to prevent overheating with consequent wheel dressing and replacement problems reduced to a minimum. Cooling is accomplished either by internally circulating water or by an external spray of water

over the electrode rollers. Since the heat input is low, very little of the welded area is hardened and, therefore, the yield point is not materially affected. The fact that very little grain growth takes place in the seam welding process is also important from the standpoint of such corrosion-resistant alloys as stainless and other chromium alloy steels whose behavior is modified by grain growth.

An unusual seam welder is one having a combination of portable welding features with special longitudinal seam welding electrodes. This adaptation permits seam welding on large assemblies where the welder must be brought to the work. See Figure 31-10.

Projection welding.   Projection welding involves the joining of parts by a resistance welding process which closely resembles spot welding. This type of welding is widely used in attaching fasteners to structural members.

The point where the welding is to be done has one or more projections which have been formed by embossing, stamping, casting, or machining. The projections serve to concentrate the welding heat at these areas and cause fusion without employing a large current. The welding process consists of placing the projections in contact with the mating fixtures and aligning them between the electrodes. See Figure 31-11. Either a single or a multitude of projections can be welded simultaneously.

There are many variables involved in projection welding, such as stock thickness, kind of material,

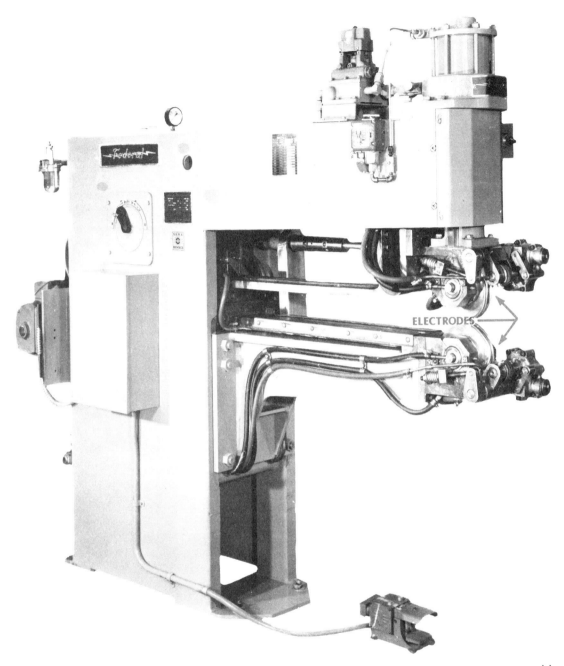

Figure 31-7. Roller electrodes with automatic controls rapidly and effectively produce a continuous weld seam of any length. (*Berkeley-Davis, Inc.*)

and number of projections, that make it impossible to predetermine the correct current setting and pressure required. Only by trial runs followed by careful inspection can proper control settings be established.

Not all metals can be projection welded. Brass and copper do not lend themselves to this method because the projections usually collapse under pres-

sure. Aluminum projection welding is limited to extruded parts (shapes formed by forcing metal through a die). Galvanized iron and tin plate, as well as most other thin-gauge steels, can be successfully projection welded.

Flash welding. In the flash welding process the two pieces of metal to be joined are clamped in dies which conduct the electric current to the work. The

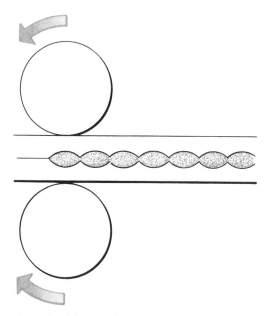

Figure 31-8. Welds must be closely spaced to provide an airtight seam.

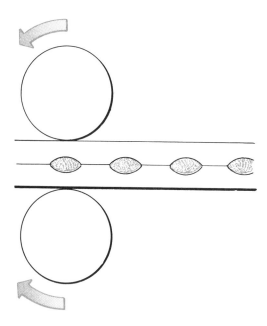

Figure 31-9. In seam welding, intermittent applications of current produce a series of spot welds spaced apart on the piece.

Figure 31-10. A portable seam welder is used for welding large structures that cannot conveniently be carried to the welding machine. (*Berkeley-Davis, Inc.*)

ends of the two metal pieces are moved together until an arc is established. The flashing action across the gap melts the metal, and as the two molten ends are forced together, fusion takes place. See Figure 31-12. The current is cut off as soon as the forging action is completed.

Flash welding is used to butt or miter-weld sheet, bar, rod, tubing, and extruded sections. It has almost unlimited application for both ferrous and nonferrous metals. It is not generally recommended for welding cast iron, lead, or zinc alloys for a variety of reasons. See Figure 31-13.

Parts to be welded are clamped by copper alloy dies shaped to fit each piece. For some operations the dies are water cooled to dissipate the heat from the welded area. The important factor in flash weld-

Figure 31-11. This projection welding machine (above) fastens leg sockets, handles, and shelf brackets to charcoal grill bowls. In projection welding, the weld area (below) has been preformed with raised points. (*Berkeley-Davis, Inc.*)

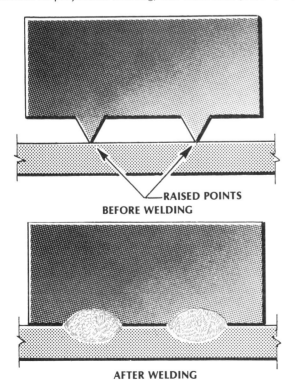

RAISED POINTS
BEFORE WELDING

AFTER WELDING

ing is the precision alignment of parts. Misalignment not only results in a poor joint but also produces uneven heat and telescoping of one piece over another.

The only serious problem met in flash welding, if grain growth is not a problem, is the resulting bulge or increased size left at the point of weld. If the finish area of the weld is important, then it becomes necessary to grind or machine the joint to the proper size after welding.

Butt welding. In butt welding the metals to be welded are brought into contact under pressure, an electric current is passed through them, and the edges are softened and fused together. See Figure 31-14. This process differs from flash welding in that constant pressure is applied during the heating process, which eliminates flashing. The heat generated at the point of contact results entirely from resistance. Although the operation and control of the butt-welding process is almost identical to flash welding, the basic difference is in the use of less current and allowing more time for the weld to be completed. See Figure 31-15.

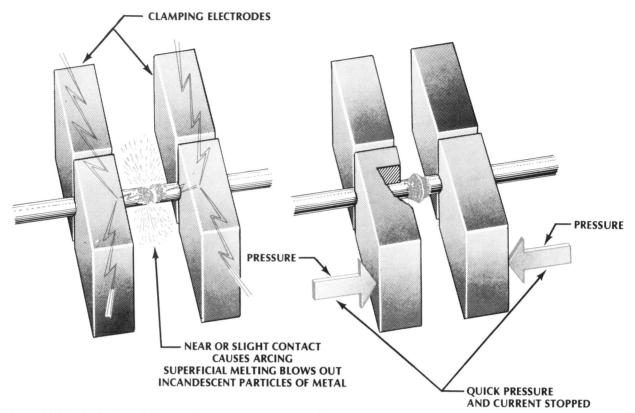

CLAMPING ELECTRODES

PRESSURE

PRESSURE

NEAR OR SLIGHT CONTACT
CAUSES ARCING
SUPERFICIAL MELTING BLOWS OUT
INCANDESCENT PARTICLES OF METAL

QUICK PRESSURE
AND CURRENT STOPPED

Figure 31-12. In flash welding, an intense arcing—caused by the electrical current flowing through the two pieces being brought together—melts the metal. (*LTV Steel*)

Figure 31-13. This flash welding machine is being used to fabricate an aluminum wheel for an automobile. Flash welding has a wide range of applications on both ferrous and nonferrous metals. (*Berkeley-Davis, Inc.*)

Gas Tungsten Arc Spot Welding

Resistance spot welding requires that pressure be applied to both sides of the pieces. The work must also be a size that will feed into the spot-welding machine. Gas tungsten arc spot welding makes it possible to produce localized fusion similar to resistance spot welding without requiring accessibility to both sides of the joint. A special tungsten arc gun is applied to one side of the joint only. Heat is generated from resistance of the work to the flow of electrical current in a circuit of which the work is part. Gas tungsten arc spot welding provides a deeper, more localized penetration when compared to a

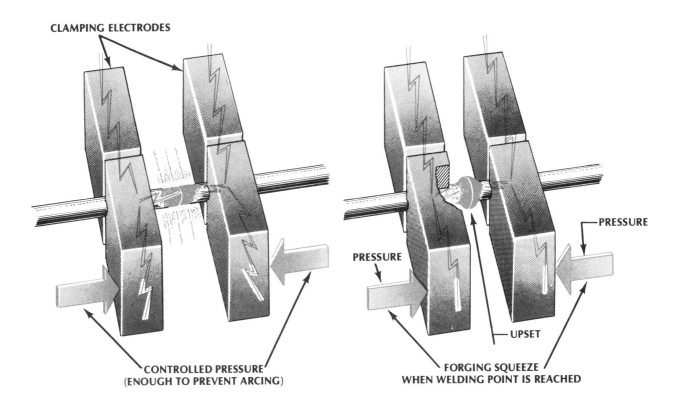

CLAMPING ELECTRODES

CONTROLLED PRESSURE
(ENOUGH TO PREVENT ARCING)

PRESSURE

PRESSURE

UPSET

FORGING SQUEEZE
WHEN WELDING POINT IS REACHED

Figure 31-14. Different from flash welding, butt welding involves passing high current through the metal pieces while continuous pressure is applied. (*LTV Steel*)

Figure 31-15. An automatic butt welding machine fuses metals in contact with a lower electric current when compared to flash welding. This eliminates the arc flash but requires more welding time. (*Berkeley-Davis, Inc.*)

conventional resistance spot weld. See Figure 31-16.

The spot-welding process has a wide range of applications in fabricating sheet-metal products involving joints which are impractical to resistance spot welding because of the location of the weld or the size of the parts or where welding can be made only from one size.

Equipment. Any DC power supply providing up to 250 amperes with a minimum open circuit voltage of 55 volts can be adapted for spot welding. The gun has a nozzle with a tungsten electrode. See Figure 31-17. Various shape nozzles are available to meet particular job requirements. See Figure 31-18. The standard nozzle can also be machined to permit access in tight corners or its diameter reduced to weld on items such as small holding clips.

For most operations a 1/8" diameter electrode is used. The end of the electrode should normally be flat and of the same diameter as the electrode. However, when working at low amperage settings (100 amperes or less) better results will be obtained if the end of the electrode is tapered slightly to provide a blunt point approximately one half the diameter of the electrode. This will prevent the arc from wandering.

Whenever the end of the electrode balls exces-

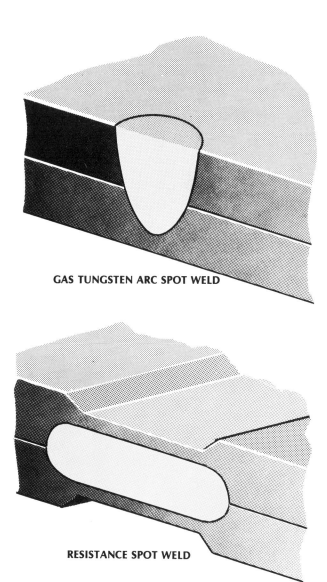

GAS TUNGSTEN ARC SPOT WELD

RESISTANCE SPOT WELD

Figure 31-16. Gas tungsten arc spot welding provides a deeper and more localized penetration compared to that obtained by conventional resistance welding. (*Airco*)

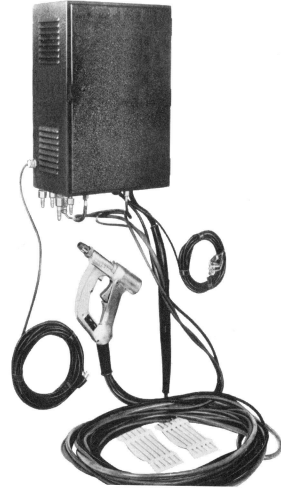

Figure 31-17. Equipment used in gas tungsten arc spot welding includes a gun with a tungsten electrode in the nozzle. (*Airco*)

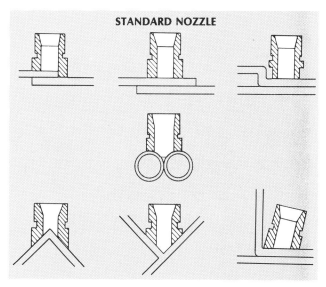

STANDARD NOZZLE

Figure 31-18. The nozzle in a gas tungsten arc spot-welding gun can be shaped for a variety of welding jobs. (*Airco*)

sively after only a few welds have been made, it is usually an indication of excessive amperage, dirty material, or insufficient shielding gas.

Making a weld. To make a spot weld, the end of the gun is placed against the work and the trigger is pulled. See Figure 31-19. Squeezing the trigger starts the flow of cooling water and shielding gas and also advances the electrode to touch the work. At the same time, the electrode automatically retracts, establishing an arc which is extinguished at the end of the present length of time. The electrode is usually preset at the factory to provide an arc length of 1/16″ which has been found to be generally satisfactory for practically all types of welding applications.

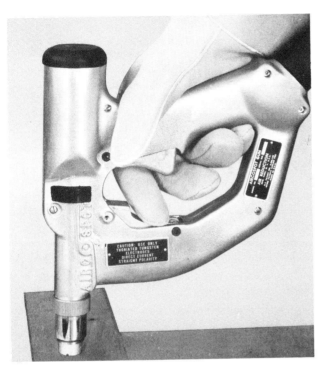

Figure 31-19. To make a spot weld, the gun is placed against the work and the trigger is pulled to activate the current. (*Airco*)

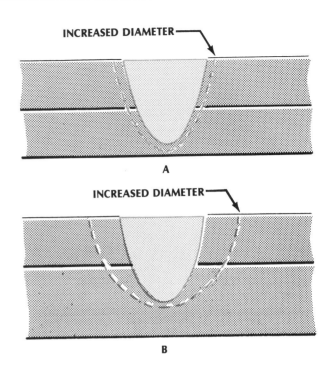

Figure 31-20 The thickness of the pieces being welded has an effect on weld diameter and penetration when the amperage is increased. (*Airco*)

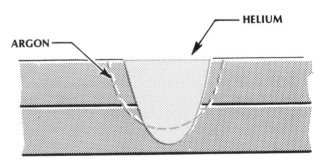

Figure 31-21. When used as a shielding gas, argon produces less penetration and a greater weld diameter compared to helium. (*Airco*)

Amperage. The amperage required for a weld will naturally be governed by the thickness of the metal to be welded. The major effect of increasing the amperage, when both pieces are approximately the same thickness, is to increase the penetration. However, it also tends to increase the weld diameter. See Figure 31-20A. Increasing the amperage, when the bottom part is considerably heavier than the top part, will result in an increase in weld diameter with little or no increase in penetration. See Figure 31-20B.

Weld time. Weld time is set on the dial in the control cabinet. The dial is calibrated in 60ths of a second and is adjustable from 0 to 6 seconds. The effect of increasing the weld time is to increase the weld diameter. But in so doing it also increases the penetration.

Shielding gas. Helium will produce greater penetration than argon, although argon will produce a larger weld diameter. See Figure 31-21. Gas flow should be set at approximately 6 cfh.

Surface condition and surface contact. Mill scale, oil, grease, dirt, paint, and other foreign materials on or between the contacting surfaces will prevent good contact and reduce the weld strength. The space between the two contacting surfaces resulting from these surface conditions or poor fit-up acts as a barrier to heat transfer and prevents the weld from breaking through into the bottom piece. Conse-

quently, good surface contact is important for sound welds. See Figure 31-22.

Backing. Although GTAW spot welding can be done from one side only, it is obvious that the bottom piece must have sufficient rigidity to permit the two parts to be brought into contact with pressure applied by the gun. If the thickness, size, or shape of the bottom part is such that it does not provide this rigidity, then some form of backing support or jigging will be required. Backing may be either of steel or copper.

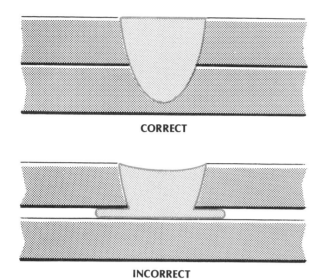

Figure 31-22. Good surface contact is important to making a sound spot weld. *(Airco)*

Gas Metal Arc Welding in Production Welding

Gas metal arc welding has been acclaimed as the most economical and effective method of joining light gauges of hard-to-weld metals such as nickel, stainless steel, aluminum, brass, copper, titanium, columbium, molybdenum, Inconel, Monel, and silver, as well as structural plates and beams. This process is performed semiautomatically (as discussed in Chapter 25) and automatically. The semiautomatic process is welding with equipment which controls only the filler metal feed. The advance of welding is manually controlled. See Figure 31-23. When the trigger is pressed, the gas, current, and wire automatically begin to flow. The operator simply has to concentrate on performing the weld in the designated area of the work piece.

Portability of welding equipment is an increasing requirement in today's construction environment. Airco's Super MIGet gun can be used to join steel

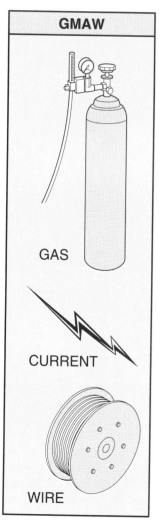

Figure 31-23. With a portable gas metal-arc welding gun, the gas, current, and wire automatically begin to flow when the trigger is pressed. *(Miller Electric Manufacturing Company)*

S-beam (formerly I-beam) sections high above ground level. See Figure 31-24. This equipment can weld steel or aluminum and may be operated up to 50 feet from the power source. Standard spools of electrode wire are housed in the canister suspended above the operator and provide many hours of welding.

The mechanized process is welding with equipment

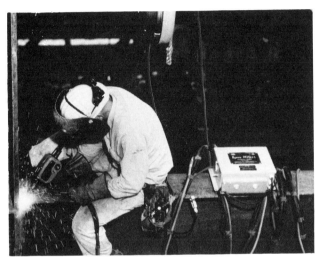

Figure 31-24. Semiautomatic portable welding guns are particularly useful when welding is performed up to 50 feet from the power source. (*Airco*)

which performs the welding operation under the constant observation and control of the operator. In the gas metal arc process the welding head is stationary instead of being portable. The head is either mounted on a carriage which travels over the work cr it is in a fixed position and the structure to be welded is moved under the unit. See Figure 31-25. In the automatic welding process welding is done with equipment which performs the entire welding operation without constant observation and adjustment of the controls by the operator. See Chapter 33.

Stud Welding[1]

Stud welding is a form of electric arc welding. Two methods have been developed, each with a different principle of operation.

The Nelson method is recognized by the use of a flux and a ceramic guide or ferrule. Equipment consists of a gun, a timing device which controls the DC welding current, the specially designed studs and ceramic ferrules. Studs are available in a wide variety of shapes, sizes and types to meet a variety of pur-

---

1. LTV Steel

poses. These studs have a recess in the welding end which contains the flux. This flux acts as an arc stabilizer and a deoxidizing agent. An individual porcelain ferrule is used with each stud when welding. It is a most vital part of the operation in that it concentrates the heat, acts (with the flux) to restrict the air from the molten weld, confines the molten metal to the weld area, shields the glare of the arc, and

Figure 31-25. In the mechanized process, the operator monitors the equipment that is performing the weld. (*Hobart Brothers Company*)

prevents charring of the material through which the stud is being welded.

In operation, a stud is loaded into the chuck of the gun and a ferrule positioned over the stud. When the trigger is depressed the current energizes a solenoid coil which lifts the stud away from the plate, causing an arc which melts the end of the stud and the area on the plate. A timing device shuts off the current at the proper time. The solenoid releases the stud and spring action plunges the stud into the molten pool and the weld is made.

Another method, the Graham method, is characterized by a small cylindrical tip on the joining face of the stud. The diameter and length of this tip vary with the diameter of the stud and the material being welded. This method operates on alternating current, and a source of about 85 pounds air pressure is also required.

The gun is air-operated with a collet (to hold the stud) attached to the end of a piston rod. Constant air pressure holds the stud away from the metal until ready to make the weld; then air pressure drives the stud against the work. When the small tip touches the workpiece, a high-amperage, low-voltage discharge results, creating an arc which melts the entire area of the stud and the corresponding area of the work. Arcing time is about one mil second (0.001); thus a weld is completed with little heat penetration, no distortion and practically no fillet. The stud is driven at a velocity of about 31 inches per second and the explosive action as it meets the workpiece cleanses the area to be welded. A minimum thickness of the workpiece of 0.02 inches is desired, particularly if no marking on the reverse side is required.

Both methods of stud welding are adaptable to welding of most ferrous and nonferrous metals, their alloys, and any combination thereof. See Figures 31-26, 31-27, and 31-28.

Electron Beam Welding

Electron beam welding is essentially a fusion welding process. Fusion is achieved by focusing a high power-density beam of electrons on the area to be joined. Upon striking the metal, the kinetic energy of the high-velocity electrons changes to thermal energy, causing the metal to melt and fuse.

The electrons are emitted from a tungsten filament heated to approximately 3630°F (2000°C). Since the filament would quickly oxidize at this temperature if it were exposed to normal atmosphere, the welding must be done in a vacuum chamber. Therefore a vacuum chamber is necessary to prevent the elec-

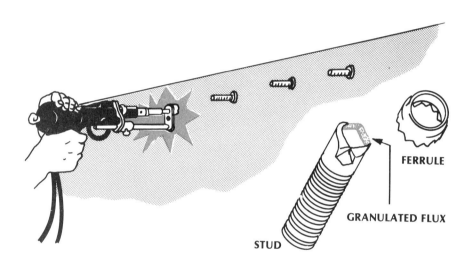

Figure 31-26. The Nelson method of stud welding utilizes granulated flux in the stud and a ceramic guide or ferrule along with the gun. (*LTV Steel*)

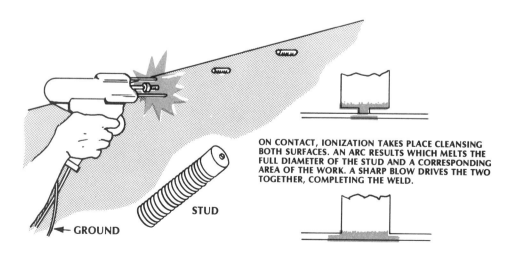

ON CONTACT, IONIZATION TAKES PLACE CLEANSING BOTH SURFACES. AN ARC RESULTS WHICH MELTS THE FULL DIAMETER OF THE STUD AND A CORRESPONDING AREA OF THE WORK. A SHARP BLOW DRIVES THE TWO TOGETHER, COMPLETING THE WELD.

Figure 31-27. The Graham method of stud welding is characterized by a small cylindrical tip on the joining face of the stud. (*LTV Steel*)

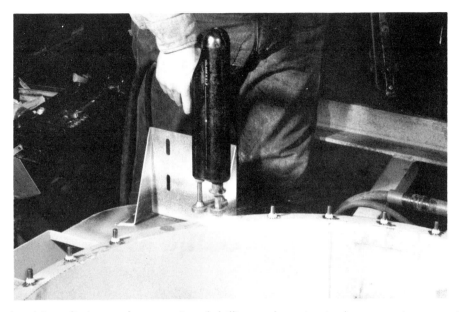

Figure 31-28. Stud welding eliminates the necessity of drilling and tapping in the cover plate assembly of this tank. (*Nelson Stud Welding Company*)

trons from colliding with molecules of air which would make the electrons scatter and lose their kinetic energy.

Electron beam welding can be used to join materials ranging from thin foil to 2″ thick. It is particularly adaptable to the welding of refractory metals such as tungsten, molybdenum, columbium, tantalum, and metals which readily oxidize, such as titanium, beryllium, and zirconium. It also has wide application in joining dissimilar metals, aluminum, standard steels, and ceramics.

Electron beam welding processes. Electron beam welding is done using either of two processes. The vacuum chamber process utilizes a controlled vacuum environment where the welding gun and the work are enclosed. The beam-in-air process uses a gun which has a vacuum chamber surrounding the area where electrons exit from the gun; welding is done in open atmosphere.

The vacuum chamber process, because of the absence of atmospheric contaminants, produces a cleaner weld without the use of a shielding gas. The weld can be more precise as the beam is much narrower in the vaccum chamber.

The beam-in-air process does not require a vacuum chamber. The welds produced by the beam-in-air process are similar to welds made by the gas tungsten-arc process. To shield the weld area from atmospheric contaminants argon is used as a shielding gas. See Figure 31-29.

Advantages: Electron beam welding has several distinct advantages. It faciiitates welding with a low total-energy input. Workpiece distortion and effects on the material properties of the workpiece are reduced to a minimum. The weld size and location can be controlled relative to the energy input. Using the beam-in-air process allows greater welding speed when compared to gas tungsten-arc welding. The process is chemically clean and facilitates welding without contamination of the workpiece.

Electron beam welding is often associated with the joining of difficult to weld metals. It is used in aerospace fabrication where new metals require more exacting joining characteristics; however, adaptation of the process to commercial applications is increasing. There is every indication that this growth will continue.

Limitations. One of the major limitations of electron beam welding using the vacuum chamber process is that the the piece must be small enough to fit into the vacuum chamber. This limitation is being reduced to some extent because larger chambers are now manufactured to accommodate a vast variety of products.

Another limitation is that when the workpiece is in the chamber in a vacuum it becomes inaccessible. It must be manipulated by some special controls.

Major components of the system. Electron beam welding equipment usually includes the following basic modules:

Electron gun. The gun consists of a filament, cathode, anode, and focusing coil. See Figure 31-30. The electrons emitted from the heated filament carry a negative charge and are repelled by the cathode and

Figure 31-29. The beam-in-air electron beam welding process does not require a vacuum chamber and produces welds similar in characteristics to the gas tungsten-arc welding process but at a much faster rate. (*Sciaky Brothers, Inc.*)

attracted by the anode. The electrons pass through an aperture in the anode and then through a magnetic field generated by the electromagnetic focusing coil. An optical viewing or numerical control system determines the path of the electron beam centerline to the weld area. See Figure 31-31. By varying the current to the focusing coil, the operator can focus the beam for gun-to-work distances ranging from a half inch to 25 inches. The electron beam can be controlled with a focusing coil to produce a spot diameter of less than 0.005".

Vacuum chamber. The chamber is usually rectangular in shape, and has heavy glass windows to permit viewing the work. A work table in the chamber is arranged so it can be operated either manually or electrically in the X and Y directions. T-slots are provided on the table to attach fixtures or workpieces for welding. See Figure 31-32.

Vacuum pumping system. The system is designed to provide a clean, dry vacuum chamber in a relatively short time. The capacity of the pump required is governed by the volume and area of the chamber and the time required to evacuate the chamber. The pumping equipment is usually completely automatic once the set-up has been completed.

Electrical controls. These include set-up controls and operating controls. The set-up controls include instruments required for the initial set-up of the welding operation, such as meters for beam voltage, beam current, focusing current, and filament current.

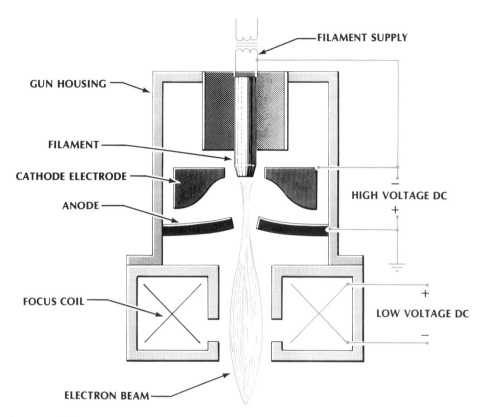

Figure 31-30. The typical electron beam gun consists of a filament, cathode, anode, and focusing coil. (*Sciaky Brothers, Inc.*)

The operating controls consist of stop-and-start sequence, high-voltage adjustment, focusing adjustment, filament activation, and work table motion.

Power unit. The power unit furnishes the main high-voltage supply up to 150 kv and a low filament-power up to 6 volts.

Operating the equipment: *Set-up procedure.* The workpiece is positioned on the work carriage in the chamber. The set-up steps are begun. The beam gun and work-to-gun distance are aligned manually and visually by use of the optical system. Work travel or gun travel (depending on the type of welding facility used) are checked and adjusted for alignment. If numerical control is used, data is entered.

The vacuum chamber is then closed. Vacuum-controls are started and the chamber is pumped down to the required vacuum (usually prescribed in a weld schedule).

Beam voltage, beam current, filament current and focusing current controls are then set from the weld schedule. This schedule is usually made by a welding engineer. When these control settings have been checked, the beam current may be switched ON and OFF for an instant for a weld spot alignment check. By viewing the weld spot through the optical system,

determination is made with regard to operation for the actual welding.

The weld or weld area is viewed by opening the shutter only when the beam current is turned off. If the shutter is opened when beam current is on, the result will be severe damage to the optical system.

Weld procedure. After everything has been checked and all switching made operative, the welding is begun by switching the sequence start switch to the ON position. The weld is made automatically.

The growth of applications for electron beam welding no longer limit it exclusively to exotic metal welding. However, electron beam welding will never entirely replace other welding processes.

Inertia Welding

Inertia or friction welding is a process where stored kinetic energy is used to generate the required heat for fusion. The two workpieces to be joined are aligned end to end. One is held stationary by means of a chuck or fixture, and the other is clamped in a rotating spindle.

The rotating member is brought up to a certain speed so as to develop sufficient energy. Then the

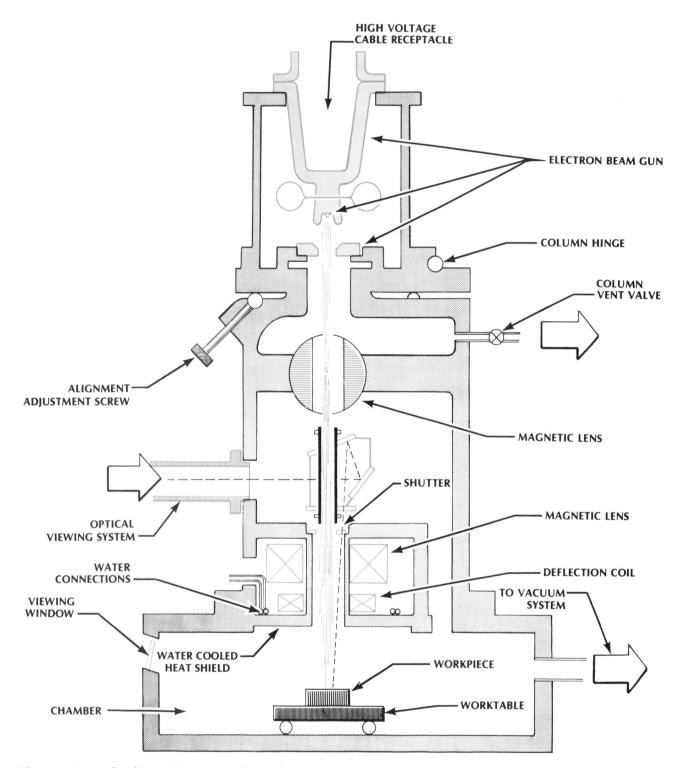

Figure 31-31. In the electron beam gun column, the electrons pass through an aperture in the anode and then through a magnetic field generated by the electromagnetic focusing coil. An optical viewing system provides a line of sight down the path of the electron beam centerline to the weld area. (*Hamilton Standard United Aircraft Corporation*)

Figure 31-32. The vacuum chamber houses the electron beam welding equipment and the work positioners. Heavy glass windows allow viewing by the operator when the door is closed during welding. (*Sciaky Brothers, Inc.*)

drive source is disconnected and the pieces brought into contact under a computed thrust load. At this point the kinetic energy contained in the rotating mass converts to frictional heat. The metal at and immediately behind the interface is softened, permitting the workpieces to be forged together. See Figures 31-33 and 31-34.

Inertia welding has several advantages over conventional flash or butt welding. It produces improved welds at higher speed and lower cost, less electrical current is required, and costly copper fixtures for hold parts are eliminated. With inertia welding there is less shortening of the components, which often occurs when flash or butt welding. Also the heat-affected zone near the weld is confined to a narrow band and therefore does not draw the temper of the surrounding area.

The inertia-welding process is applicable for welding many dissimilar or exotic metals as well as similar metals. Weld strength is normally equal to that of

the original metals.

Laser Welding

Laser welding is like welding with a white-hot needle. Fusion is achieved by directing a highly concentrated beam to a spot about the diameter of a human hair. The highly concentrated beam generates a power intensity of one billion or more watts per square centimeter at its point of focus. Because of its excellent control of heat input, the laser can fuse metal next to glass or weld near varnish-coated wires without damaging the insulating properties of the varnish.

Since the heat input to the workpiece is extremely small in comparison to other welding processes, the size of the heat-affected zone and the thermal damage to the adjacent parts of the weld are minimized. It is possible to weld heat-treated alloys without affecting their heat-treated condition, and the weld-

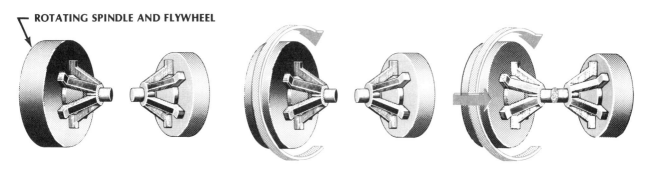

1. PIECES ARE ALIGNED AND CLAMPED   **2. FLYWHEEL IS ROTATED BY AN EXTERNAL ENERGY SOURCE**   **3. MEMBERS BROUGHT INTO CONTACT**

Figure 31-33. In inertia welding, heat resulting from stored kinetic energy is used to forge the pieces together. (*Caterpillar Tractor Company*)

ment can be held in the hand immediately after the weld is completed.

The laser can be used to join dissimilar metals such as copper, nickel, tungsten, aluminum, stainless steel, titanium and columbium. Furthermore, the laser beam can pass through transparent substances without affecting them, thereby making it possible to weld metals that are sealed in glass or plastic. Because of the fact that the heat source is a light beam, atmospheric contamination on the weld joint is not a problem.

The current application of laser welding is largely in aerospace and electronic industries where extreme control in weldments is required. Its major limitation is the shallow penetration. Present day equipment restricts it to metals not over 0.020″ thick.

The duration of the beam is usually about 0.002 seconds, with a pulse rate of one to ten times per second. As each point of the beam hits the metal, a spot is melted but solidifies in microseconds. The line of weld thus consists of a series of round, solid puddles each overlapping the other. The workpiece

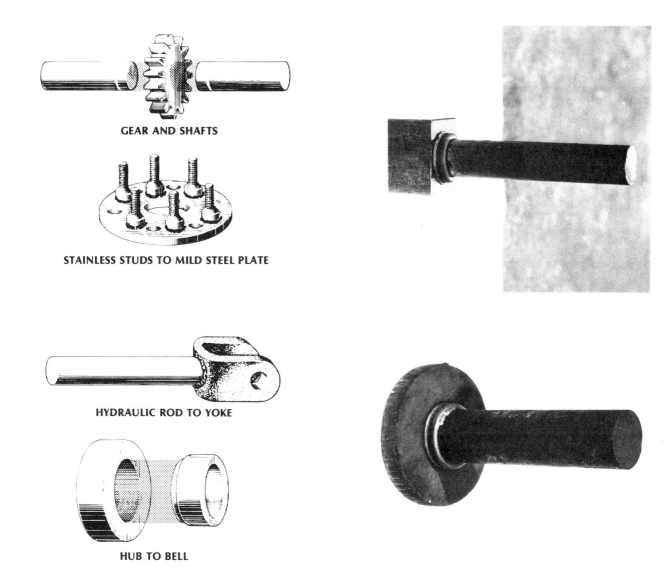

GEAR AND SHAFTS

STAINLESS STUDS TO MILD STEEL PLATE

HYDRAULIC ROD TO YOKE

HUB TO BELL

Figure 31-34. In inertia-welded parts, the heat-affected zone is confined to a narrow band near the weld. (*Caterpillar Tractor Company*)

is either moved beneath the beam or the energy source is moved across the line of weld.

Focusing the beam onto the workpiece is accomplished with an optical system and the actual control of the welding energy by means of a switch. See Figure 31-35.

Theory of the laser beam[2]. Atoms have been made to generate energy by exciting them in such common devices as fluorescent lights and television tubes. Fluorescence refers to the ability of certain atoms to emit light when they are exposed to external radiation of shorter wave lengths.

In the laser welder, the atoms that are excited to produce the laser light beam are produced in a man-made ruby rod $3/8''$ in diameter. See Figure 31-36.

The ruby is identical to a natural ruby but has a more perfect crystal structure. About 0.05 percent of its weight is chromium oxide.

The chromium atoms give the ruby its red color because they absorb green light from external light sources. When the atoms absorb this light energy, some of their electrons are excited. Thus, green light is said to pump the chromium atoms to a higher energy state.

The atoms eventually return to their original state. In doing so, they give up a portion of the extra energy they previously absorbed (as green light) in the form of red fluorescent light.

When the red light emitted by one excited atom hits another excited atom, the second atom gives off red light which is in phase with the colliding red light wave. In other words, the red light from the first atom

___

2. Linde Division, Union Carbide Corp.

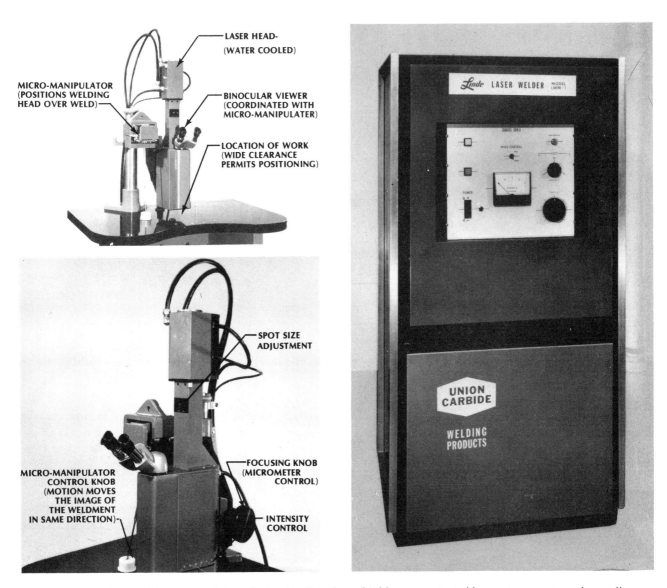

Figure 31-35. A laser welding unit achieves fusion by directing a highly concentrated beam to an extremely small area. (*Linde Company*)

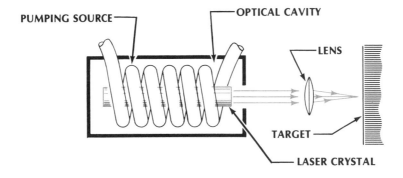

Figure 31-36. The laser welder's highly concentrated beam is focused on the workpiece with an optical system. (*Linde Company*)

is amplified because more red light exactly like it is produced.

By using a very intense green light to excite the chromium atoms in the ruby rod, a larger number of its atoms can be excited and the chances of collisions are increased. To further enhance this effect, the parallel ends and the sides of the rod are mirrored to bounce the red light back and forth within the rod. When a certain critical intensity of pumping is reached (the so-called threshold energy), the chain reaction collisions become numerous enough to cause a burst of red light. The mirror at the front end of the rod is only a partial reflector, allowing the burst of light to escape through it.

Plasma-Arc Welding

Plasma-arc welding is a process which utilizes a central core of extreme temperature surrounded by a sheath of cool gas. The required heat for fusion is generated by an electric arc which has been highly intensified by the injection of a gas into the arc stream. The superheated columnar arc is concentrated into a narrow stream and when directed on metal makes possible butt welds up to one-half inch or more in thickness in a single pass without filler rods or edge preparation. See Figure 31-37.

In some respects plasma-arc welding may be considered as an extension of the conventional gas tungsten arc welding. The main difference is that in plasma-arc welding the arc column is constricted and it is this constriction that produces the much higher heat transfer rate.

The arc plasma actually becomes a jet of high current-density. The arc gas upon striking the metal cuts or keyholes entirely through the piece producing a small hole which is carried along the weld seam. See Figure 31-38. During this cutting action, the melted metal in front of the arc flows around the arc column, then is drawn together immediately behind the hole by surface tension forces and reforms in a weld bead.

The specially designed torch for plasma-arc welding can be hand held or mounted for stationary or mechanized applications. See Figures 31-39 and 31-40. The process can be used to weld stainless steels, carbon steels, Monel, Inconel, titanium, aluminum, copper and brass alloys. See Figure 31-41. Although for many fusion welds no filler rod is needed, a continuous filler wire can be added for various fillet types of weld joints.

Equipment. A regular heavy duty DC rectifier is used as the source of power for plasma welding. A special control console is required to provide the necessary operating controls. A water cooling pump is usually needed to assure a controlled flow of cooling water to the torch at a regulated pressure. Proper cooling prolongs the life of the electrode and nozzle. See Figure 31-42.

Gas supply is either argon or helium. In some applications, argon is used as the plasma gas and helium as the shielding gas. However, in many operations argon is used for shielding and generating the plasma arc.

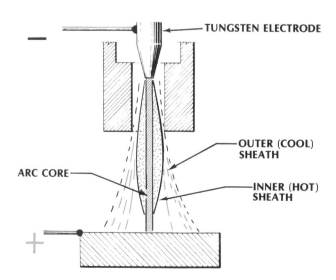

Figure 31-37. Plasma-arc welding process uses a central core of extreme temperature surrounded by a sheath of cool gas. (*Thermal Dynamics Corporation*)

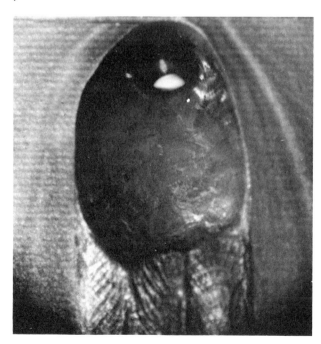

Figure 31-38. The plasma-arc welding process was used to obtain this penetration keyhole in a 1/2″ thick plate of titanium. (*Thermal Dynamics Corporation*)

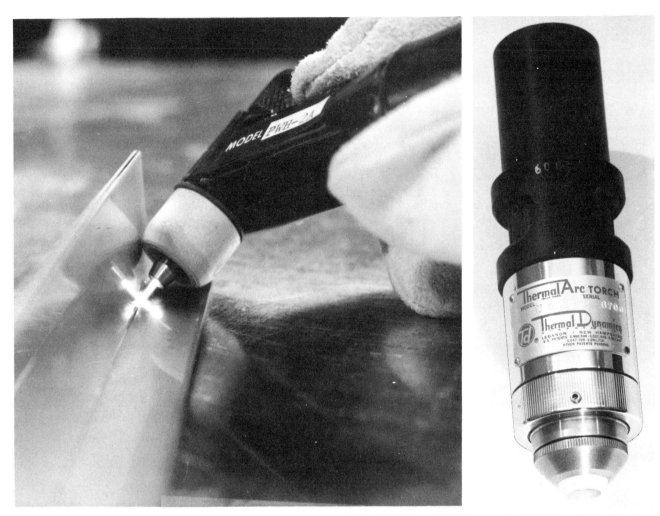

Figure 31-39. The specially designed torches for plasma-arc welding can be hand held or mounted on a machine. (*Thermal Dynamics Corporation*)

No matter what means are used to smooth surfaces, they will still possess peaks and valleys as measured by a microscope. As a result only the peaks of two workpieces which come into close contact will unite, leaving the countless valleys without producing a bond. Furthermore, smooth surfaces are never actually clean. Oxygen molecules from the atmosphere react with the metal to form oxides. These oxides attract water vapor, forming a film of moisture on the oxidized metal surface. Both the moisture and oxide film also act as barriers to prevent intimate contact.

In the ultrasonic welding process, these three existing barriers are broken down by plastically deforming the interface between the workpieces. This is done by means of vibratory energy which disperses the moisture, oxide, and irregular surface to bring the areas of both pieces into close contact and form a solid bond. Vibratory energy is generated by a transducer. See Figure 31-43.

The welding equipment consists of two units: a power source or frequency converter, which converts 60 cycle line power into high-frequency electrical power; and a transducer, which changes the high-frequency electrical power into vibratory energy.

The components to be joined are simply clamped between a welding tip and supporting anvil with just enough pressure to hold them in close contact. The high-frequency vibratory energy is then transmitted to the joint for the required period of time. The bonding is accomplished without applying external heat, or adding filler rod or melting metal. Either spot welds or continuous-seam welds can be made on a variety of metals ranging in thickness from 0.00017″ (aluminum foil) to 0.10″. Thicker sheet and plate can be welded if the machine is specifically designed for them. High-strength bonds are possible both in similar and dissimilar metal combinations.

Ultrasonic welding is particularly adaptable for joining electrical and electronic components, her-

Figure 31-40. The mounted plasma-arc welding torch can be used for stationary or mechanized applications. (*Thermal Dynamics Corporation*)

Ultrasonic Welding

If two metal pieces with perfectly smooth surfaces are brought into close contact, the metal atoms of one piece will theoretically unite with the atoms of the other piece to form a permanent bond. Regardless of how smooth such surfaces are, a sound metallurgical bond normally will not occur because it is impossible to prepare surfaces that are absolutely smooth.

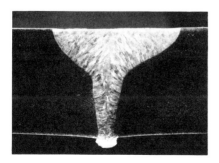

Figure 31-41. Plasma arc welding can produce several types of welding penetration on different types of metal. (*Thermal Dynamics Corporation*)

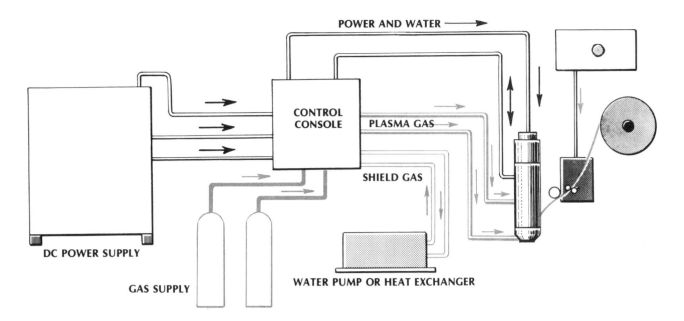

Figure 31-42. Plasma-arc welding equipment includes a heavy duty DC rectifier, control console, water cooling pump, and gas supply. (*Thermal Dynamics Corporation*)

metic sealing of materials and devices, splicing metallic foil, welding aluminum wire and sheet, and fabricating nuclear fuel elements. Because of the low heat input of the ultrasonic welding process, exotic metals may be joined without changing their original properties. See Figure 31-44.

Welding variables such as power, clamping force, weld time for spot welds or welding rate for continuous-seam welds can be preset and the cycle completed automatically. A switch lowers the welding head, applies the clamping force and starts the flow of ultrasonic energy.

Successful ultrasonic welding depends on the proper relationship between these welding variables which is usually determined experimentally for a specific application. Thus clamping force may vary from a few grams for very light materials to several thousand pounds for heavy pieces. Weld time may

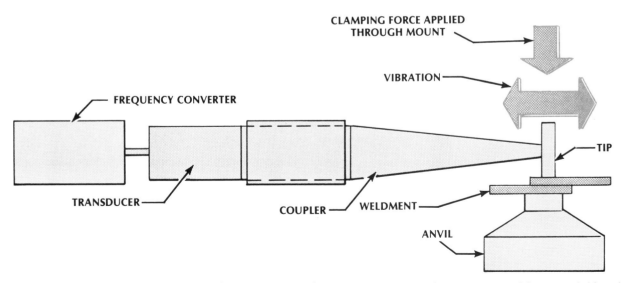

Figure 31-43. An ultrasonic welder used vibratory energy to bring two pieces into close contact and form a solid bond. (*Sonobond Ultrasonics*)

Figure 31-44. An ultrasonic continuous-seam welder using a rotary tip and anvil is often used for complete sealing of components used in electronics. *(Sonobond Ultrasonics)*

range from 0.005 to 1.0 seconds for spot welding and a few feet per minute to 400 fpm for continuous-seam welding. The high-frequency electrical input to the transducer may very from a fraction of a watt to several kilowatts.

Figure 31-45. Electro-gas welding is designed for single-pass welding of vertical joints. *(Airco)*

Electro-Gas Welding

Electro-gas welding is a process which uses a gas-shielded metal arc and is designed for single-pass welding of vertical joints in steel plates ranging in thickness from 3/8″ to 1½″. See Figure 31-45.

The welding head is suspended from an elevator mechanism which provides automatic control of the vertical travel speed during welding. This mechanism raises the welding head automatically at the same rate as the advancing weld metal. The welding head is self-aligning and can follow any alignment irregularity in plate or joint.

Once the equipment is positioned on the joint, welding is completely automatic. Wire feed and current are constant. At the end of the weld, the process automatically stops.

This welding technique is especially adaptable for ship building and fabrication of storage tanks and large diameter pipe.

Adhesive Bonding

Adhesive bonding is the joining of parts with an adhesive placed between faying (mating) surfaces. Adhesive bonding is useful for joining dissimilar metals, plastics and composites in manufacturing and repair operations. Adhesive bonding can be used to reduce the number of fasteners required, and strengthen joints prone to failure from vibration. See Figure 31-46.

Figure 31-46. Adhesive bonding is used to join dissimilar materials and strengthen joints prone to failure from vibration. *(Fel-Pro Chemical Products)*

Thin parts subject to heat distortion can be joined with adhesives. For example, auto body panels joined with adhesives do not have depressions caused by resistance welding heat. Joint member dimensions do not affect bonding strength. Thin parts can be joined with thick parts. Adhesives fill voids between parts without breaking surface contours. Flexibility of adhesives also allows distortion without failure. Joint types for adhesive bonding require large contact areas for adhesion as in brazing and soldering.

Adhesive bonding requires proper surface preparation, application, and curing procedures. The faying surfaces must be clean and free of foreign matter. Application processes of adhesives include manual, semiautomatic, mechanized, automatic, and robotic depending on the equipment available. Adhesives are cured by chemical action using catalyst cure (two parts), evaporation, ultraviolet (UV) light, heat, pressure, or both heat and pressure. Equipment required for adhesive bonding varies depending on application and curing methods.

Adhesives are available in various viscosities. *Viscosity* is the resistance of a substance to flow in a fluid or semi-fluid state. Low viscosity adhesives are liquid in form, and flow readily into small spaces. High viscosity adhesives range from gel to plastic-like forms. In some applications, an adhesive functions as a sealant. A sealant is a product used to seal, fill voids, and waterproof parts. Adhesive selection is based on the material and application of the parts joined. Adhesives can be broadly classified by chemical content or base as acrylic, anaerobic, cyanoacrylate, epoxy, hot melt, polyurethane, polysulfide, solvent-base, or water-base adhesives. See Figure 31-47.

An acrylic adhesive is a one-part UV or heat cure, or two-part adhesive that can be used on a variety of materials. It has a fast setting time and excellent flexibility. An anaerobic adhesive is one-part adhesive or sealant that cures by the absence of air displaced between mated parts. Low viscosity anaerobic adhesives are commonly used for locking metal parts together such as screws, nuts, and other fasteners. High viscosity anaerobic adhesives are used for joining parts with large gaps between faying surfaces.

A cyanoacrylate adhesive is a one-part adhesive that cures instantly by reacting to trace surface moisture to bond mated parts. Cyanoacrylate adhesives are commonly called super glue or instant glue and have a low resistance to high temperatures, moisture, vibration, and shock. An epoxy adhesive is a two-part adhesive that cures when resin and hardener are combined. Some epoxy adhesives are heat cured.

ADHESIVE BONDING

Adhesive	Components	Cure Time	Viscosity	Void-Filling	Flexibility	Head Resistance	Cold Resistance	Thermal Resistance	Water Resistance	Metal Bonding
Acrylic	two-part one-part (UV or heat cure)	medium to fast	medium	good	good	good	good	good	good	good
Anaerobic	one-part	medium	low	poor to fair	good	good	good	good	good	fair
Cyanoacrylate	one-part	fast	low	poor to fair	poor to fair	fair	fair	good	fair	good
Epoxy	two-part one-part (heat cure)	slow to medium	medium to high	excellent	fair	good	fair	good	good	good
Hot Melt	one-part	fast	high	excellent	fair to good	poor to fair	fair	fair	good	fair
Polyurethane	one-part two-part	medium	medium	good	good	fair	good	good	fair	good
Polysulfide	one-part two-part	medium	high	excellent	good	good	good	excellent	good	good
Silicone	one-part two-part	medium	high	excellent	excellent	excellent	excellent	excellent	excellent	fair
Solvent-Base	one-part	medium	low to medium	poor to fair	good	good	good	good	good	good
Water-Base	one-part	medium	low to medium	poor to fair	poor to fair	fair	fair	poor	poor	poor to fair

Figure 31-47. Adhesive selection is determined by the material and application of the parts joined.

A hot melt adhesive is thermoplastic material applied in a molten state that cures to a solid state when cooled. It is not as strong as epoxy but is very fast setting. A polyurethane adhesive is a one- or two-part adhesive with excellent flexibility that cures by evaporation, catalyst, or heat. A polysulfide adhesive is a one- or two-part adhesive or sealant that cures by evaporation or catalyst. It is commonly used in the aerospace and building materials industry. Silicone is a one- or two-part adhesive or sealant that cures by evaporation or catalyst. It has high temperature resistance and excellent sealing characteristics. A solvent-base adhesive is a one-part adhesive with a rubber or plastic base that cures by solvent evaporation. It is commonly used as contact cement for bonding large surface areas and lamination applications. A water-base adhesive is a one-part adhesive that cures by water evaporation. It has low flexibility and is primarily used for wood and paper products.

Points to Remember

1 Spot welding is a form of resistance welding with wide applications in industry.

2 Spot welders are available to produce single spot welds or multiple spot welds.

3 In pulsation welding, the current is regulated to go on and off a number of times during the welding process.

4 Seam welding produces a series of overlapping spot welds, thereby making a continuous-weld seam.

5 Projection welding is widely used in attaching fasteners to structural members.

6 When gas tungsten arc spot welding, set the amperage to suit the thickness of the metal to be spot welded.

7 Electron beam welding is a fusion process where a high power-density beam of electrons is focused on the area to be joined.

8 In inertia welding heat, resulting from the parts being rotated together, is used to forge the pieces.

9 Laser welding is a fusion process where a high concentrated beam is focused on a tiny spot.

10 Plasma welding uses an electric arc which is highly intensified by the injection of gas into the arc stream and results in a jet of high current-density.

11 Ultrasonic welding is a process where vibratory energy disperses the moisture, oxide, and surface irregularities between the pieces, thereby bringing the surfaces into close contact to form a permanent bond.

QUESTIONS FOR STUDY AND DISCUSSION

1 What is the basic principle of resistance welding?

2 What is a seam welder?

3 What is roller spot-welding?

4 What is projection welding?

5 How does flash welding differ from butt welding?

6 What is meant by pulsation welding?

7 What is the advantage of gas tungsten arc spot-welding over the conventional resistance spot welding?

8 What inert gases are used for spot-welding?

9 How does the stud-welding gun operate?

10 What are some advantages, and limitations, of electron beam welding?

11 How does plasma welding differ from regular gas tungsten arc welding?

12 What is the function of the transducer in ultrasonic welding?

13 How is fusion of metal accomplished in ultrasonic welding?

14 In laser-beam welding, how is the high-intensity laser light beam generated?

15 What is the principle of inertia welding?

16 What is electro-gas welding and what is it designed for?

17 What are some advantages of adhesive bonding?

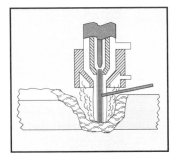

PLASTIC WELDING

Special Welding Processes

In the fabrication of many plastic products a welding process is often used to fasten parts together. Joining plastic edges by welding is a common technique in assembling such products as storage tanks, boxes, and other containers. Installation of pipe and duct work by welding has become common. The manufacture of many other custom products of plastic are possible by welding techniques. See Figure 32-1.

Plastic welding is very much like welding metal where localized heat is used to produce fusion. The same material preparation such as fit-up, root gap, joint design and beveling are required in plastic welding as in metal welding. However, there is one significant difference. In metal welding a sharply defined melting point develops and both the metal and welding rod melt and flow together to form the weld joint. Plastics, on the other hand, are poor heat conductors and consequently do not melt and flow; they simply soften. To achieve a permanent bond, a soft (heated) filler rod has to be forced into the softened surface of the joint. Thus plastic welders have to develop the skill of working within narrower temperature ranges than those doing metal welding.

Some plastics are easier to weld than others and one group of plastics cannot be welded at all. The basic plastic welding techniques are hot gas, heated tools, induction, and friction.

TYPES OF PLASTICS

Most plastics are known by trade names or by the principal substance from which they are made. The two main plastic families are *thermosetting* and *thermoplastics.*

The thermosetting plastics will soften only once when exposed to heat. After they are molded into a particular form and become hard no subsequent heating will soften them. These plastics are not weldable. Typical thermosetting plastics are ureas, phenolics, melamines, polyesters, silicones, epoxies, and urethanes.

Thermoplastics will repeatedly soften whenever heat is applied. These plastics can easily be welded. There are many kinds of thermoplastics, such as acrylics, polystyrenes, polyamides, polyfluorides, hydrocarbons, and vinyls. Generally the more common thermoplastics used where welding is involved are the polyethylenes, polyvinyl chlorides (PVC) and polypropylenes. Welding these plastics will produce seams that are as strong or stronger than the materials being bonded.

PLASTIC WELDING TECHNIQUES

Hot Gas Welding

Hot gas welding is accomplished with a special design gun containing an electrical heating unit. A stream of compressed air or inert gas (nitrogen) is directed over the heated element which then flows out of the nozzle onto the surface of the material being bonded. The gun permits the use of several tips for different welding operations. The basic tips are round for general purpose welding and welding in tight corners, flat for straight welding, and V-shaped for corner welding. A special tip is also available for high-speed welding. The increased speed is achieved by the design of the tip which holds the filler rod and applies the needed pressure as the weld is made. A tacker tip is sometimes used for tack welding. See Figure 32-2.

Joint preparation: The type of joints used in plastic welding are the same as those in metal welding—butt, lap, T, corner, and edge. The edges of the joint are beveled to have a sufficient area so a good bond can be formed. The beveled edges should have an included angle of 60°, with a root gap between ¹⁄₆₄″ to ¹⁄₁₆″. See Figure 32-3.

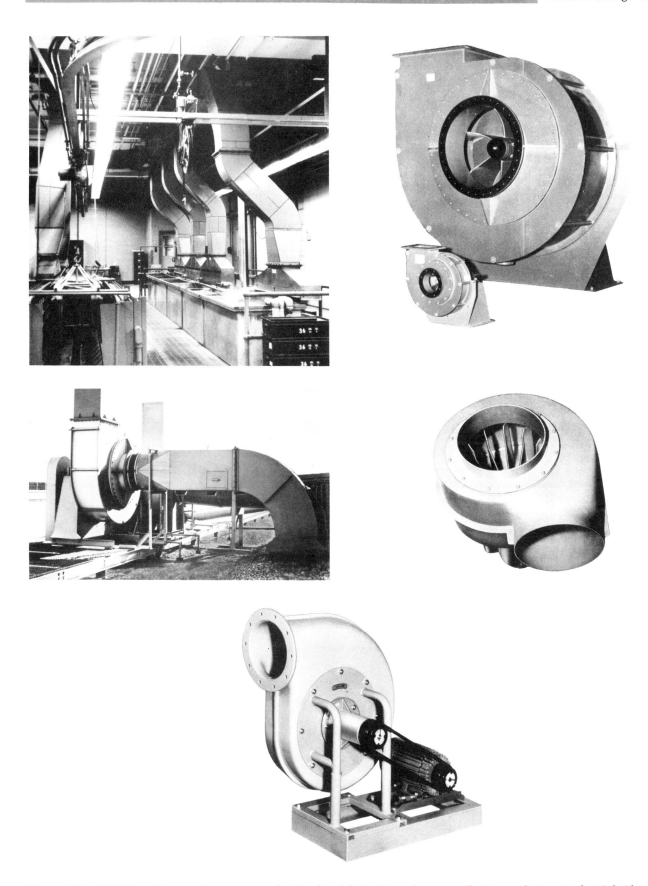

Figure 32-1. Welding plays an important role in the fabrication of many plastic products. (*Industrial Plastic Fabricators, Inc.*)

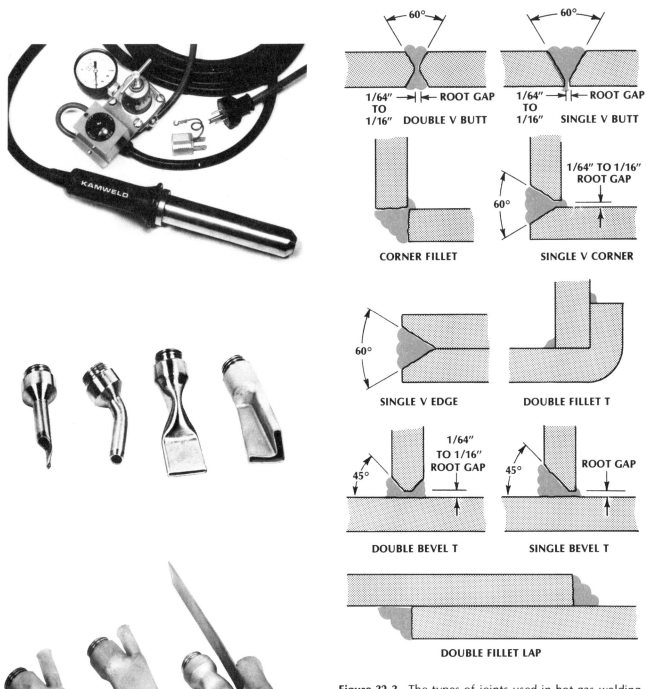

Figure 32-2. There are several types of tips available for hot gas welding of plastic. (*Kamweld Products Company*)

Figure 32-3. The types of joints used in hot gas welding of plastic are the same as those used in welding metal: butt, lap, T, corner, and edge.

Welding procedure. The technique for welding plastic is very similar to oxyacetylene welding of metals. The gun is held in one hand and the filler rod in the other. Welding goggles or helmet are not necessary. In general the following procedure is used:

Select the correct shape tip and insert in the gun. Normally guns should be able to supply a temperature varying from 400°F to 600°F (204 to 316°C) or more (up to 925°F or 496°C). Different materials and different thicknesses require different degrees of heat.

1. *Set the air or gas pressure according to the manufacturer's recommendations.* Although the wattage of the heating element determines the range of heat, the air or gas pressure determines the actual amount of heat at the tip. See Table 32-1.

Compressed air is best for welding PVC and several other types of plastics. However, better results are obtained when welding oxygen-sensitive plastics such as polyethylene and polypropylene by using nitrogen gas. See Table 32-2. Both the gas and compressed air should be controlled by regulators to provide the correct pressure flow. With some installations a gas flow meter on the nitrogen tank is equipped with a Y or by-pass valve to permit shutting off the nitrogen and switching to compressed air. This allows the use of compressed air to keep the weld at the proper heat when welding is temporarily stopped, to conserve nitrogen.

2. *Select the correct filler rod and cut its end at a 60° angle.* The rod should be of the same basic composition as the parent plastic. Either flat, round or triangular rods may be used. Triangular rods are particularly advantageous in V or fillet welds since the area can be filled with one pass. This reduces welding time and minimizes chances of porosity which may occur with multiple passes of round rods. See Figure 32-4.

TABLE 32-1. AIR PRESSURE RECOMMENDATIONS.

ELEMENT* (watts)	AIR PRESSURE (pounds)	TEMPERATURE (F° at 3/16" from tip)
320	2-3	400
340*	2-3	410
350	2 1/2-3 1/2	430
450	3-4	540
460*	3-4	600
550	4-5	700
650	4 1/2-5 1/2	800
750	5-6	860
800*	5-6	900

*Note: Three-heat unit with a rotary heat selector switch: Low—340 w, Medium—460 w, High—800 w.

TABLE 32-2. THERMOPLASTIC WELDING CHART.

THERMO-PLASTIC	PVC	POLY-ETHYLENE	POLY-PROPYLENE	PENTON	ABS	PLEXIGLASS
Welding temp	525	550	575	600	500	575
Welding gas	Air	Nitrogen	Nitrogen	Air	Nitrogen	Air

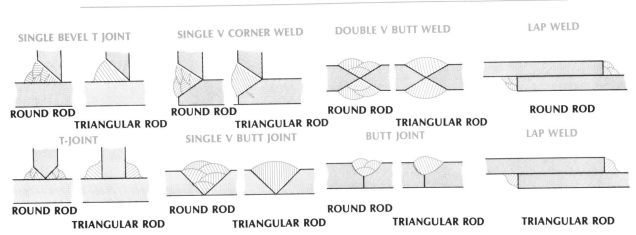

Figure 32-4. Triangular filler rods produce better welds in less time. (*Kamweld Products Company*)

3. *Hold the tip of the gun about ³/₁₆″ to ½″ away from where the weld is to be started and begin a fanning motion.* Place the rod in a vertical position so the heat from the gun is directed both on the rod and base material. When the base material and rod become tacky, press the rod firmly into the joint and bend it back at a slant with the point away from the direction of welding. See Figure 32-5. As the gun is moved along the seam continue to exert pressure on the rod to force it into the groove. Maintain a constant fanning motion at a 45° angle so both the rod and joint area are heated equally. On heavy gauge sheet with light rod most of the heat should be directed on the joint.

During the welding cycle it is important not to exert too much pressure on the rod since it will cause the rod to stretch excessively. The length of the rod used should be no more nor less than the length of the weld. Equally important is to avoid overheating because it will cause the rod and base material to char and discolor. Overheating will weaken the weld and cause cracks to radiate into the sheet from the weld. Underheating is also objectionable since it produces a cold weld which has poor tensile strength.

4. *Check weld.* Bend a test weld 90°. If the weld is made properly, the weld beads will not separate from the base material nor will it be possible to pry the rod out of the weld when cooled. Cutting a cross-section through a test weld will disclose whether or not there is correct penetration. See Figure 32-6.

WARNING: Some plastic materials produce obnoxious odors and poisonous fumes (polyvinyl chloride). Precautions must be taken to avoid inhaling these fumes. If necessary a respirator should be used.

In any case plastic welding should always be done in a well ventilated area. Follow the manufacturer's recommendations for safe practices when welding specific types of plastics.

Tack welding.   Tack welding is simply a means of fusing pieces together prior to welding them in order to eliminate the use of clamps or fixtures. See Figure 32-7. A tacker tip is used for this purpose which is quickly drawn at intervals along the joint. The gun is held at an angle of about 80° with the point of the tip touching the material. Tacks should be about ½″ long. When a tack welding tip is not available, tacking can be done with a regular round tip and filler rod.

High-speed welding.   The speed of making welds can be substantially increased with the use of a special tip that holds the welding rod in the correct position. Flat strips are often used for this operation although round or triangular rods will also serve the same purpose. Since the rods are supplied in roll

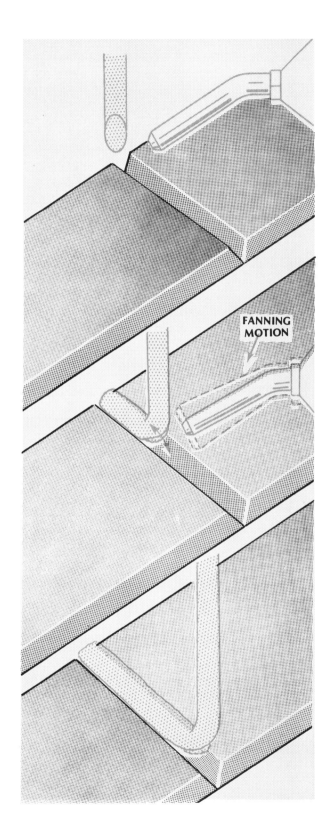

Figure 32-5. When the base material and rod become tacky, press the rod firmly into the joint and bend it back at a slant away from the direction of welding. (*Seelye Plastic, Inc.*)

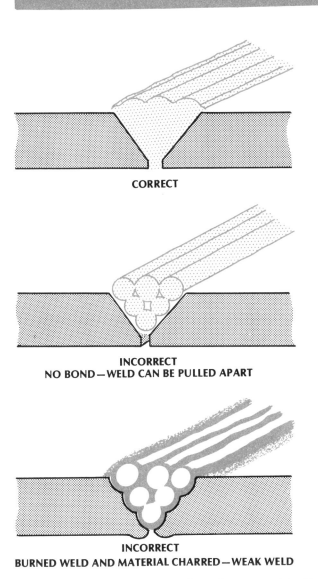

Figure 32-6. A cross-section cut through a test weld will show the amount of penetration. (*Seelye Plastic, Inc.*)

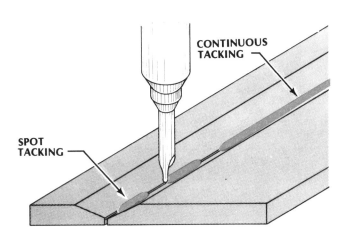

Figure 32-7. Tack welding fuses pieces together so that clamps or fixtures are not necessary during welding. (*Kamweld Products Company*)

form they first must be cut into required lengths with one or two inches allowed for trimming. Welding procedure should be as follows:

1. *Insert the rod into the feeder tube.* Start the weld by tamping the broad shoe of the tip on the surface until the first inch of the rod adheres firmly to the base material. Hold the welding gun at a 90° angle to the work and press the end of the rod into the weld. Feed the weld rod manually until the weld bead is well started. See Figure 32-8.

2. *Move the welder forward along the seam,* dropping the angle of the welder to 45°. Once the welding operation is well underway the rod will feed automatically into the preheated tube by simply exerting firm downward pressure on the gun with the wrist.

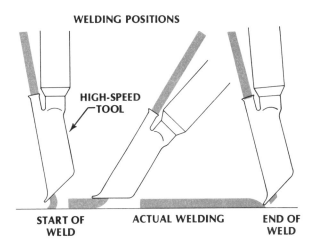

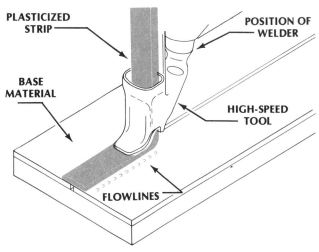

Figure 32-8. High-speed welding involves the use of a special tip that holds the welding rod in the correct position. (*Kamweld Products Company*)

3. *Keep the gun moving at sufficient speed.* Correct speed can be observed by the formation of flow lines on both sides of the rod. Insufficient speed will cause the rod to stretch because of the built-up excessive heat. This condition can be corrected by a quick tamping motion of the shoe as used in starting the weld.

To stop the weld, lift the welder and allow the remaining rod to pull through the feeder tube. Then cut the rod with a special curved knife. Never allow the filler rod to remain in the feeder tube because it will clog the opening.

When a weld is to be made from one corner to another such as in a tank, start the weld in one corner and proceed only halfway. Then begin the weld from the other corner and overlap the scarfed end of the first weld.

Heated-Tool Welding

In the heated-tool process the edges to be joined are heated to a fusing temperature and then brought into contact and allowed to cool under pressure. The edges of the plastic sheet are softened with some heat producing unit such as an electrical strip or bar heater, hot plate, or resistance-coil heater. The heaters should be of the aluminum or nickel plate type since hot steel and copper units have a tendency to decompose plastic. This welding technique is frequently used in joining sections of pipe and tubing and in the assembly of many molded articles.

Heat is applied by holding the edges in contact with the heating unit until the surface is softened. When the material is sufficiently in a molten state it is removed from the heater and the edges quickly pressed together. The pressure between the pieces should be enough to force out air bubbles and form a solid contact. Normally, pressures of 5 to 15 psi will produce good bonded joints. Pressure can be applied by hand or in production work by jigs. The pressure must be maintained until the weld is cooled.

The most important factor in securing sound welds by the heated-tool technique, outside of proper softening of materials and firm contact, is the lapsed time between removing the pieces from the heating unit and joining them together. This interval should be as short as possible to prevent any degree of solidification before the edges come in contact.

Induction Welding

In induction welding, heat is generated by causing a high-frequency current to flow into a metal insert placed between the areas to be joined. Although induction welding is one of the fastest methods of joining plastic, its greatest limitation is that the metal insert must remain in the weld.

The metal inserts usually consist of metallic foil, a coil of wire, wire screen, metallic conducting particles or any other configurations of conductive metal. They must be placed in the interface so they are not exposed to the air; otherwise rapid heating is induced which may cause the inserts to disintegrate. Fusion occurs only in an area immediately near the insert. When the edges become soft, uniform pressure is applied to bond them together. As a rule welds made by the induction process are not as strong as those obtained by other heating methods.

Friction Welding

Friction or spin-welding consists of rubbing the surfaces of the parts to be joined until sufficient heat is developed to bring them up to a fusing temperature. Pressure is then applied and maintained until the unit is cooled. Usually in friction welding one piece is held stationary and the other is rotated. When sufficient melt occurs the spinning is stopped and the pressure increased to squeeze out air bubbles and distribute the softened plastic uniformly between the surfaces.

The principal advantages of spin-welding are the speed and simplicity of the process itself. However, this technique is limited to circular areas. Sometimes spin-welding produces a flashing out of soft material beyond the weld area, but usually the excess flashing can be directed to the interior of the part if the weld is properly designed. Excess flashing can also be avoided by preventing the parts from overheating and by maintaining the proper pressure.

Spin-welding has wide applications in fastening instrument knobs, pipe fittings, container caps, handles and other units where parts can readily be spun.

Points to Remember

1 Weld plastics only in a well ventilated area. Refer to manufacturer's specifications for safe procedures.

2 Bevel all edges to secure a proper weld joint.

3 Use a filler rod of the same composition as the parent material.

Points to Remember, cont.

4 Set the air or gas pressure to provide the proper flow as prescribed by the manufacturer.

5 Use a fanning motion to insure uniform heat distribution over the rod and edges of the joint.

6 Avoid exerting too much pressure on the filler rod as it will stretch.

7 Do not allow the surface to char or discolor.

8 Triangular shape filler rods usually produce the best results in making fillet welds.

QUESTIONS FOR STUDY AND DISCUSSION

1 What is the main difference between plastic welding and metal welding?

2 Why are thermosetting plastics not weldable?

3 At what range of temperatures are plastics generally welded?

4 What governs the degree of heat which is to be used in plastic welding?

5 What is the particular advantage of using a triangular shape filler rod over a round rod?

6 How far away from the surface should the gun be held when welding plastics?

7 Why is a fanning motion necessary in manipulating the gun over the weld joint?

8 Why should excessive pressure on the rod be avoided?

9 How can you tell if the heat is too great?

10 What happens if insufficient heat is used when a welder is making a plastic weld?

11 How can you check to see if a weld is made properly?

12 What precautions should be taken when welding plastics?

13 How does a high-speed plastics welding technique differ from the regular hot gas welding technique?

14 When using a high-speed welding tip, why should the filler rod not be allowed to remain in the feeder tube?

15 How is the heated-tool welding technique accomplished?

16 What is one of the main limitations of induction plastic welding?

17 How are plastic joints bonded by spin-welding?

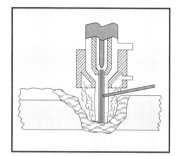

ROBOTICS AND WELDING

Special Welding Processes

In a production setting, consistent, rapidly repeated welds are necessary for greater productivity. Machine and automatic welding applications are used to maintain the speed and accuracy necessary for efficient fabrication. There are instances, however, where the position of the weld requires manual or semiautomatic welding. When using manual and semiautomatic applications, accessibility to the workpiece is limited only by the operator's ability to position the welding torch, gun, or electrode. See Figure 33-1.

An automatic welding machine can perform the welding operation without operator supervision, but the variety of positions is limited. The weld is executed as directed mechanically or as programmed by a computer. See Figure 33-2.

The use of industrial robots (robotics) with welding machines has increased the capabilities of automatic welding applications. By combining the versatility of manual and semiautomatic applications with the speed and accuracy of automatic welding, robot welding has become a growing manufacturing process. The advantages of robotics in welding operations include higher productivity, consistent quality, adaptability to different weld locations, elimination of costly jigs, and the ability to maintain a high rate of welding speed.

According to a recent study by the American Welding Society (AWS), the average arc-on time for a welding machine operator is 30%. The remainder of the time is spent moving the part or adjusting the equipment. With the welding machine operator and the robot working together, arc-on time is in excess of 70%. Thus, use of a robot minimizes production costs. See Figure 33-3.

Robot welding machines and their supporting components are complex and expensive. In the past, their applications were limited to long repetitive production runs. Given the advances in programming, however, low-volume production plants are using robot welding systems in increasing numbers.

Figure 33-1. Semiautomatic welding applications allow an operator to use a variety of welding positions. (*Hobart Brothers Company*)

Figure 33-2. After the controls are programmed or set, the automatic welding machine performs welding operations without the supervision of an operator. (*Miller Electric Company*)

Figure 33-3. The robot welding machine and the operator are capable of higher productivity and reduced manufacturing costs. (*Hobart Brothers Company*)

Components of a Robot Welding System

A typical robot welding system includes several key components. See Figure 33-4.

1. Operator controls
2. Robot controller
3. Robot
4. Automatic welding equipment
5. Positioner

Operator controls. The operator controls start, execute the program, and stop the robot welding cycle. See Figure 33-5. A portable control pendant is used for on-site programming and remote program editing away from the welding system. See Figure 33-6. When used in the teaching (programming) mode, the control pendant can be used to program specific weld functions by moving the robot through each operation while entering necessary data into its memory. The robot then duplicates the movements when recalled from its memory.

Robot controller. The robot controller translates and relays commands from the operator controls to the robot. Coordination between the robot and the positioner is also processed through the robot controller.

Robot. The robot, or manipulator, is classified by the joints or axes which facilitate movement. A five-axes robot has five joints which can be programmed to position the welding equipment in the correct location by moving any combination of the

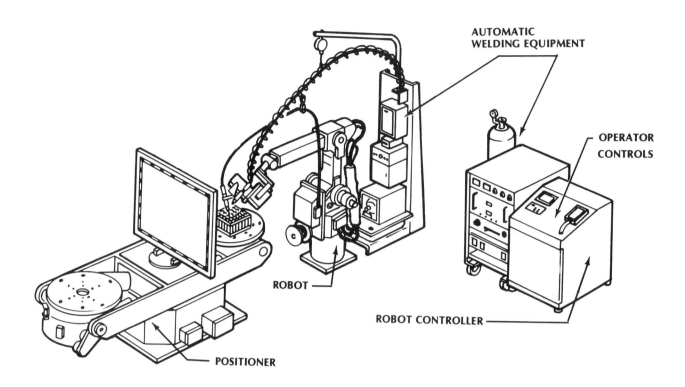

Figure 33-4. The major components of a robot welding system include operator controls, robot controller, robot, automatic welding equipment, and positioner. (*Advanced Robotics Corporation*)

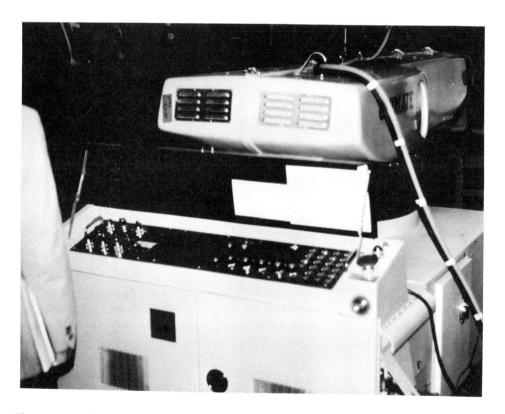

Figure 33-5. The operator monitors welding operations at the control console. (*Unimation Inc.*)

Figure 33-6. A portable control pendant is used to program the welding operation. *(Motoman, Inc.)*

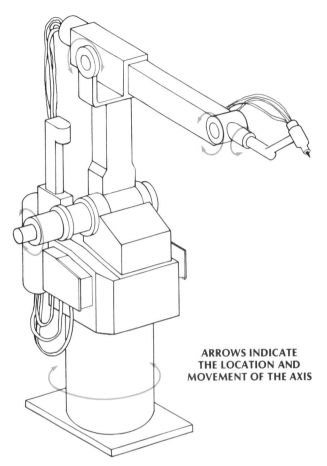

**ARROWS INDICATE
THE LOCATION AND
MOVEMENT OF THE AXIS**

Figure 33-7. The robot's axes and mechanical joints facilitate specific movements. A five-axes robot is shown. *(Advanced Robotics Corporation)*

five axes. See Figure 33-7. Depending on the type of robot, axis movement can be accomplished hydraulically or electrically.

Specific applications determine the type of robot to be used. The articulating-type robots are particularly suited for intricate maneuvering in and around the work. These robots provide a flexible working volume using limited floor space. Working volume refers to the maximum range of movement capability of the robot. See Figure 33-8.

The rectilinear-type robot is used primarily for welding larger pieces. A larger working volume is possible with an extendable horizontal axis. See Figure 33-9. If necessary, the rectilinear-type robot can be used at work stations on both sides of the horizontal axis.

Because the robot is a very complex piece of machinery, correct maintenance is critical for optimum performance.

Automatic welding equipment. The automatic welding equipment used for robotic welding is basically the same equipment used for semiautomatic and machine applications. Automatic welding equipment typically includes a power source, wire feeder, a torch or gun, and a shielding gas (if necessary). See Figure 33-10.

Robot welding systems use a variety of welding processes. These processes include gas metal-arc welding (GMAW), gas tungsten-arc welding (GTAW), flux-cored arc welding (FCAW), plasma-arc welding (PAW), and resistance spot welding (RSW).

Positioner. The positioner is used to locate the workpiece in a predetermined position. See Figure 33-11. The accessibility of the work is increased by rotating the workpiece in a two-axes positioner. Movement of the positioner is synchronized with the robot through the robot controller. A fixed-position jig can also be used to hold the work securely. See Figure 33-12.

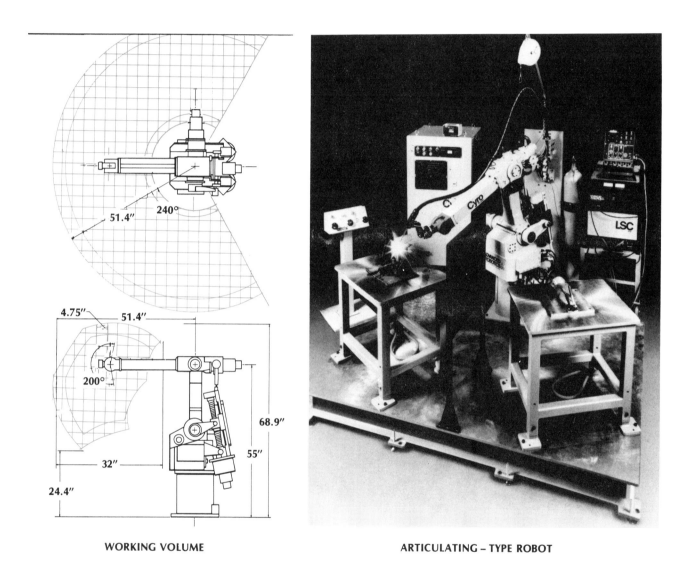

WORKING VOLUME

ARTICULATING – TYPE ROBOT

Figure 33-8. The articulating-type robot is capable of intricate maneuvering in and around the work. (*Advanced Robotics Corporation*)

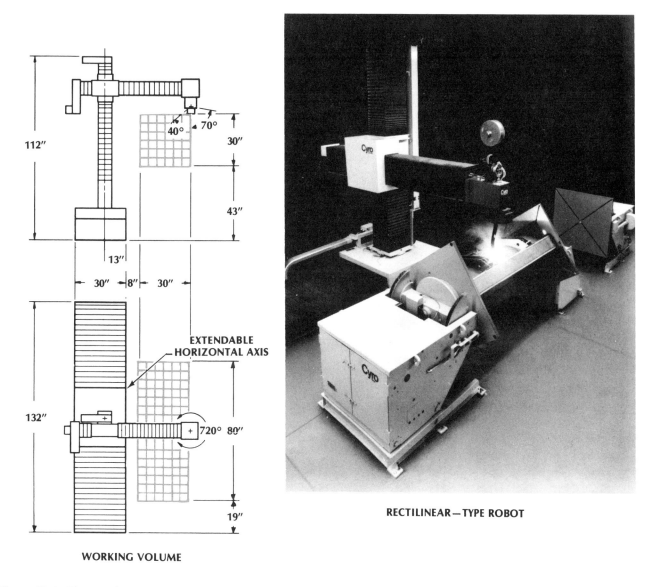

Figure 33-9. The rectilinear-type robot is used primarily for welding larger pieces. (*Advanced Robotics Corporation*)

Figure 33-10. Robotic welding systems use basically the same equipment used for semiautomatic and machine applications. *(GM — North American Operations)*

Figure 33-11. The movement of the positioner is coordinated with the robot to increase accessibility to the weld joint. *(Cincinnati Milacron)*

Figure 33-12. A jig can be used to hold the work securely. (*Berkley-Davis, Inc.*)

Points to Remember

1 Stand clear of any automatic welding machine in operation to avoid possible injury.

2 Robotics, when used with welding machines, are a type of automatic welding process.

3 Robot welding machines are expensive and complex pieces of equipment which require regular maintenance.

4 Two commonly used robots are the articulating type and the rectilinear type.

5 The basic components of a typical robot welding system include operator controls, robot controller, robot, automatic welding equipment, and positioner.

6 Robot welding systems use many different welding processes.

QUESTIONS FOR STUDY AND DISCUSSION

1 What are the advantages of using industrial robots (robotics) with welding machines?

2 How does welding with robotics differ from semiautomatic welding?

3 What components are found in the typical robot welding system?

4 What is meant by the robot's "working volume"?

5 What determines what type of robot should be used for welding?

6 What is meant by a five-axes robot?

7 What automatic welding equipment is necessary for robotic welding?

8 What welding processes are commonly used in a robot welding system?

9 What is the function of a positioner?

10 What are two types of commonly used robots?

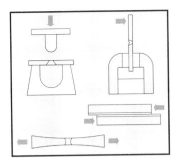

In the fabrication of any welded product, tests are generally employed to determine the soundness of welds. The nature of the test is governed to a great extent by the service requirements of the finished product.

Various types of tests have been devised, each for a specific purpose. These tests may be broadly classified as visual, destructive, and nondestructive.

VISUAL EXAMINATION

Visual examination consists of looking at a weld, or examination through a magnifying glass. A thorough examination of the weldment may disclose such surface defects as cracks, shrinkage cavities, undercuts, inadequate penetration, lack of fusion, overlaps, and crater deficiencies. Very often weld gauges are used to check for proper weld bead size and contour.

The limitation of any visual examination is that there is no way of knowing if *internal defects* exist in the welded area. The outer appearance of a weld may be satisfactory, yet cracks, porosity, slag inclusions, or excessive grain growth can be present which are not visible.

DESTRUCTIVE TESTING

Destructive testing involves the use of sample portions of a welded structure and subjecting them to loads until they fail. The most common types of destructive tests are known as *tensile, shear, weld uniformity, etching,* and *impact.*

Tensile Testing

Tensile testing involves the placement of a weld specimen in a tensile testing machine and pulling the piece until it breaks. See Figure 34-1. The specimen is cut either from an all-weld area or from a welded butt joint for plate and pipe.

The specimen for an all-weld area should conform to specific dimensions. See Figure 34-2. It should be cut from the welded section so its reduced area contains only weld metal.

For a plate and pipe welded butt joint, the specimen should be similar to the ones illustrated in Figures 34-3 and 34-4.

Before the specimen is placed in the tensile machine an accurate measurement should be taken of the gauge length so the percent of elongation can be determined.

The actual tensile strength is found by dividing the maximum load needed to break the piece by the cross-sectional area of the specimen. The cross-sectional area is determined by multiplying the width of the bar by its thickness. For example, assume that the specimen is 1½″ wide and ¼″ thick. The computation is carried out as follows:

Cross-sectional area = 1½ x ¼ = ⅜ sq in
Pull to break the bar = 24,500 lb
Tensile strength = 24,500 ÷ ⅜ = 65,333 psi

The percent of elongation is found by fitting the broken ends of the two pieces and measuring the new gauge length. The percent of elongation is a good indicator of the plasticity of the weld and is calculated with this formula:

$$\frac{FGL - OGL}{OGL} \times 100$$

where: FGL = Final gauge length
OGL = Original gauge length

Shearing Strength

The shearing strength of a weld can apply either to a transverse or a longitudinal weld.

Transverse shearing strength. To check the shearing strength of a transverse weld, a specimen

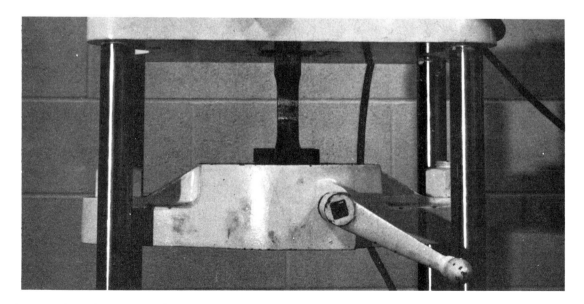

Figure 34-1. A universal testing machine determines the strength of a weld. (*Tinius Olsen*)

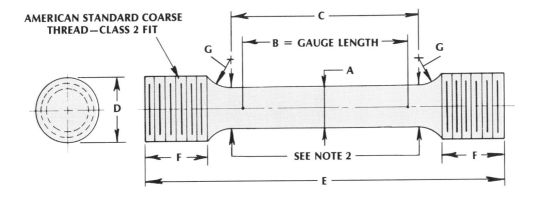

SPECIMEN DIMENSIONS OF SPECIMEN (INCHES MINIMUM)

	A	B	C	D	E	F	G
C-1	0.500 ± 0.01	2	2 1/4	3/4	4 1/4	3/4	3/8
C-2	0.437 ± 0.01	1 3/4	2	5/8	4	3/4	3/8
C-3	0.357 ± 0.007	1.4	1 3/4	1/2	3 1/2	5/8	3/8
C-4	0.252 ± 0.005	1.0	1 1/4	3/8	2 1/2	1/2	1/4
C-5	0.126 ± 0.003	0.5	3/4	1/4	1 3/4	3/8	1/8

Note 1: Dimension A, B and C shall be as shown, but alternate shapes of ends may be used as allowed by ASTM specification E-8.
Note 2: It is desirable to have the diameter of the specimen within the gauge length slightly smaller at the center than at the ends. The difference shall not exceed 1 percent of the diameter.

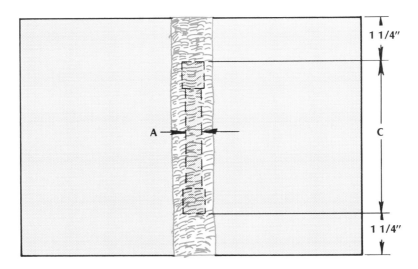

FOR GAS WELDING BACKING STRIP SHALL BE OMITTED
DOTTED LINES SHOW POSITION FROM WHICH SPECIMEN SHALL BE MACHINED

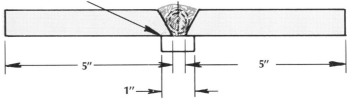

Figure 34-2. Tensile specimen for an all-weld metal.

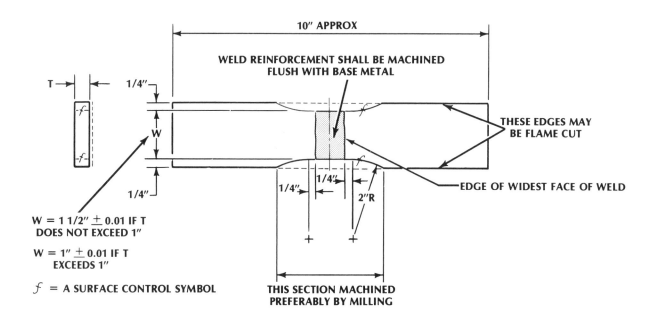

Figure 34-3. Tensile specimen for flat plate butt weld.

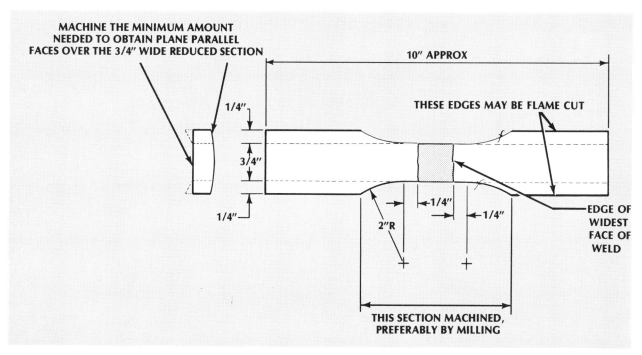

Figure 34-4. Tensile specimen for pipe butt weld.

is prepared similar to the one shown in Figure 34-5. The specimen is then placed in a tensile testing machine and pulled until it breaks. Dividing the maximum load in pounds by twice the width of the specimen will indicate the shearing strength in pounds per linear inch. If the shearing strength in pounds per square inch (psi) is desired, the shearing strength in pounds per linear inch is divided by the throat dimension of the weld. See Figure 34-6. Expressed as formulas, these relationships are shown as:

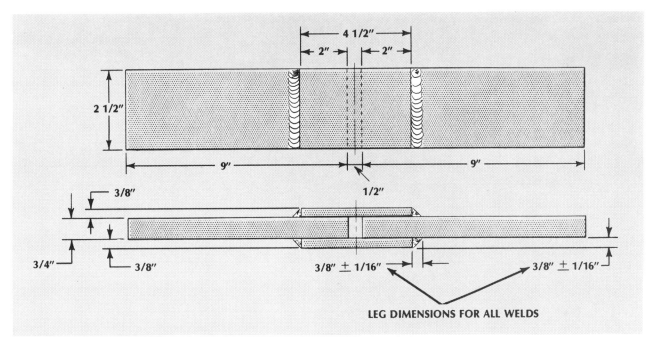

Figure 34-5. Specimen for determining the shear strength of a transverse weld.

Shearing strength (lb/in) =

$$\frac{\text{Maximum load}}{2 \text{ x width of specimen}}$$

or

Shearing strength (psi) =

$$\frac{\text{Shearing strength (lb/in)}}{\text{Throat dimension of weld}}$$

Longitudinal shearing strength. To determine the shearing strength of a longitudinal weld, a specimen is prepared. See Figure 34-7. The length of each weld is then measured and the piece fractured in a tensile testing machine. The shearing strength in pounds per linear inch is found by dividing the maximum load by the length of the ruptured weld. Expressed as a formula:

Shearing strength (lb/linear inch) =

$$\frac{\text{Maximum Load}}{\text{Length of ruptured weld}}$$

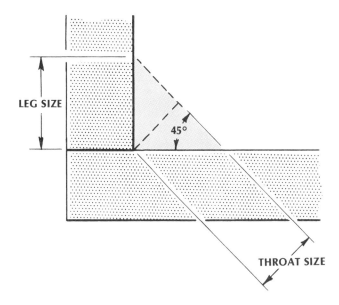

Figure 34-6. Fillet weld throat and leg.

Weld Uniformity Test

The soundness or uniformity and ductility of a weld can be ascertained by the nick-break test, the free-bend test, and the guided-bend test.

Nick-break test. A test specimen as illustrated in Figure 34-8 is prepared and placed on supporting members as shown in Figure 34-9. A load is applied on this specimen until it breaks. The surface of the fracture is then examined for porosity, gas pockets, slag inclusions, overlaps, penetration, and grain size. For a more accurate check of the weld, the fractured pieces should be subjected to an etch test as described in the paragraph on etching testing.

Free-bend test. The free-bend test is used to determine the ductility of the welded specimen. For this test, cut the test piece from the plate so as to include the weld. See Figure 34-10. Grind or ma-

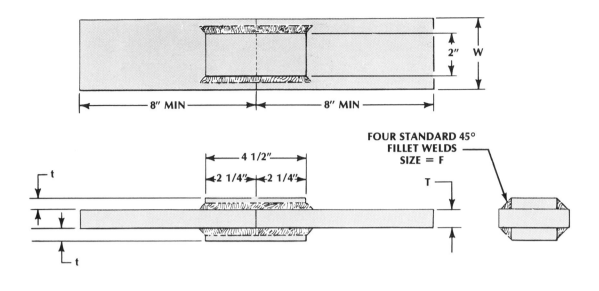

DIMENSIONS			
SIZE OF WELD F, INCHES	1/8	1/2	1/4
THICKNESS T, INCHES MIN	3/8	1/2	1
THICKNESS T, INCHES MIN	3/8	3/4	1 1/4
WIDTH W, INCHES	3	3	3 1/2

LONGITUDINAL FILLET-WELD SHEARING SPECIMEN AFTER WELDING

Figure 34-7. Specimen for determining the shear strength of a longitudinal weld.

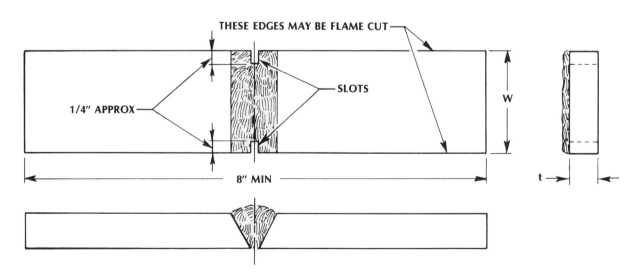

NICK BREAK SPECIMEN FOR BUTT JOINTS IN PLATE

Figure 34-8. Specimen for a nick-break test.

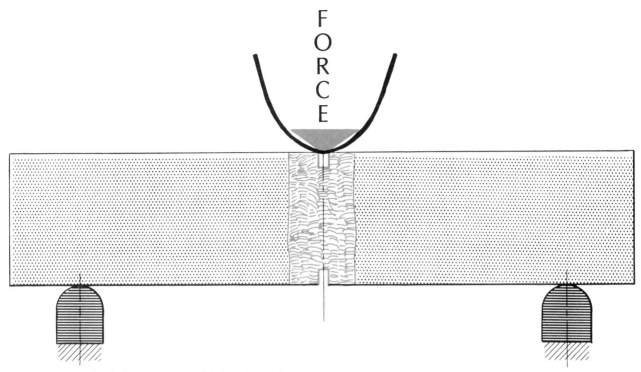

Figure 34-9. Method of rupturing a nick-break specimen.

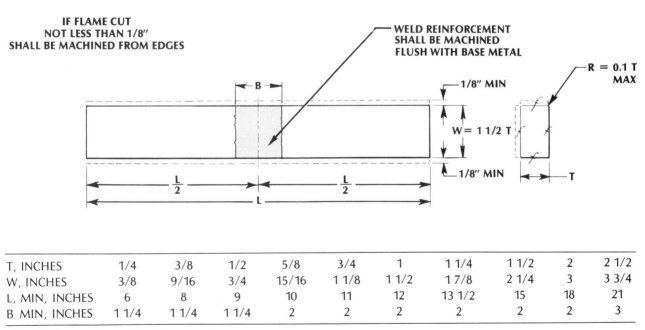

**IF FLAME CUT
NOT LESS THAN 1/8″
SHALL BE MACHINED FROM EDGES**

**WELD REINFORCEMENT
SHALL BE MACHINED
FLUSH WITH BASE METAL**

R = 0.1 T
MAX

T, INCHES	1/4	3/8	1/2	5/8	3/4	1	1 1/4	1 1/2	2	2 1/2
W, INCHES	3/8	9/16	3/4	15/16	1 1/8	1 1/2	1 7/8	2 1/4	3	3 3/4
L, MIN, INCHES	6	8	9	10	11	12	13 1/2	15	18	21
B MIN, INCHES	1 1/4	1 1/4	1 1/4	2	2	2	2	2	2	3

Note: The length L is suggestive only, not mandatory

Figure 34-10. Specimen for a free-bend test. *(AWS)*

chine the top of the weld so it is flush with the base metal surface. The scratches produced by grinding should run across the weld in the direction of the bend. See Figure 34-11. If the scratches extend along the weld they might cause premature failure and give incorrect results. Now measure the distance across the weld and mark it with prickpunch marks. The measured length (between gauge lines) must be about

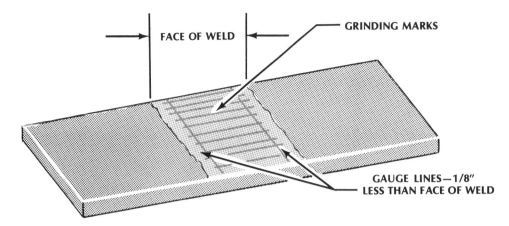

Figure 34-11. Gauge marks are located on the weld face when preparing the specimen for a free bend test.

⅛″ less than the width of the face of the weld as shown by the gauge marks. See Figure 34-11.

Bend the specimen by a steady force, with the face containing the gauge lines on the outside of the bend. Start the initial bend by placing the piece in a device as illustrated in Figure 34-12. After the specimen is given a permanent set, make the final bend in a vise. See Figure 34-13.

Continue the bend until a crack or depression appears, and then immediately remove the load. Measure the distance between the prickpunch marks or gauge lines with a flexible rule graduated in hundredths of an inch. The elongation is measured between the gauge lines along the convex surface of the weld to the nearest 0.01″. The percent of elongation is obtained by dividing the elongation by the initial gauge length and multiplying by 100. For example, suppose the initial gauge lines measure 2″, and after the bend the elongation measured 2.5″. The calculation then would be:

Increase in elongation $= 2.5″ - 2″ = 0.5″$

Percent elongation $= (0.5 \div 2) \times 100 = 25$

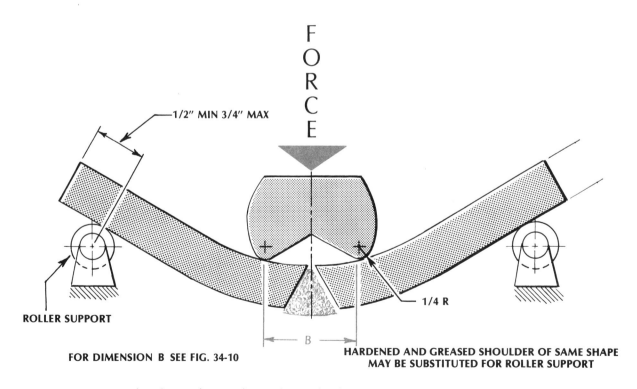

Figure 34-12. Here is another device that can be used to make the initial bend. (*AWS*)

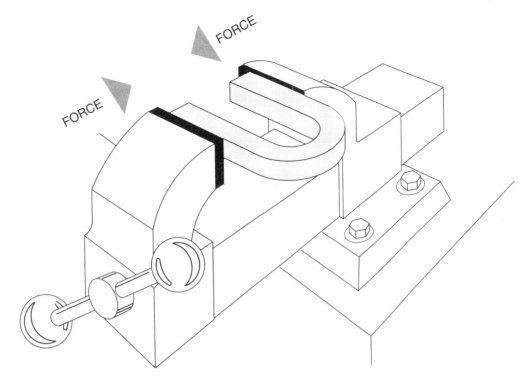

Figure 34-13. For the final bend, the specimen must be bent in a vise.

The ductility of this specimen is therefore considered to be 25 percent in 2″.

Guided-bend test. For this test, two specimens are required. See Figure 34-14. One piece, referred to as the *face-bend specimen,* is used to check the quality of fusion; that is, whether the weld is free of defects such as porosity, inclusions, etc. The second piece, referred to as the *root-bend specimen,* is used to check the degree of weld penetration.

To perform the face-bend test, place the specimen in the guided-bend jig face down and depress the plunger until the piece becomes U-shaped in the die. See Figure 34-15. If upon examination, cracks greater than ⅛″ appear in any direction, the weld is considered to have failed.

In the root-bend test, place the specimen in the jig with the root down or in just the reverse position of the face-bend piece. The results must also show no cracks to be acceptable.

Fillet Weld Joint Test

This test is used to ascertain the soundness of fillet welds. *Soundness* refers to the degree of freedom a weld has from defects discernable by visual inspection of any exposed surface of weld metal. These defects include penetrations, inclusions, and gas pockets. Prepare a test specimen. See Figure 34-16. Apply force on point A until a break in the specimen

occurs. See Figure 34-17. The force may be applied by a press, a testing machine, or hammer blows.

In addition to checking the fractured weld for soundness, the weld specimen should be etched so that the weld can be examined for cracks.

Etch Test

The etch test is used to determine the soundness of a weld and to make visible the boundary between the weld metal and the base metal.

To make such a test, cut a specimen from the welded joint so it displays a complete transverse section of the weld. The piece may be cut either by sawing or flame cutting. File the face of the cut and polish it with grade 00 abrasive cloth. Then place the specimen in one of the following etching solutions:

Hydrochloric acid. This solution should contain equal parts by volume of concentrated hydrochloric (muriatic) acid and water. Immerse the weld in the boiling reagent. Hydrochloric acid will etch unpolished surfaces. It will usually enlarge gas pockets and dissolve slag inclusions, enlarging the resulting cavities.

Ammonium persulfate. Mix one part of ammonium persulfate (solid) to nine parts of water by weight. Vigorously rub the surface of the weld with cotton saturated with this reagent at room temper-

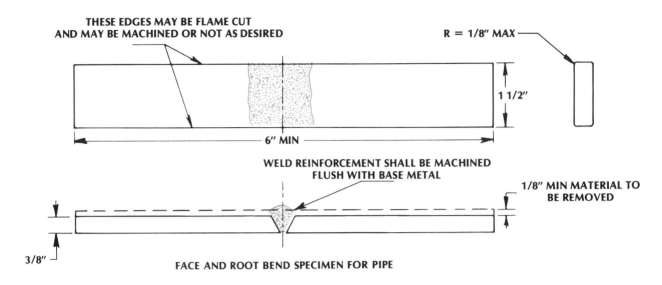

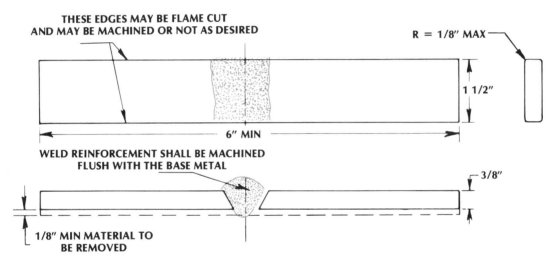

Figure 34-14. Specimen for a guided-bend test. (*AWS*)

ature.

Iodine and potassium iodide. This solution is obtained by mixing one part of powdered iodine (solid) to twelve parts of a solution of potassium iodide by weight. The latter solution should consist of one part potassium iodide to five parts water by weight. Brush the surface of the weld with this reagent at room temperature.

Nitric acid. Mix one part of concentrated nitric acid to three parts of water by volume.

WARNING: Always add acid to water when diluting. Nitric acid causes bad stains and severe burns. Wash instantly with water if on skin.

Either apply this reagent to the surface of the weld with a glass stirring rod at room temperature, or immerse the weld in a boiling reagent provided the room is well ventilated. Nitric acid etches rapidly. It should be used on polished surfaces only, and will show the refined zone as well as the metal zone.

After etching, wash the weld immediately in clear water, preferably hot water; remove the excess water; dip the etched surface in ethyl alcohol; and then remove and dry it in a steady blast of warm air.

Impact Testing

Impact testing is concerned with the ability of a weld to absorb energy under impact without fracturing. This is a dynamic test in which a specimen is broken by a single blow and the energy absorbed in breaking the piece is measured in foot pounds. The purpose of the test is to compare the toughness of

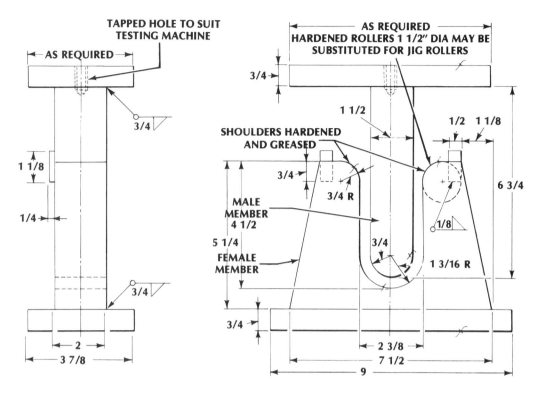

Figure 34-15. A guided bend test can be performed using a jig (top), or a guided bend tester (bottom).

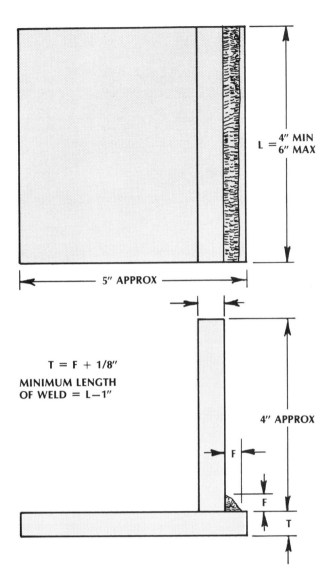

$$L = \frac{4'' \text{ MIN}}{6'' \text{ MAX}}$$

5" APPROX

T = F + 1/8"

**MINIMUM LENGTH
OF WELD = L−1"**

4" APPROX

Figure 34-16. Specimen for a fillet weld test. (*AWS*)

Figure 34-17. Method of rupturing fillet weld specimen. (*AWS*)

NONDESTRUCTIVE TESTING

Nondestructive testing is used to evaluate a structure without destroying it or impairing its actual usefulness. Tests of this nature will disclose all of the common surface and internal defects that normally occur with improper welding procedures or practices.

Currently a variety of testing devices are available which do provide effective data concerning the reliability of a weldment. These devices are often more convenient to use than other regular destructive testing techniques particularly on large and costly welded units.

Magnetic Particle Inspection

The magnetic particle inspection method uses a strong magnetizing current and a finely divided powder suspended in a liquid to detect lack of fusion, very fine cracks, and inclusions or internal flaws which are slightly below the surface in weldments.

In this test the piece to be examined is subjected to a very strong magnetizing current and the areas of inspection are covered with suspended powder. Any impurities or discontinuities in the magnetized material will interrupt the lines of magnetic force causing the particles of suspended powder to concentrate at the defect showing its size, shape, and location. Surface cracks of all kinds are detected by

the weld metal with the base metal. It is especially significant in finding if any of the mechanical properties of the base metal have been destroyed due to welding.

The two types of specimens used for impact testing are known as Charpy and Izod. See Figure 34-18. Both specimens are broken in an impact testing machine. The difference is simply in the manner in which the specimens are anchored. The *Charpy* piece is supported horizontally between two anvils and the pendulum allowed to strike opposite the notch as shown in Figure 34-19A. The *Izod* specimen is supported as a vertical cantilever beam and struck on the free end projecting above the holding vise. See Figure 34-19B.

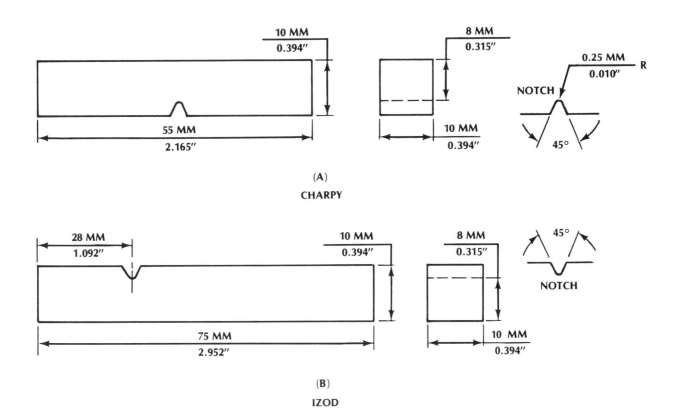

Figure 34-18. Specimens for impact testing.

this method. It is one of the most reliable techniques for nondestructive testing. See Figure 34-20.

Dye Penetrant Inspection

In dye penetrant inspection, surface defects are found by means of proprietary dyes suspended in liquids having high fluidity. These liquids are readily drawn into all surface defects by capillary action. Application of a suitable developer brings out the dye and outlines the defect.

In this test, the surface of the weldment, which must be clean and dry, is coated with a thin film of a penetrant. After allowing a small amount of time for the penetrant to flow into the defects, the part is wiped clean. Only the penetrant in the defects remains. An absorbent material, called a developer, is put on the weldment and allowed to remain until the liquid from the imperfection flows into the developer. The dye now clearly outlines the defects.

Some of the penetrants used contain a fluorescent dye. The method of applying and developing are the same as for the previously mentioned dye penetrants; however, the fluorescent penetrant must be viewed under ultraviolet light, commonly referred to

as *black light.* This light causes the penetrants to fluoresce (glow) to a yellow-green color which is more clearly defined than regular dye penetrants. The dye penetrant methods are particularly useful for bringing out defects in nonferrous materials such as aluminum. These materials are nonmagnetic so magnetic particle tests cannot be used on them. See Figure 34-21.

Eddy Current Testing

Eddy current testing uses electromagnetic energy to detect discontinuities in weld deposits and is effective in testing both ferrous and nonferrous materials for porosity, slag inclusions, internal cracks, external cracks, and lack of fusion.

Whenever a coil carrying a high-frequency alternating current is brought close to a metal, it produces a current in the metal by induction. The induced current is called an *eddy current.*

The part to be tested is subjected to electromagnetic energy by being placed in or near high-frequency alternating current coils. Differences in metal in the weld deposit change the impedance of the coil. The change in impedance is indicated on elec-

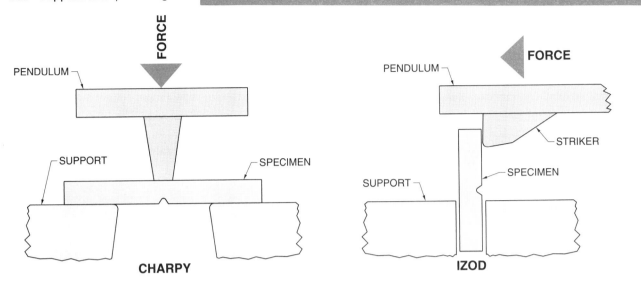

Figure 34-19. Performing impact tests.

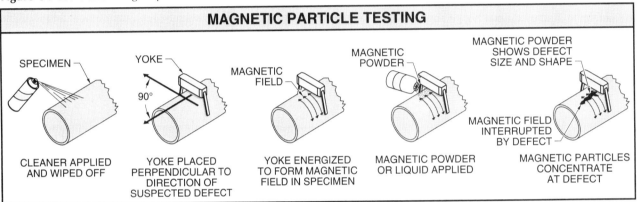

Figure 34-20. Magnetic particle testing uses a strong magnetic field and iron particles in a powder or liquid to locate defects.

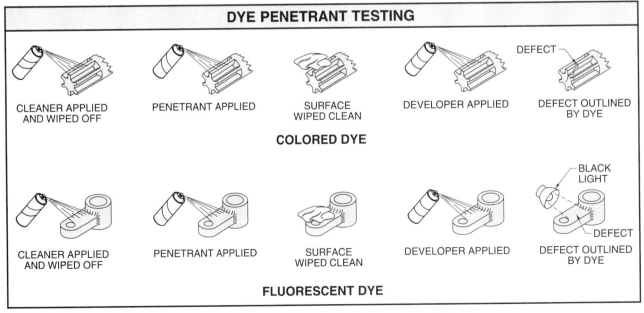

Figure 34-21. Dye penetrant testing uses colored or fluorescent dye to locate defects.

tronic measuring instruments, and the size of the defect is shown by the amount of this change. See Figure 34-22.

Radiographic Inspection

Radiographic inspection is a method of determining the soundness of a weldment by means of rays which are capable of penetrating through the entire weldment. X-rays and gamma rays are two types of electro-magnetic waves used to penetrate opaque materials. A permanent record of the internal structure is obtained by placing a sensitized film in direct contact with the back of the weldment. When these rays pass through a weldment of uniform thickness and structure, they fall upon the sensitized film and produce a negative of uniform density. If the weldment contains gas pockets, slag inclusions, cracks, or lacks penetration, more rays will pass through the less dense areas and will register on the film as dark areas, clearly outlining the defects and showing their size, shape and location.

X-rays are produced by electrons travelling at high speed which are suddenly stopped by impact with a tungsten anode. Gamma rays are given off by radium or by other radioactive substances. Gamma rays are of shorter wave length than x-rays. See Figure 34-23.

Ultrasonic Testing

In ultrasonic testing, high-frequency vibrations or waves are used to locate and measure defects in both ferrous and nonferrous materials. This method is very sensitive, and is capable of locating very fine surface and subsurface cracks, as well as other internal defects. All types of joints can be evaluated and the exact size and location of defects measured.

Ultrasonic testing utilizes high-frequency vibratory impulses to ascertain the soundness of a weld. If a high-frequency vibration is sent through a sound piece of metal, a signal will travel through it to the other side of the metal and be reflected back and shown on a calibrated screen of an oscilloscope. Discontinuities of structure interrupt the signal and reflect it back sooner than the signal of the sound material. This is shown on the oscilloscope screen and indicates the depth of the defect. Only one side of the weldment needs to be exposed for testing purposes. See Figure 34-24.

Hardness Testing

Hardness testing is often used in preference to the

Figure 34-22. Eddy current testing uses electromagnetic energy to detect defects.

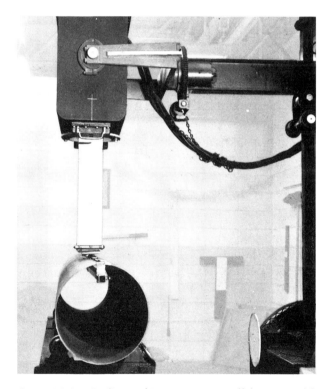

Figure 34-23. Radiographic inspection will locate weld defects by passing x-rays through the welded structure.

more expensive tensile testing methods since comparable results are obtained.

Hardness tests are effective in determining the relative hardness of the weld area as compared with the base metal. This hardness is indicated by values

Figure 34-24. Ultrasonic inspection is capable of locating very fine surface and subsurface cracks.

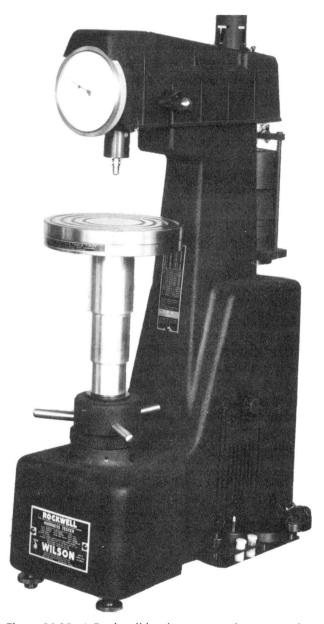

Figure 34-25. A Rockwell hardness tester determines the relative hardness of the weld area as compared with the base metal.

obtained from various hardness testing machines. Hardness numbers represent the resistance offered by the metal to the penetration of an indenter. The standard hardness machines are known as *Brinell* and *Rockwell*. See Figure 34-25.

In the Brinell test, a 10 mm diameter ball is forced into the surface of a metal by a load of 3000 kg. The load must remain on the specimen 15 seconds for ferrous materials, and 30 seconds for nonferrous materials. Sufficient time is required for adequate flow of the material being tested otherwise the readings will be in error. Brinell hardness numbers are calculated by dividing the applied load by the area of the surface indentation. The diameter of the indentation is read from a calibrated microscope and this number is then converted to a Brinell hardness number from a chart.

The Rockwell hardness tester employs a variety of loads and indenters; consequently different scales can be used. These scales are designated by letters. For example, R_c 60 represents a Rockwell scale with a diamond penetrater and a 150 kg load. Since hardness numbers give relative or comparative hardness values of materials, the scale must always be specified.

Points to Remember

1. The soundness of a weld can be determined by visual inspection or destructive and nondestructive testing.

2. Visual inspection will not show if there are any internal defects in a weld.

3. The true strength of a weld can be determined by tensile, shear, and by impact testing procedures.

Points to Remember, cont.

4   Tensile testing involves placing a weld specimen in a tensile testing machine and pulling the piece until it breaks.

5   Weld uniformity can be checked by applying a load on a weld specimen until it breaks.

6   A free-bend test is used to determine the ductility of a weld.

7   A guided-bend test will disclose the quality of fusion and degree of penetration.

8   An etch test will show the boundary lines between the weld metal and the base metal.

9   When preparing the mixture for an etch test, always add the acid to the water.

10   Impact testing is used to determine the toughness of the weld metal.

11   In nondestructive testing, special ultrasonic, radiographic, eddy current or magnetic particle equipment is used to evaluate a weld.

12   Nondestructive testing is used when it is impractical to check weld quality by other kinds of test methods.

QUESTIONS FOR STUDY AND DISCUSSION

1   What is the limitation of any visual inspection method for testing a weld?

2   What is the difference between destructive and nondestructive testing methods?

3   What will a tensile test show?

4   In conducting a tensile test, what is the value of finding the percent of elongation?

5   What is the function of a shearing strength test?

6   What is meant by a longitudinal and transverse weld?

7   What is the function of a nick-break test?

8   When is a free-bend test used?

9   How is a guided-bend test conducted?

10   How can the soundness of fillet welds be determined?

11   What will etching tests disclose?

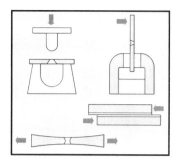

Supplementary Welding Data

In the fabrication of metal products, the welder usually has to work from a print which shows in detail exactly how the structure is to be made. On the print will be specified where the welds are to be located, the type of joint to be used, as well as the correct size and amount of weld to be deposited at the designated joints. This information is indicated by a set of symbols which have been standardized by the American Welding Society (AWS).

Some of the more common symbols for weldments are included in this chapter.

Welding Symbols

A weld symbol is made up of a reference line, to which are attached instructions as to the type of weld required, the location of the weld, whether it is a field weld or a shop weld, and other reference data which are necessary to do a complete weld job. While these symbols may be very complex and carry a large amount of data, they may also be quite simple. The welding student must study the various examples and learn to read the symbols before he or she can

qualify for apprenticeship or ask for a full-time job as a welder.

Reference line. The main foundation of the weld symbol is a reference line with an arrow at one end, as shown in Figure 35-1. Notice that above or below the reference line the type of weld is indicated, whether it is a fillet, groove, flange, plug, spot, or seam weld. Also included is such information as surface contour of a weld, size of a weld bead, length of a weld, how beads are to be finished, and often what type welding process is to be used.

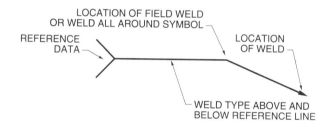

Figure 35-1. The foundation of the weld symbol is a reference line with an arrow at one end.

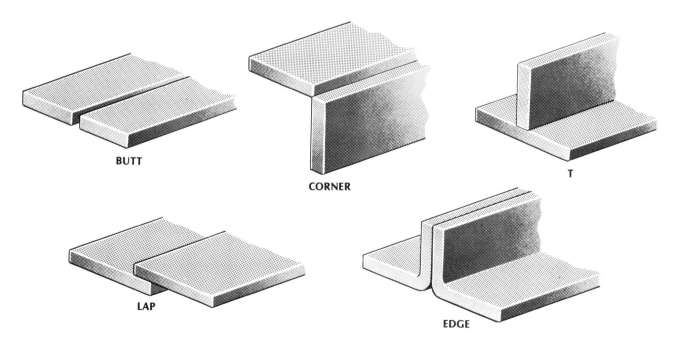

Figure 35-2. The five basic types of joints are the butt, corner, T, lap, and edge.

ing process is to be used. All of this data is indicated by geometric figures, numerical values, and abbreviations.

Designating types of welds.   The most important factor in a welding symbol is the type of weld. The type of weld is used in relationship to five basic types of joints: butt, corner, lap, T, and edge. These are represented in Figure 35-2.

Welds are classed as fillet, plug, spot, seam, or groove. Groove welds are further divided and classified according to the particular shape of the grooved joint. See Chapter 4.

Each weld has its own specific symbol. For example, a fillet weld is designated by a right triangle, and a plug weld by a rectangle. See Figure 35-3. (All of the weld symbols are included in Figure 35-28, in the Summary section of this chapter.)[1]

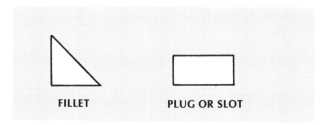

FILLET   **PLUG OR SLOT**

Figure 35-3. The symbol for a fillet weld is a triangle; a rectangle represents a plug or slot weld.

Location of symbols.   Another requirement in understanding weld symbols is the method which is used to specify on what side of a joint a weld is to be made. A weld is said to be either on the *arrow* or *other* side of a joint. The arrow side is the surface that is in direct line of vision, while the other side is the opposite surface of the joint. See Figure 35-4.

Weld location is designated by running the arrowhead of the reference line to the joint. The direction of the arrow is not important; that is, it can run on either side of a joint and extend upward or downward. See Figure 35-5. If the weld is to be made on the *arrow side,* the appropriate weld symbol is placed *below* the reference line. If the weld is to be located on the *other side* of the joint, the weld symbol is placed *above* the reference line. When both sides of the joint are to be welded, the same weld symbol

---

1. A more complete treatment of symbols as they apply to all forms of manual and automatic machine welding will be found in the pamphlet A2.4-79 *Standard Welding Symbols,* published by the American Welding Society.

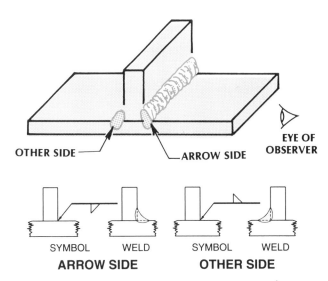

OTHER SIDE   ARROW SIDE   **EYE OF OBSERVER**

SYMBOL   WELD   SYMBOL   WELD
ARROW SIDE   **OTHER SIDE**

Figure 35-4. The arrow side of the weld is in the observer's direct line of vision. The opposite surface of the joint is called the other side.

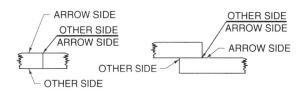

ARROW SIDE   OTHER SIDE
OTHER SIDE   ARROW SIDE
ARROW SIDE   ARROW SIDE
          OTHER SIDE
OTHER SIDE

ARROW LOCATION

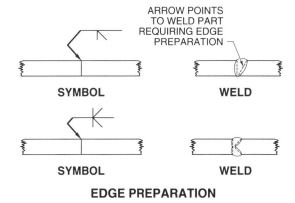

ARROW POINTS TO WELD PART REQUIRING EDGE PREPARATION

SYMBOL   **WELD**

SYMBOL   **WELD**

EDGE PREPARATION

Figure 35-5. The arrow specifies weld location and edge preparation.

appears above and below the reference line. See Figure 35-6.

The only exception to this practice of indicating weld location is in seam and spot welding. With seam or spot welds, the arrowhead is simply run to the centerline of the weld seam and the appropriate weld symbol centered above or below the reference line. See Figure 35-6. If no arrow side or other side is important, the symbol is placed astride the reference

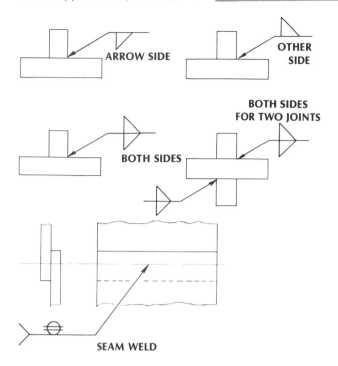

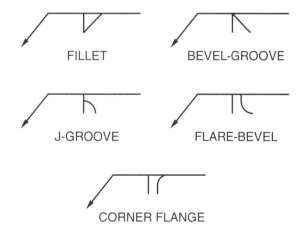

Figure 35-6. The weld symbol designates weld location by its placement on the reference line.

line to indicate this condition.

On beveled joints it is often necessary to show which member is to be beveled. In such cases, the arrow points with a definite break toward the member to be beveled. See Figure 35-5.

Information on weld symbols is placed to read from left to right along the reference line in accordance with the conventions of drafting.

Fillet, bevel and J-groove, flare-bevel groove, and corner-flange weld symbols are shown with the *perpendicular leg* always to the left of the weld symbol. See Figure 35-7.

FILLET   BEVEL-GROOVE

J-GROOVE   FLARE-BEVEL

CORNER FLANGE

Figure 35-7. For these joints, the perpendicular leg always appears to the left of the weld symbol.

Combining Weld Symbols

In the fabrication of a product, there are occasions when more than one type of weld is to be made on a joint. Thus a joint may require both a fillet and double-bevel groove weld. When this happens, a symbol is shown for each weld. See Figure 35-8.

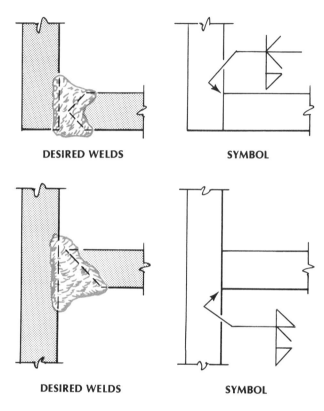

DESIRED WELDS   SYMBOL

DESIRED WELDS   SYMBOL

Figure 35-8. A joint that requires more than one type of weld is represented by a combined weld symbol.

Size of fillet welds. The width of a fillet weld is shown to the left of the weld symbol and is expressed in fractions, decimals, or metric units (mm). See Figure 35-9. When both sides of a fillet are to be welded and both welds have the same dimensions, both are dimensioned. If the welds differ in dimensions, both are dimensioned. Where a note appears on a drawing that governs the size of a fillet weld, no dimensions are usually shown on the symbol.

The length of the weld is shown to the right of the weld symbol by numerical values representing the actual required length.

When a fillet weld with unequal legs is required, the size of the legs is placed to the left of the weld symbol.

Intermittent fillet welds. The length and pitch increments of intermittent welds are shown to the

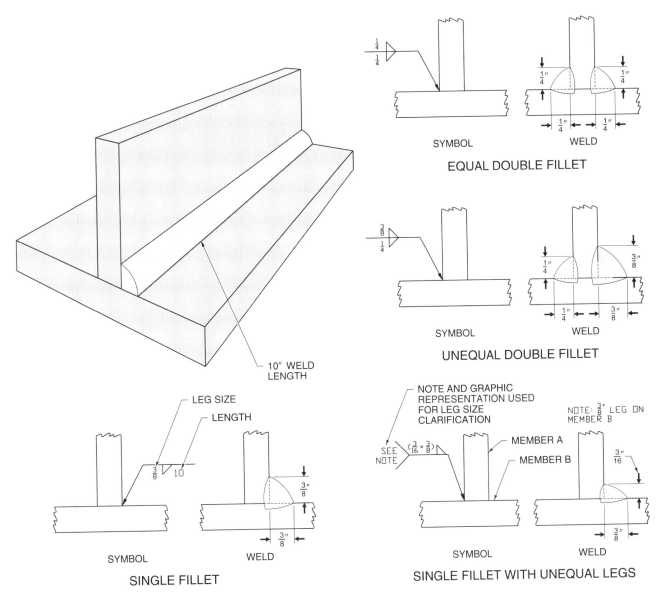

Figure 35-9. The size and length of fillet welds are indicated in the various weld symbols.

right of the weld symbol. The first figure represents the length of the weld section and the second figure the pitch (center-to-center spacing) between the welds. See Figure 35-10.

Size of groove welds. There are several types of groove welds. Their sizes (effective throat in fractions, decimals or millimeters) are shown as follows:

1. For single-groove and symmetrical double-groove welds which extend completely through the members being joined, no size is included on the weld symbol. See Figure 35-11.

2. For groove welds which extend only partly through members being joined or on nonsymmet-

rical double-groove joints, weld size (effective throat) is shown in parentheses to the left of the weld symbol. See Figure 35-12.

3. A dimension not in parentheses when placed to the left of the weld symbol indicates the depth of the bevel only. See Figure 35-13. When both the effective throat and bevel depth are indicated, the groove bevel depth is located to the left of the effective throat size as shown in Figure 35-13.

4. Root opening of groove joints is shown inside the weld symbol. The included angle of the bevel is placed below or above the weld symbol. See Figure 35-14.

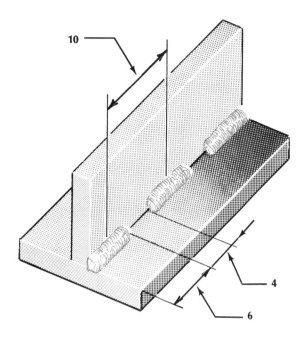

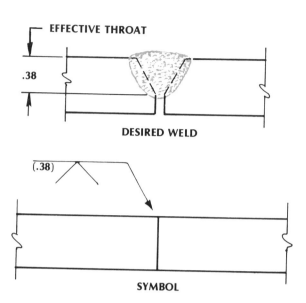

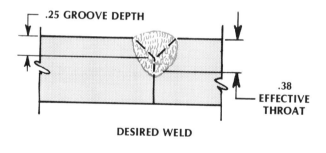

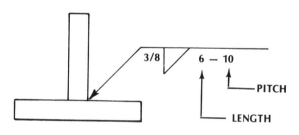

Figure 35-10. The length and pitch of intermittent fillet welds are indicated in the weld symbol.

Figure 35-12. For grooved welds with partial penetration, weld size is shown in parentheses to the left of the weld symbol.

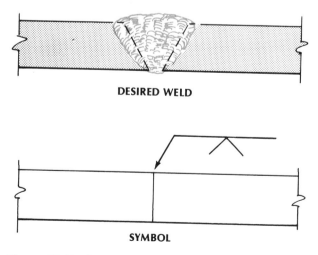

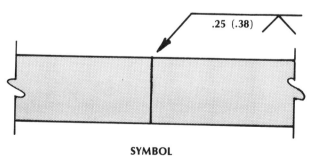

Figure 35-11. Size is not shown for single and symmetrical double-groove welds with a complete penetration.

Figure 35-13. Groove bevel depth is shown to the left of the effective throat size, shown in parentheses, when both dimensions are indicated. When groove depth is shown by itself, the dimension appears without parentheses.

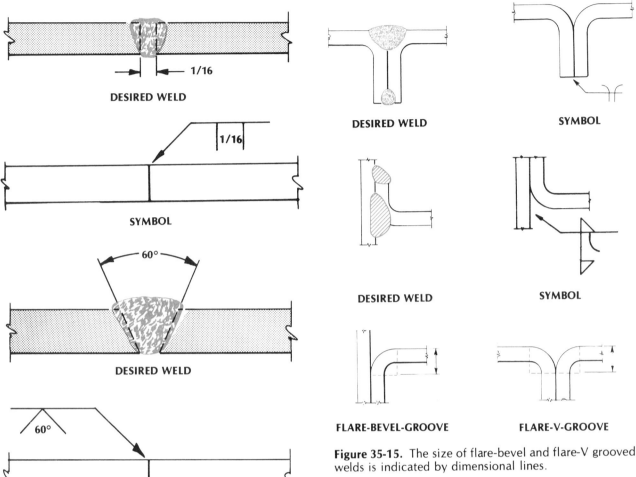

Figure 35-14. For grooved welds, the included angle of the bevel is shown above or below the weld symbol. The root opening of a square butt joint is indicated inside the weld symbol.

5. The size of the flare-groove welds is considered as extending only to the tangent points as indicated by dimensional lines. See Figure 35-15.

Size of flange welds. The radius and height of the flange is separated by a plus mark and placed to the left of the weld symbol. The size of the weld is shown by a dimension located outward of the flange dimensions. See Figure 35-16.

Size of plug welds. The size of plug welds is shown to the left of the weld symbol, the depth, when less than full, on the inside of the weld symbol, the center-to-center spacing (pitch) to the right of the symbol, and the included angle of countersink below the symbol. See Figure 35-17.

Size of slot welds. Length, width, spacing, angle of countersink, and location of slot welds are not

Figure 35-15. The size of flare-bevel and flare-V grooved welds is indicated by dimensional lines.

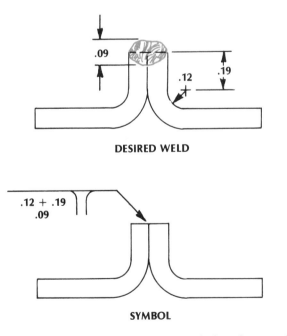

Figure 35-16. The flange weld symbol indicates the radius and height of the flange as well as the size of the weld.

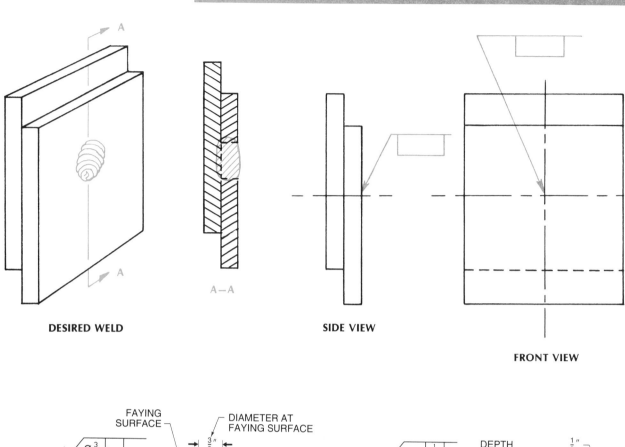

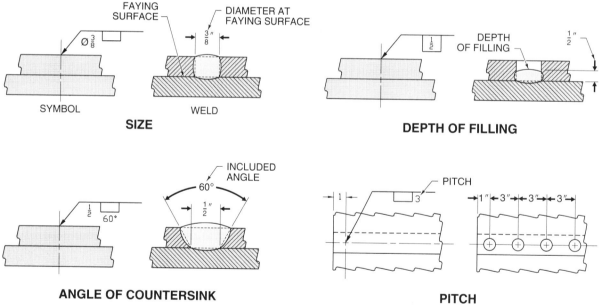

Figure 35-17. The dimensions of plug welds are shown as indicated.

shown on the symbol. This data is included by showing a special detail on the print. If the slots are partly filled, the depth of filling is shown inside. See Figure 35-18.

Size of spot welds. Spot welds are dimensioned either by size or strength. Size is designated as the

diameter of the weld expressed in fractions, decimals or millimeters and placed to the left of the symbol. The strength is also placed to the left of the symbol and expresses the required minimum shear strength in pounds or neutrons per spot. The spacing of spot welds is shown to the right of the symbol. When a

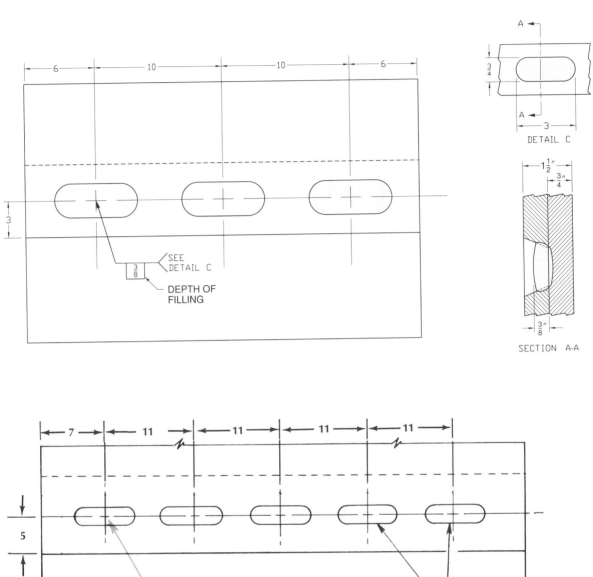

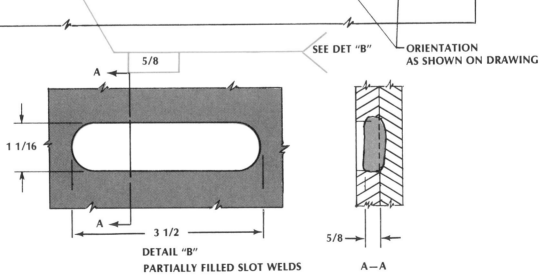

Figure 35-18. The dimensions of slot welds are shown as indicated.

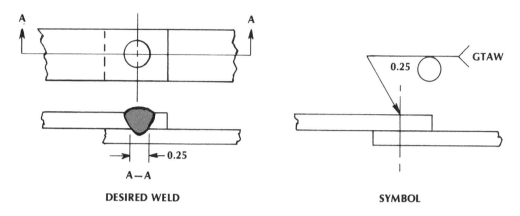

DESIRED WELD   SYMBOL

DIAMETER OF SPOT WELDS
(GAS TUNGSTEN-ARC SPOT)

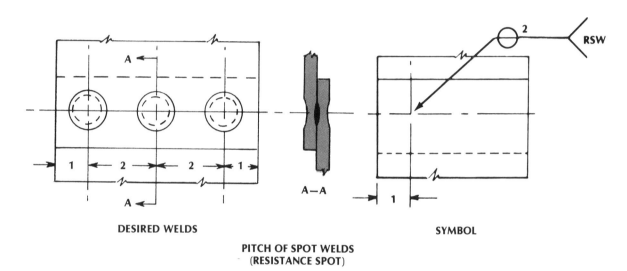

DESIRED WELDS   SYMBOL

PITCH OF SPOT WELDS
(RESISTANCE SPOT)

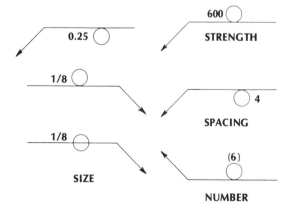

Figure 35-19. Spot weld designations include size, strength, spacing, and number of spot welds.

definite number of spot welds are needed in a joint, this number is indicated in parentheses either above or below the weld symbol. See Figure 35-19.

Size of seam welds. Seam welds are dimensioned either by size or strength. Size is designated as the width of the weld in fractions, or decimals or millimeters and shown to the left of the weld symbol. The length of the weld seam is placed to the right of the weld symbol. The pitch of intermittent seam welds is shown to the right of the length dimension. See Figure 35-20.

The strength of the weld, when used, is located to the left of the symbol, and is expressed as the minimum acceptable shear strength in pounds per linear inch or metric units.

Weld-all-around symbol. When a weld is to extend completely around a joint, a small circle is placed

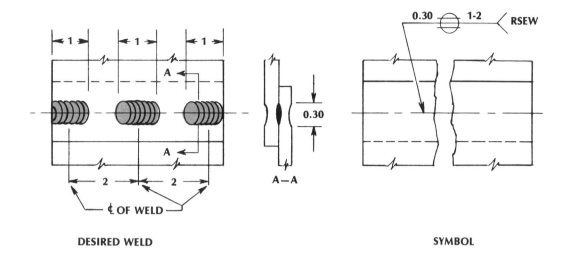

DESIRED WELD

SYMBOL

A—A

SIZE, LENGTH AND PITCH OF INTERMITTENT SEAM WELDS
(RESISTANCE SEAM)

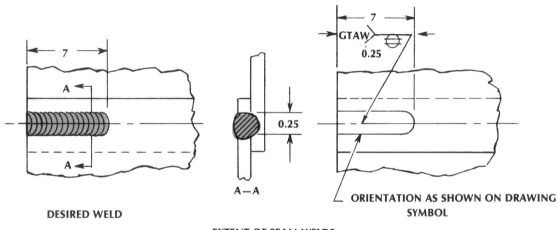

DESIRED WELD

A—A

ORIENTATION AS SHOWN ON DRAWING
SYMBOL

EXTENT OF SEAM WELDS
(GAS TUNGSTEN-ARC SEAM)

where the arrow connects the reference line. See Figure 35-21.

Field weld symbol. Welds to be made in the field (not in a shop or at the place of initial construction) are shown by a darkened triangular flag at the juncture of the reference line and arrow. The flag always points toward the tail of the arrow. See Figure 35-22.

Reference tail. The tail is included only when some definite welding specification, procedure, reference, weld or cutting process needs to be called out; otherwise it is omitted. This data is often in the form of symbols. See Figure 35-23 and Table 35-1. Abbreviations in the tail may also call out some specifications that are included on some other part of the print.

Surface contour of welds. When bead contour

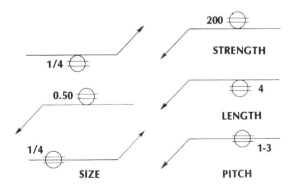

Figure 35-20. Seam weld designations include size, strength, length of weld seam, and pitch of intermittent seam welds.

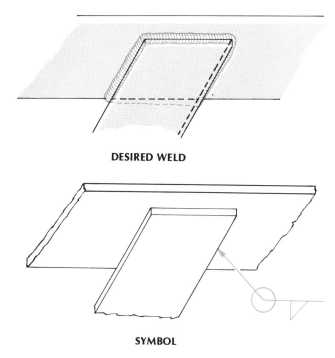

DESIRED WELD

SYMBOL

Figure 35-21. In the weld-all-around symbol, a small circle appears where the arrow connects the reference line.

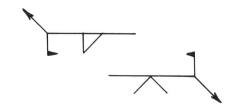

Figure 35-22. The field weld symbol is placed at a right angle to the reference line at the junction with the arrow.

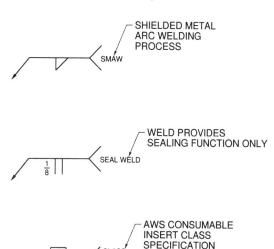

SHIELDED METAL
ARC WELDING
PROCESS

SMAW

WELD PROVIDES
SEALING FUNCTION ONLY

SEAL WELD

$\frac{1}{8}$

AWS CONSUMABLE
INSERT CLASS
SPECIFICATION

CLASS
3
INSERT

Figure 35-23. The tail is used to indicate some specific detail or weld process. (*AWS*)

is important, a special flat, concave, or convex contour symbol is added to the weld symbol. Welds that are to be mechanically finished also carry a finish symbol along with the contour symbols. See Figure 35-24.

Back or backing welds. Back or backing welds refer to the weld made on the opposite side of the regular weld. Back welds are occasionally specified to ensure adequate penetration and provide additional strength to a joint. This particular symbol is included opposite the weld symbol. No dimensions of back or backing welds except height of reinforcement are shown on the weld symbol. See Figure 35-25.

Melt-through welds. When complete joint penetration of the weld through the material is required in welds made from one side only, a special melt-through weld symbol is placed opposite the regular weld symbol. No dimension of melt-through, except height of reinforcement, is shown on the weld symbol. See Figure 35-26.

Surfacing welds. Welds whose surfaces must be built up by single or multiple-pass welding are provided with a surfacing weld symbol. The height of the built-up surface is indicated by a dimension placed to the left of the surfacing symbol. See Figure 35-27. The extent, location, and orientation of the area to be built up is normally indicated on the drawing.

Nondestructive Examination (NDE) Symbols

Nondestructive examination (NDE) symbols are symbols that specify examination methods and requirements to verify weld quality. The method of examination required can be specified on a separate reference line of the welding symbol or as a separate NDE symbol.

When specified on a separate reference line, the order of operation is the same as for multiple welding operations. The reference line furthest from the arrowhead indicates the last operation to be performed. The operation on the reference line nearest the arrowhead is performed first. When used separately, NDE symbols include an arrow, reference line, examination letter designation, dimensions, areas, number of examinations, supplementary symbols, tail, and specifications and other references. See Figure 35-28. See Appendix.

Summary

A more complete listing of basic arc and gas welding symbols and some supplementary symbols are listed in Figure 35-29. The welding symbol is designed so that specific information has designated locations on the symbol. See Figure 35-30.

TABLE 35-1. MASTER CHART OF WELDING AND ALLIED PROCESSES.

Gas metal arc welding............GMAW
—electrogas............GMAW-EG
—pulsed arc............GMAW-P
—short circuiting arc............GMAW-S
gas tungsten arc welding............GTAW
—pulsed arc............GTAW-P
plasma arc welding............PAW
shielded metal arc welding............SMAW
stud arc welding............SW
submerged arc welding............SAW
—series............SAW-S

arc brazing............AB
diffusion brazing............DFB
dip brazing............DB
furnace brazing............FB
induction brazing............IB
infrared brazing............IRB
resistance brazing............RB
torch brazing............TB
twin carbon arc brazing............TCAB

electron beam welding............EBW
electroslag welding............ESW
induction welding............IW
laser beam welding............LBW
thermit welding............TW

oxyacetylene welding............OAW
oxyhydrogen welding............OHW
pressure gas welding............PGW

air carbon arc cutting............AAC
carbon arc cutting............CAC
gas metal arc cutting............GMAC
gas tungsten arc cutting............GTAC
metal arc cutting............MAC
plasma arc cutting............PAC
shielded metal arc cutting............SMAC

electron beam cutting............EBC
laser beam cutting............LBC

carbon arc welding............CAW
—shielded............CAW-S
—twin............CAW-T
flux cored arc welding............FCAW
—electrogas............FCAW-EG

cold welding............CW
diffusion welding............DFW
explosion welding............EXW
forge welding............FOW
friction welding............FRW
hot pressure welding............HPW
roll welding............ROW
ultrasonic welding............USW

dip soldering............DS
furnace soldering............FS
induction soldering............IS
infrared soldering............IRS
iron soldering............INS
resistance soldering............RS
torch soldering............TS
wave soldering............WS

flash welding............FW
high frequency resistance welding............HFRW
percussion welding............PEW
projection welding............RPW
resistance seam welding............RSEW
resistance spot welding............RSW
upset welding............UW

electric arc spraying............EASP
flame spraying............FLSP
plasma spraying............PSP

chemical flux cutting............FOC
metal powder cutting............POC
oxyfuel gas cutting............OFC
—oxyacetylene cutting............OFC-A
—oxyhydrogen cutting............OFC-H
—oxynatural gas cutting............OFC-N
—oxypropane cutting............OFC-P
oxygen arc cutting............AOC
oxygen lance cutting............LOC

ARC WELDING (AW) · BRAZING (B) · OTHER WELDING · OXYFUEL GAS WELDING (OFW) · ADHESIVE BONDING (ABD) · ARC CUTTING (AC) · WELDING PROCESSES · ALLIED PROCESSES · THERMAL CUTTING (TC) · OTHER CUTTING · SOLID STATE WELDING (SSW) · SOLDERING (S) · RESISTANCE WELDING (RW) · THERMAL SPRAYING* (THSP) · OXYGEN CUTTING (OC)

*Sometimes a welding process.

AWS

FINISHING METHODS

LETTER	MECHANICAL METHOD	SYMBOL		
		FLAT	CONVEX	CONCAVE
C	CHIPPING			
H	HAMMERING			
G	GRINDING			
M	MACHINING			
R	ROLLING			
U	UNSPECIFIED			

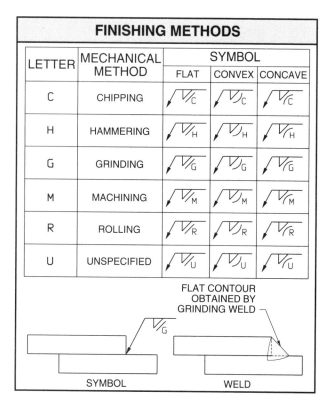

FLAT CONTOUR
OBTAINED BY
GRINDING WELD

SYMBOL   WELD

Figure 35-24. A special flat, concave, or convex symbol added to the weld symbol indicates that bead contour is important.

BACK OR BACKING WELDS

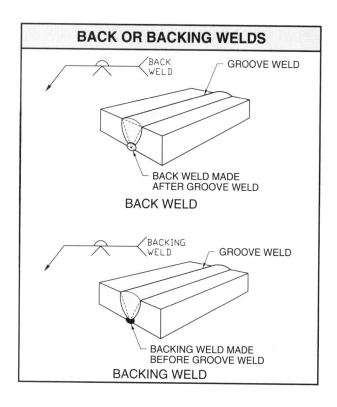

Figure 35-25. The back weld symbol is included opposite the weld symbol.

MELT-THROUGH WELDS

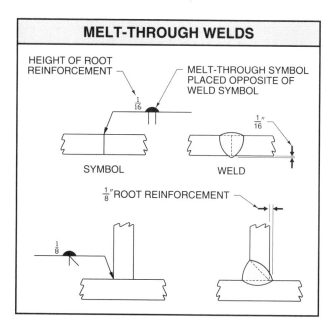

Figure 35-26. A special melt-through symbol, placed opposite the regular weld symbol, indicates that complete joint penetration of the weld through the material is required from one side only.

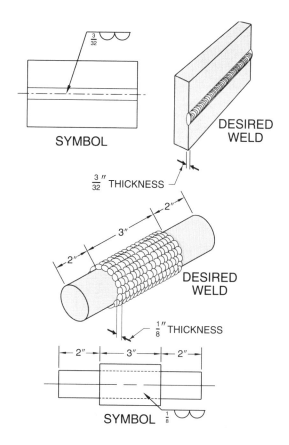

Figure 35-27. A surfacing weld symbol, with a dimension placed to the left, indicates surfaces to be built up by welding.

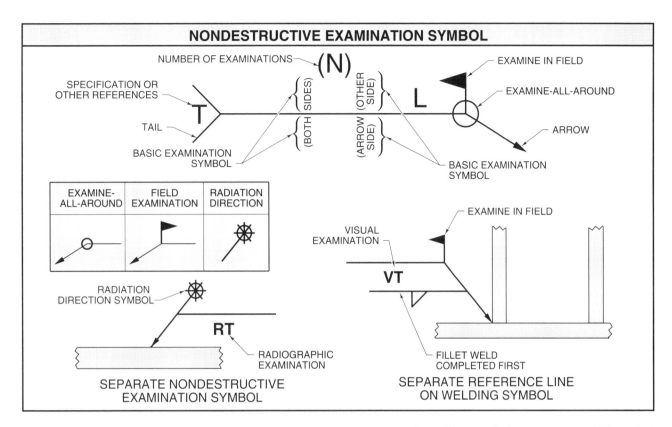

Figure 35-28. Nondestructive examination (NDE) symbols can be part of a welding symbol or a separate NDE symbol.

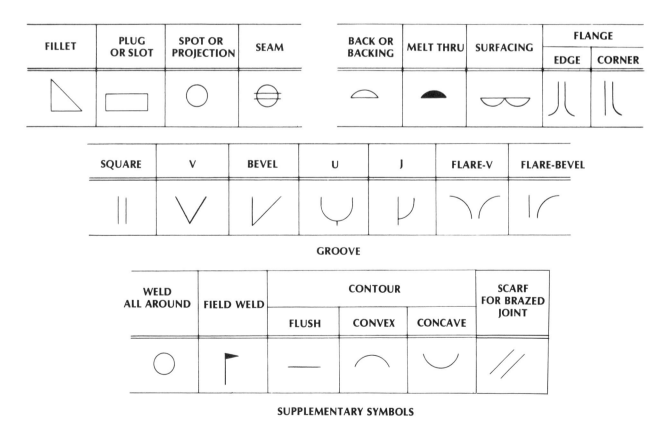

Figure 35-29. Weld symbols that are commonly used in basic arc and gas welding.

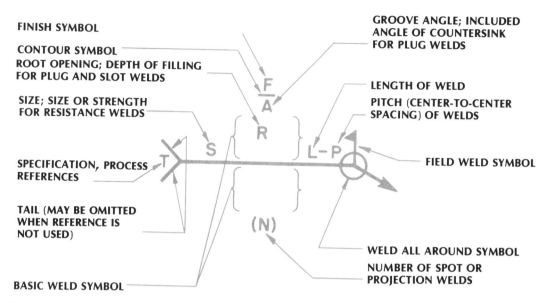

FINISH SYMBOL

CONTOUR SYMBOL

ROOT OPENING; DEPTH OF FILLING FOR PLUG AND SLOT WELDS

SIZE; SIZE OR STRENGTH FOR RESISTANCE WELDS

SPECIFICATION, PROCESS REFERENCES

TAIL (MAY BE OMITTED WHEN REFERENCE IS NOT USED)

BASIC WELD SYMBOL

GROOVE ANGLE; INCLUDED ANGLE OF COUNTERSINK FOR PLUG WELDS

LENGTH OF WELD

PITCH (CENTER-TO-CENTER SPACING) OF WELDS

FIELD WELD SYMBOL

WELD ALL AROUND SYMBOL

NUMBER OF SPOT OR PROJECTION WELDS

Figure 35-30. This master symbol combines the major elements discussed throughout the chapter.

QUESTIONS FOR STUDY AND DISCUSSION

1 What is meant by the arrow side of the weld symbol?

2 What is meant by the other side of the weld symbol?

3 Indicate the meaning of the following weld symbols.

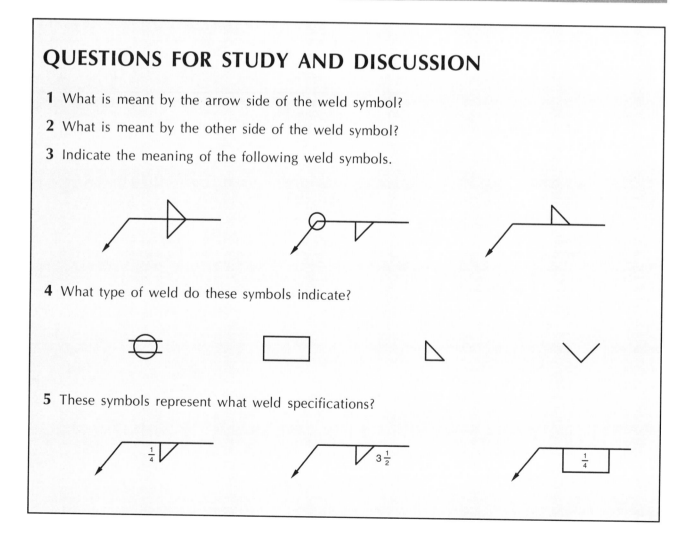

4 What type of weld do these symbols indicate?

5 These symbols represent what weld specifications?

6 These symbols represent what weld specifications?

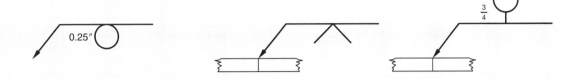

7 What do welds designated with the following symbols represent?

8 Draw completed weld symbols, including necessary information, to describe the welds.

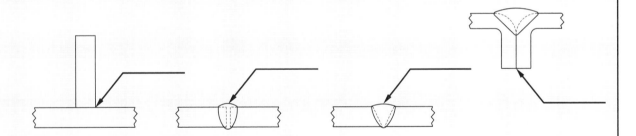

9 What do these symbols mean?

10 What do these symbols represent?

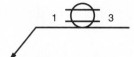   

11 What do these symbols mean?

   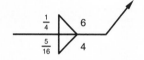   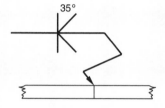

12 Identify the parts of the master weld symbol shown below.

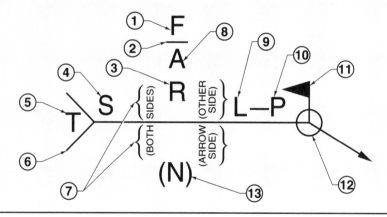

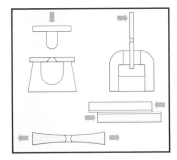

CERTIFICATION OF WELDERS

Supplementary Welding Data

To protect human lives and to make certain that products or structures fabricated by welding will function safely and effectively, certain safeguards are generally established to cover the quality of welding that must be done. The safeguards are usually stipulated in some document which clearly defines the nature and conditions of the required work.

The listing of specifications governing quality is not always as simple as it might appear because of the diversity of products and structures involving a great variety of welding processes and a wide gradation of welding skills. State and local laws sometimes specify in great detail what these requirements must be. At other times there may be only limited regulations, as established by the manufacturer, especially if high performance standards are not required. In all instances safeguards relate directly to the performance skills of the welder. Degrees of welding competences may be expressed in the form of codes, standards or specifications.

Code

A code consists of a set of regulations covering permissible materials, service limitations, fabrication, inspection, testing procedures and qualifications of welding operators. These codes have been established by a number of nationally recognized agencies such as:

American Welding Society (AWS)
American Society of Mechanical Engineers (ASME)
American Petroleum Institute (API)
American National Standards Institute (ANSI)

Codes usually deal with a specific field of work such as ship building, piping, boiler work, building construction, tanks, aircraft and many others. A typical example of what a code may include is the AWS Structural Welding Code D1.1. Some of the specifications listed in this code are:

Section I —General Provisions
Section II —Design of Welded Connections
Section III—Workmanship
Section IV—Techniques
Section V —Qualifications
Section VI—Inspection

A code is sometimes enacted into law and consequently is often the most enforceable of any safety regulation. A properly worded code is written in mandatory language, using imperative words such as *shall* or *must*. Inclusion of other words like *should* or *it is desirable* would raise the question of their enforceability and therefore are excluded.

Standards

Standards are specific regulations which cover the quality of a particular product to be fabricated by welding. By and large standards deal with work quality rather than work procedure. Thus standards may cover type of material to be used, test strength of required welds, characteristics of filler metals, preheating and postheating temperatures and other essentials which have a direct bearing on the quality of the finished product.

Standards are usually developed by the manufacturer and apply only to its own welding personnel and work or product to be produced. The stringency of the standards depends on the nature of the work or product and the demands of the consumer. Thus for some jobs the required competency of the welders may not be unduly high, whereas for other tasks the performance requirements could be extremely critical. On many occasions the established standards are based on a nationally recognized code and may be supplemented by other demands which are dictated by the parties for whom the work is to be done.

Specifications

Specifications are specific descriptions of fabricating procedure. Among other manufacturing instruction they include such welding data as location of welds, welding process to be used and method of testing the soundness of welds. These specifications are usually formulated by the design engineer and are included on production prints or on separate specification sheets. The nature of the specifications is governed by established standards and quality of work required. Welders must then produce the type of welds indicated by these specifications.

Certification Requirements

There is no one set of certification requirements dealing with all segments of the welding trade. Each area whether it involves welding pipe, aircraft parts, building structures, boilers or ships will have its own certification requirements.

Although certification requirements may vary somewhat from one classification of welders to another, in general, they all specify comparable tests which welders have to take. Most tests involve one or more of the following:

1. Tension test to establish the strength of a weld.
2. Guided bend tests to determine the ductility of the weld bead.
3. A fillet weld test to check the lack of fusion or cracks and proper weld contour.
4. A radiographic test or other testing techniques to detect porosity, cracks, inclusions and penetration.

These tests may have to be carried out in one or several positions, such as flat, horizontal, vertical, or overhead.

Qualifying tests are performed on specimens sometimes called *coupons*, of the same material to be used in the product involved. Each test weld is done on certain size pieces and then subjected to some destructive or nondestructive tests. See Chapter 34. A welder is considered qualified only if his test specimens meet the required standards of quality.

Certifying Agency

Generally manufacturers have their own testing programs for qualifying welders. In such a program someone in the plant is designated as the certifying agent. A welder then reports to the agent and performs certain welding tasks that will demonstrate the ability to meet the requirements of the established standards.

The results of the test are analyzed and if found satisfactory are so stated and recorded by the company. The welder is then said to be qualified to work on the contracted job. See Figure 36.1.

Certification Procedure

The question often arises among welders, "How can I become certified?" First of all one must re-

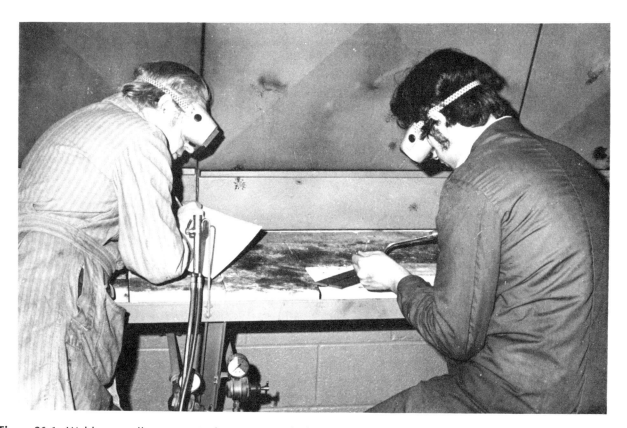

Figure 36-1. Welders usually are required to pass specified tests in order to be certified to work on a job. A certifying agent watches as a welder demonstrates the ability to meet the requirements of established standards.

member that there are no standardized certification requirements which will lead to a general permanent certificate. Secondly, certification of welders is assumed by the manufacturer, or supplier who has contracted to provide certain types of products or perform special kinds of welding jobs. Consequently, welders cannot go to a single agency and apply for permanent certification. Each time a welder applies for a welding job, or if already employed and is designated to work on a different welding assignment, a qualifying examination is taken. If the designated tests are passed, the welder is then certified for that particular job. If moved to some other assignment, then the welder must be certified again. For example, qualification to weld pipe does not automatically certify the welder to weld boilers. Each time the welder must be re-examined even though some of the actual tests may be similar to those that have been taken previously.

Qualification Requirements for Structural Welding*

The qualification requirements established for structural welding apply to any steel welded structure. They are applicable to groove and fillet welds performed with manual shielded metal-arc or gas metal-arc welding processes.

Weld profiles. Fillet and groove weld profiles must meet minimum AWS standards for size and shape. See Figure 36-2. Fillet weld convexity is based on measurement of weld face and leg size. A weld fillet gage can be used to measure weld profile. Groove weld reinforcement must not exceed ⅛″.

Weld sizes. Minimum fillet weld sizes and minimum groove weld sizes are shown in Figures 36-3 and 36-4.

Electrode types and sizes. Either E60xx or E70xx type electrodes should be used to weld test specimens with the shielded metal arc and E70S electrodes when welding with gas metal arc.

Maximum electrode diameters should be as follows:

Shielded Metal Arc 5/16″—flat position
¼″—horizontal position
5/32″—vertical and overhead positions

Gas Metal Arc 5/32″—flat and horizontal positions
3/32″—vertical position
5/64″—overhead position

Types of tests. The seven tests used for structural welding qualifying purpose are:
1. Reduce section—for tensile strength
2. Root bend—for weld soundness
3. Face bend—for weld soundness
4. Free bend—for weld ductility
5. All weld metal—for weld soundness
6. Impact—for weld toughness
7. Fillet weld—for weld toughness

The results of these tests not only serve to ascertain the competency of a welder but also to substantiate the limits of acceptability of a particular work standard. Thus evidence must often be provided to show that the prescribed strength, safety or endurance requirements for a product or job are being met. The type and number of tests administered depends on the nature of the job or product involved. The method of conducting these tests is described in Chapter 34.

Preparation of test specimens. The general requirement for preparing test specimens is that the length of the weld and dimensions of the base metal should be such as to produce sufficient material to adequately conduct the test. The removal of test specimens from a welded plate is shown in Figure 36-5 and 36-6. Actual shape and sizes of specimens are included in Chapter 34.

Joint design of groove weld test plate of limited thickness should consist of a single-vee groove with a 45° included angle and a ¼″ root opening with backing. The backing must be at least ⅜″ x 1″ for mechanical testing or ⅜″ x 3″ if x-ray is used for testing without removal of the backing. See Figure 36-7.

Position of test welds. The American Welding Society has established the following four qualification designations to represent levels of welding competencies:

1G—flat position for groove welds
2G—horizontal position for groove welds
3G—vertical position for groove welds
4G—overhead position for groove welds

Qualification Requirements for Pipe Welding*

The following three qualification levels have been established for pipe and tubing welding:

*For a more complete coverage see AWS D1.1.

*For a more complete coverage see AWS D10.9.

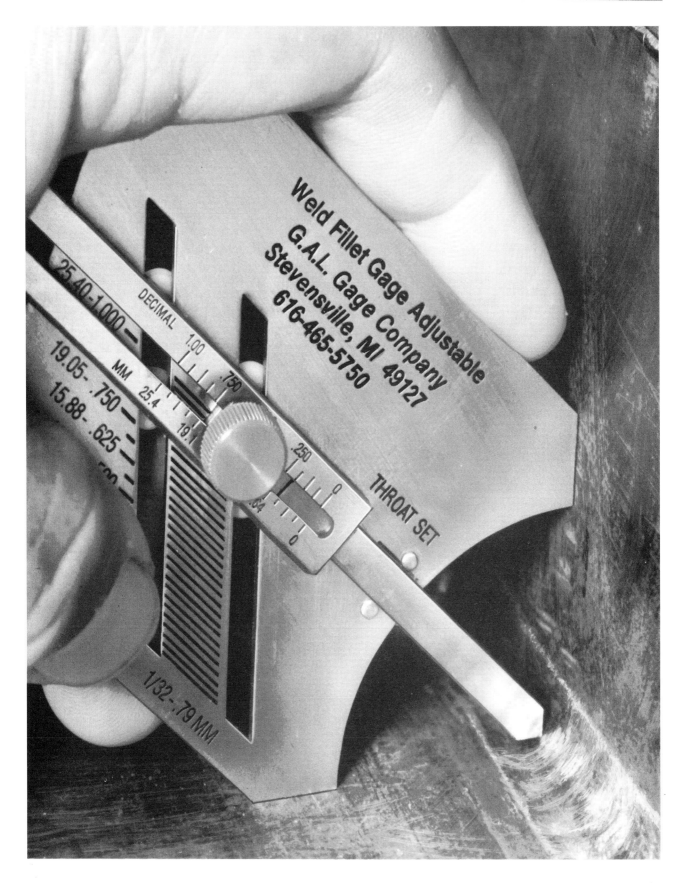

Figure 36-2. Acceptable fillet and groove weld profiles must have the proper size and shape to assure weld strength. (*G.A.L. Gage Company*)

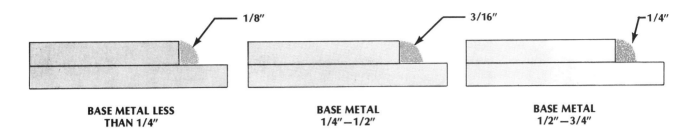

Figure 36-3. These illustrations show minimum fillet weld sizes.

AR-1—This level applies to systems where the highest level of quality is required. Use of the AR-1 level of quality is intended to provide the confidence required for certain lines which may be found in nuclear energy, space, high pressure or high temperature, chemical or gas systems.

AR-2—This level applies to systems where a high degree of weld quality is required. Use of the AR-2 level of quality is intended to provide the confidence necessary for some lines which may be found in nuclear energy,

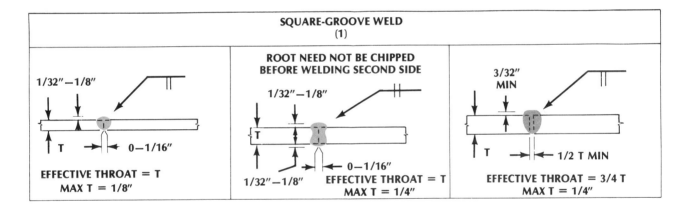

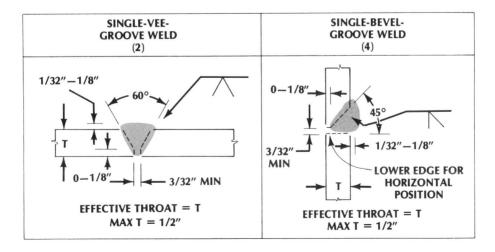

Figure 36-4. Minimum groove sizes for structural welding.

steam, water, petroleum, gas or chemical systems.

AR-3—This level applies to systems where a nominal degree of weld quality is required. Use of the AR-3 level of quality is intended to provide the confidence adequate for lines such as low-pressure heating, air conditioning, sanitary water and some gases or chemicals.

Types of tests. The two basic destructive qualifying tests for pipe and tubing are tensile and guided bend. See Chapter 34. Specimens for these two tests are to be removed from a welded test section. See Figure 36-8.

Test positions. Pipe and tubing systems often require welding in several positions. Four standard positions are defined as part of the procedure qualification. These are 1G, 2G, 5G and 6G. See Figure 36-9. Their applications for qualifying purposes are as follows:

1G—qualifies for welding in this position only

2G—qualifies for welding in the 1G and 2G positions

5G—qualifies for welding in all positions

6G—qualifies for welding in other than the four standard positions

DIRECTION OF ROLLING

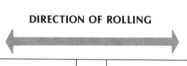

DISCARD	THIS PIECE
REDUCED-SECTION	TENSILE SPECIMEN
FREE-BEND ROOT-BEND	SPECIMEN
ALL WELD FACE-BEND	SPECIMEN
IMPACT ROOT-BEND	SPECIMEN
FACE-BEND	SPECIMEN
REDUCED-SECTION	TENSILE SPECIMEN
DISCARD	THIS PIECE

Figure 36-5. This chart shows the order of removal of test specimens from welded test plate 3/8" thick—procedure qualification.

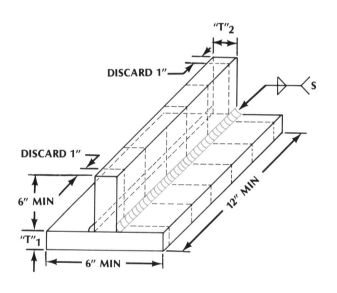

Weld size	T_1 Inches	T_2 Inches-Min.
3/16	1/2	3/16
1/4	3/4	1/4
5/16	1 1/2	5/16
3/8	2 1/4	3/8
1/2	3	1/2
5/8	3	5/8
3/4	3	3/4

Figure 36-6. A test of fillet weld soundness is required for procedure qualification.

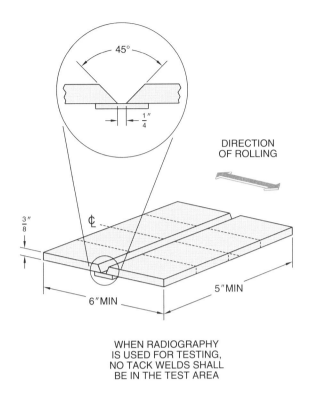

Figure 36-7. A test plate for limited thickness in all positions has a single groove with a 45° included angle and a ¼″ root opening with backing.

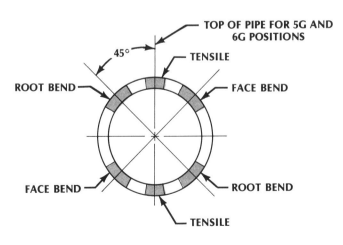

Figure 36-8. This chart shows the order of removal of test specimens from welded test pipe or tubing.

Testing and Certification Documentation

Weld testing is required to verify the strength of a given weld. Weld testing is conducted in accordance with Welding Procedure Specifications (WPS) established for the weld.

Weld testing can be conducted using destructive and nondestructive testing methods. *Destructive testing* is any type of testing that damages the test part (specimen). It is used primarily in the qualification of welders. *Nondestructive testing* is any type of testing that leaves the specimen undamaged. It is used to determine weld quality.

The WPS specifies the base metal, preheating and postheating specifications, and filler metal for a given weld. The Procedure Qualification Record (PQR) documents the specific welding variables and procedures used to complete an acceptable test weld and the results of the required weld tests.

Welder certification is not governed by a central agency. Efforts have been made to standardize a nationwide certification program. However, because of the many requirements of the welding industry, these efforts have not been successful. Individual companies and agencies have preferred to certify welders in their own controlled environments based on the applicable WPS. WPSs and PQRs must be filed for each weld procedure and welder qualified. This allows testing on specific fabrications produced and/or regulated by the agency.

The WPS specifies welds to be completed in a specific test weld position. Test weld positions are based on AWS groove, fillet, and stud weld test positions. In groove and fillet weld tests:

1 = flat
2 = horizontal
3 = vertical
4 = overhead

WPSs and PQRs are designed to record specific information relative to the welding procedure and the quality of the test results. Care should be taken when completing WPSs and PQRs. Always complete all applicable blanks in a neat and professional manner.

Information regarding test specifications and procedures are detailed in ANSI/AWS B2.1 *Standard for Welding Procedure and Performance Qualification.* As part of the procedure for qualification, WPS and PQR forms are completed which specify all welding directives and requirements (Welding Procedure Specifications), and specific welding variables and procedures used to complete an acceptable test weld (Procedure Qualification Record). See Appendix.

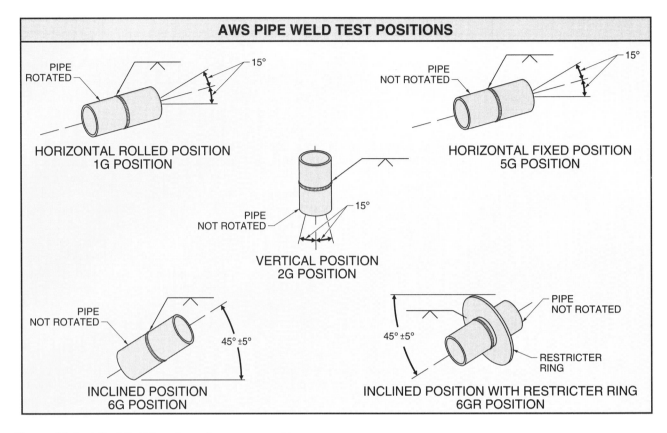

AWS PIPE WELD TEST POSITIONS

HORIZONTAL ROLLED POSITION
1G POSITION

PIPE ROTATED

15°

HORIZONTAL FIXED POSITION
5G POSITION

PIPE NOT ROTATED

15°

VERTICAL POSITION
2G POSITION

PIPE NOT ROTATED

15°

INCLINED POSITION
6G POSITION

PIPE NOT ROTATED

45° ±5°

INCLINED POSITION WITH RESTRICTER RING
6GR POSITION

PIPE NOT ROTATED

45° ±5°

RESTRICTER RING

Figure 36-9. 1G, 2G, 5G, 6G, and 6GR are welding test positions that are part of the procedure qualification for pipe and tubing.

QUESTIONS FOR STUDY AND DISCUSSION

1 Why are strict regulations often established for welders?

2 What is the difference between a *code, standards,* and *specifications?*

3 What are some of the agencies that prescribe welding codes?

4 Why are codes designated for only one category of work?

5 Who is generally responsible for certifying welders?

6 What are some of the tests which welders must take to be certified?

7 Why isn't it possible to apply for permanent certification status?

8 How often do you have to be certified?

APPENDIX

page

WELD JOINTS AND POSITIONS

	BUTT	LAP	T	EDGE	CORNER
FLAT					
HORIZONTAL					
VERTICAL					
OVERHEAD					

WELDING SYMBOL

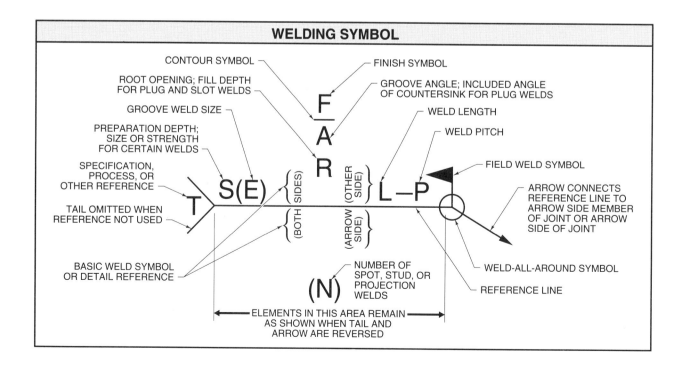

CONTOUR SYMBOL — FINISH SYMBOL

ROOT OPENING; FILL DEPTH FOR PLUG AND SLOT WELDS — GROOVE ANGLE; INCLUDED ANGLE OF COUNTERSINK FOR PLUG WELDS

GROOVE WELD SIZE — WELD LENGTH

PREPARATION DEPTH; SIZE OR STRENGTH FOR CERTAIN WELDS — WELD PITCH

SPECIFICATION, PROCESS, OR OTHER REFERENCE — FIELD WELD SYMBOL

TAIL OMITTED WHEN REFERENCE NOT USED — ARROW CONNECTS REFERENCE LINE TO ARROW SIDE MEMBER OF JOINT OR ARROW SIDE OF JOINT

F / A / R — S(E) — T — L–P

(SIDES) (OTHER SIDE) (ARROW SIDE) (BOTH)

BASIC WELD SYMBOL OR DETAIL REFERENCE

NUMBER OF SPOT, STUD, OR PROJECTION WELDS

(N)

WELD-ALL-AROUND SYMBOL

REFERENCE LINE

ELEMENTS IN THIS AREA REMAIN AS SHOWN WHEN TAIL AND ARROW ARE REVERSED

WELD JOINTS AND TYPES						
APPLICABLE WELDS	WELD SYMBOL	BUTT	LAP	T	EDGE	CORNER
SQUARE-GROOVE			—			
BEVEL-GROOVE						
V-GROOVE			—	—		
U-GROOVE			—	—		
J-GROOVE						
FLARE-BEVEL-GROOVE						
FLARE-V-GROOVE			—	—		
FILLET		—			—	
PLUG		—			—	
SLOT		—			—	
EDGE-FLANGE			—	—		—
CORNER-FLANGE		—	—	—		
SPOT		—			—	
PROJECTION		—			—	
SEAM		—				
BRAZE	BRAZE				—	

STRUCTURAL STEEL SHAPES

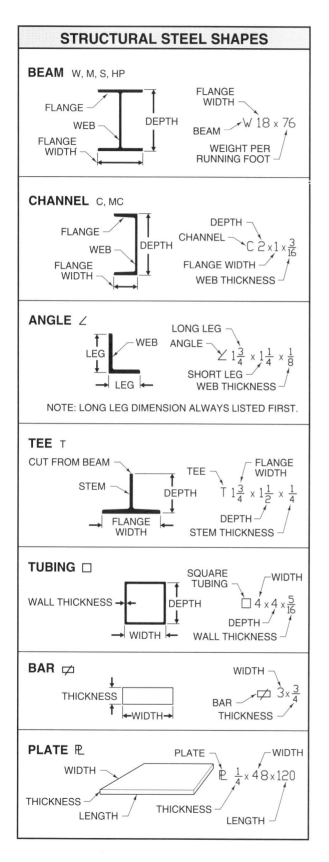

BEAM W, M, S, HP

FLANGE
WEB
FLANGE WIDTH
DEPTH
FLANGE WIDTH
BEAM
W 18 x 76
WEIGHT PER RUNNING FOOT

CHANNEL C, MC

FLANGE
WEB
FLANGE WIDTH
DEPTH
DEPTH
CHANNEL
C 2 x 1 x $\frac{3}{16}$
FLANGE WIDTH
WEB THICKNESS

ANGLE ∠

LEG
WEB
LEG
LONG LEG
ANGLE
∠ 1$\frac{3}{4}$ x 1$\frac{1}{4}$ x $\frac{1}{8}$
SHORT LEG
WEB THICKNESS

NOTE: LONG LEG DIMENSION ALWAYS LISTED FIRST.

TEE T

CUT FROM BEAM
STEM
DEPTH
FLANGE WIDTH
TEE
FLANGE WIDTH
T 1$\frac{3}{4}$ x 1$\frac{1}{2}$ x $\frac{1}{4}$
DEPTH
STEM THICKNESS

TUBING □

WALL THICKNESS
SQUARE TUBING
DEPTH
WIDTH
WIDTH
□ 4 x 4 x $\frac{5}{16}$
DEPTH
WALL THICKNESS

BAR

THICKNESS
WIDTH
WIDTH
3 x $\frac{3}{4}$
BAR
THICKNESS

PLATE ℙ

WIDTH
THICKNESS
LENGTH
PLATE
WIDTH
ℙ $\frac{1}{4}$ x 48 x 120
THICKNESS
LENGTH

GREEK SYMBOLS

Alpha	A α	Nu	N ν
Beta	B β	Xi	Ξ ξ
Gamma	Γ γ	Omicron	O o
Delta	Δ δ	Pi	Π π
Epsilon	E ε	Rho	P ρ
Zeta	Z ζ	Sigma	Σ σ
Eta	H η	Tau	T τ
Theta	Θ θ	Upsilon	Υ υ
Iota	I ι	Phi	Φ φ
Kappa	K κ	Chi	X χ
Lambda	Λ λ	Psi	Ψ ψ
Mu	M μ	Omega	Ω ω

METRIC/ENGLISH CONVERSIONS

LENGTH

Units	m	mm	ft	in.
1 m	1	1000	3.281	39.37
1 mm	0.001	1	$3.281(10^{-3})$	0.03937
1 ft	0.3048	304.8	1	12
1 in.	0.0254	25.4	0.00694	1

AREA

Units	m^2	cm^2	ft^2	in^2
1 m^2	1	104	10.764	1549.9
1 cm^2	10^{-4}	1	0.00108	0.155
1 ft^2	0.0929	929	1	144
1 in^2	$6.452(10^{-4})$	6.452	0.00694	1

VOLUME

Units	m^3	cm^3	ft^3	in^3
1 m^3	1	10^6	35.31	61.023
1 cm^3	107	10^4	$3.531(10^{-5})$	0.061023
1 ft^3	0.028317	1	1	1,728

UNIT PREFIXES

PREFIX	UNIT	SYMBOL	NUMBER
			Other larger multiples
Mega	Million	M	$1,000,000 = 10^6$
Kilo	Thousand	k	$1,000 = 10^3$
Hecto	Hundred	h	$100 = 10^2$
Deka	Ten	d	$10 = 10^1$
			Unit $1 = 10^0$
Deci	Tenth	d	$0.1 = 10^{-1}$
Centi	Hundreth	c	$0.01 = 10^{-2}$
Milli	Thousandth	m	$0.001 = 10^{-3}$
Micro	Millionth	μ	$0.000001 = 10^{-6}$
	Other smaller multiples		

MILLIMETER AND DECIMAL INCH EQUIVALENTS*

mm	in.	mm	in.	mm	in.	mm	in.	mm	in.	mm	in.
$\frac{1}{50}$ = .00079		$\frac{25}{50}$ = .01969		1 = .03937		26 = 1.02362		51 = 2.00787		76 = 2.99212	
$\frac{2}{50}$ = .00157		$\frac{26}{50}$ = .02047		2 = .07874		27 = 1.06299		52 = 2.04724		77 = 3.03149	
$\frac{3}{50}$ = .00236		$\frac{27}{50}$ = .02126		3 = .11811		28 = 1.10236		53 = 2.08661		78 = 3.07086	
$\frac{4}{50}$ = .00315		$\frac{28}{50}$ = .02205		4 = .15748		29 = 1.14173		54 = 2.12598		79 = 3.11023	
		$\frac{29}{50}$ = .02283									
$\frac{5}{50}$ = .00394		$\frac{30}{50}$ = .02362		5 = .19685		30 = 1.18110		55 = 2.16535		80 = 3.14960	
$\frac{6}{50}$ = .00472		$\frac{31}{50}$ = .02441		6 = .23622		31 = 1.22047		56 = 2.20472		81 = 3.18897	
$\frac{7}{50}$ = .00551		$\frac{32}{50}$ = .02520		7 = .27559		32 = 1.25984		57 = 2.24409		82 = 3.22834	
$\frac{8}{50}$ = .00630		$\frac{33}{50}$ = .02598		8 = .31496		33 = 1.29921		58 = 2.28346		83 = 3.26771	
$\frac{9}{50}$ = .00709		$\frac{34}{50}$ = .02677		9 = .35433		34 = 1.33858		59 = 2.32283		84 = 3.30708	
$\frac{10}{50}$ = .00787		$\frac{35}{50}$ = .02756		10 = .39370		35 = 1.37795		60 = 2.36220		85 = 3.34645	
$\frac{11}{50}$ = .00866		$\frac{36}{50}$ = .02835		11 = .43307		36 = 1.41732		61 = 2.40157		86 = 3.38582	
$\frac{12}{50}$ = .00945		$\frac{37}{50}$ = .02913		12 = .47244		37 = 1.45669		62 = 2.44094		87 = 3.42519	
$\frac{13}{50}$ = .01024		$\frac{38}{50}$ = .02992		13 = .51181		38 = 1.49606		63 = 2.48031		88 = 3.46456	
$\frac{14}{50}$ = .01102		$\frac{39}{50}$ = .03071		14 = .55118		39 = 1.53543		64 = 2.51968		89 = 3.50393	
$\frac{15}{50}$ = .01181		$\frac{40}{50}$ = .03150		15 = .59055		40 = 1.57480		65 = 2.55905		90 = 3.54330	
$\frac{16}{50}$ = .01260		$\frac{41}{50}$ = .03228		16 = .62992		41 = 1.61417		66 = 2.59842		91 = 3.58267	
$\frac{17}{50}$ = .01339		$\frac{42}{50}$ = .03307		17 = .66929		42 = 1.65354		67 = 2.63779		92 = 3.62204	
$\frac{18}{50}$ = .01417		$\frac{43}{50}$ = .03386		18 = .70866		43 = 1.69291		68 = 2.67716		93 = 3.66141	
$\frac{19}{50}$ = .01496		$\frac{44}{50}$ = .03465		19 = .74803		44 = 1.73228		69 = 2.71653		94 = 3.70078	
$\frac{20}{50}$ = .01575		$\frac{45}{50}$ = .03543		20 = .78740		45 = 1.77165		70 = 2.75590		95 = 3.74015	
$\frac{21}{50}$ = .01654		$\frac{46}{50}$ = .03622		21 = .82677		46 = 1.81102		71 = 2.79527		96 = 3.77952	
$\frac{22}{50}$ = .01732		$\frac{47}{50}$ = .03701		22 = .86614		47 = 1.85039		72 = 2.83464		97 = 3.81889	
$\frac{23}{50}$ = .01811		$\frac{48}{50}$ = .03780		23 = .90551		48 = 1.88976		73 = 2.87401		98 = 3.85826	
$\frac{24}{50}$ = .01890		$\frac{49}{50}$ = .03858		24 = .94488		49 = 1.92913		74 = 2.91338		99 = 3.89763	
				25 = .98425		50 = 1.96850		75 = 2.95275		100 = 3.93700	

*Based on $\frac{1}{100}$ mm = .003973″ 10 mm = 1 centmeter = 0.3937″ 25.4 mm = 1″
10 cm = 1 decimeter = 3.937″ 10 dm = 1 meter = 39.37″

DECIMAL EQUIVALENTS OF AN INCH

Fraction	Decimal	Fraction	Decimal	Fraction	Decimal	Fraction	Decimal
$\frac{1}{64}$	0.015625	$\frac{17}{64}$	0.265625	$\frac{33}{64}$	0.515625	$\frac{47}{64}$	0.765625
$\frac{1}{32}$	0.03125	$\frac{9}{32}$	0.28125	$\frac{17}{32}$	0.53125	$\frac{25}{32}$	0.78125
$\frac{3}{64}$	0.046875	$\frac{19}{64}$	0.296875	$\frac{35}{64}$	0.546875	$\frac{51}{64}$	0.796875
$\frac{1}{16}$	0.0625	$\frac{5}{16}$	0.3125	$\frac{9}{16}$	0.5625	$\frac{13}{16}$	0.8125
$\frac{5}{64}$	0.078125	$\frac{21}{64}$	0.328125	$\frac{37}{64}$	0.578125	$\frac{53}{64}$	0.828125
$\frac{3}{32}$	0.09375	$\frac{11}{32}$	0.34375	$\frac{19}{32}$	0.59375	$\frac{27}{32}$	0.84375
$\frac{7}{64}$	0.109375	$\frac{23}{64}$	0.359375	$\frac{39}{64}$	0.609375	$\frac{55}{64}$	0.859375
$\frac{1}{8}$	0.125	$\frac{3}{8}$	0.375	$\frac{5}{8}$	0.625	$\frac{7}{8}$	0.875
$\frac{9}{64}$	0.140625	$\frac{25}{64}$	0.390625	$\frac{41}{64}$	0.640625	$\frac{57}{64}$	0.890625
$\frac{5}{32}$	0.15625	$\frac{13}{32}$	0.40625	$\frac{21}{32}$	0.65625	$\frac{29}{32}$	0.90625
$\frac{11}{64}$	0.171875	$\frac{27}{64}$	0.421875	$\frac{43}{64}$	0.671875	$\frac{59}{64}$	0.921875
$\frac{3}{16}$	0.1875	$\frac{7}{16}$	0.4375	$\frac{11}{16}$	0.6875	$\frac{15}{16}$	0.9375
$\frac{13}{64}$	0.203125	$\frac{29}{64}$	0.453125	$\frac{45}{64}$	0.703125	$\frac{61}{64}$	0.953125
$\frac{7}{32}$	0.21875	$\frac{15}{32}$	0.46875	$\frac{23}{32}$	0.71875	$\frac{31}{32}$	0.96875
$\frac{15}{64}$	0.234375	$\frac{31}{64}$	0.484375	$\frac{47}{64}$	0.734375	$\frac{63}{64}$	0.984375

PREFERRED METRIC SCREW THREADS

COARSE (GENERAL PURPOSE)				FINE			
Nominal Size and Thd Pitch	Tap Drill Diameter*	Nominal Size and Thd Pitch	Tap Drill Diameter*	Nominal Size and Thd Pitch	Tap Drill Diameter*	Nominal Size and Thd Pitch	Tap Drill Diameter*
M1.6 x 0.35	1.25	M20 x 2.5	17.5	—	—	M20 x 1.5	18.5
M2 x 0.4	1.6	M24 x 3	21.0	—	—	M24 x 2	22.0
M2.5 x 0.45	2.05	M30 x 3.5	26.5	—	—	M30 x 2	28.0
M3 x 0.5	2.5	M36 x 4	32.0	—	—	M36 x 2	33.0
M4 x 0.7	3.3	M42 x 4.5	37.5	—	—	M42 x 2	39.0
M5 x 0.8	4.2	M48 x 5	43.0	—	—	M48 x 2	45.0
M6 x 1	5.0	M56 x 5.5	50.5	—	—	M56 x 2	52.0
M8 x 1.25	6.8	M64 x 6	58.0	M8 x 1	7.0	M64 x 2	60.0
M10 x 1.5	8.5	M72 x 6	66.0	M10 x 1.25	8.75	M72 x 2	68.0
M12 x 1.75	10.30	M80 x 6	74.0	M12 x 1.25	10.5	M80 x 2	76.0
M16 x 2	14.00	M90 x 6	84.0	M16 x 1.5	14.5	M90 x 2	86.0
—	—	M100 x 6	94.0	—	—	M100 x 2	96.0

*in mm

STANDARD SERIES THREADS — GRADED PITCHES

NOMINAL DIAMETER	UNC		UNF		UNEF	
	TPI	TAP DRILL	TPI	TAP DRILL	TPI	TAP DRILL
0 (.0600)			80	$3/64$		
1 (.0730)	64	No. 53	72	No. 53		
2 (.0860)	56	No. 50	64	No. 50		
3 (.0990)	48	No. 47	56	No. 45		
4 (.1120)	40	No. 43	48	No. 42		
5 (.1250)	40	No. 38	44	No. 37		
6 (.1380)	32	No. 36	40	No. 33		
8 (.1640)	32	No. 29	36	No. 29		
10 (.1900)	24	No. 25	32	No. 21		
12 (.2160)	24	No. 16	28	No. 14	32	No.13
$1/4$ (.2500)	20	No. 7	28	No. 3	32	$7/32$
$5/16$ (.3125)	18	F	24	I	32	$9/32$
$3/8$ (.3750)	16	$5/16$	24	Q	32	$11/32$
$7/16$ (.4375)	14	U	20	$25/64$	28	$13/32$
$1/2$ (.5000)	13	$27/64$	20	$29/64$	28	$15/32$
$9/16$ (.5625)	12	$31/64$	18	$33/64$	24	$33/64$
$5/8$ (.6250)	11	$17/32$	18	$37/64$	24	$37/64$
$11/16$ (.6875)					24	$41/64$
$3/4$ (.7500)	10	$21/32$	16	$11/16$	20	$45/64$
$13/16$ (.8125)					20	$49/64$
$7/8$ (.8750)	9	$49/64$	14	$13/16$	20	$53/64$
$15/16$ (.9375)					20	$57/64$
1 (1.000)	8	$7/8$	12	$59/64$	20	$61/64$

DRILLED HOLES

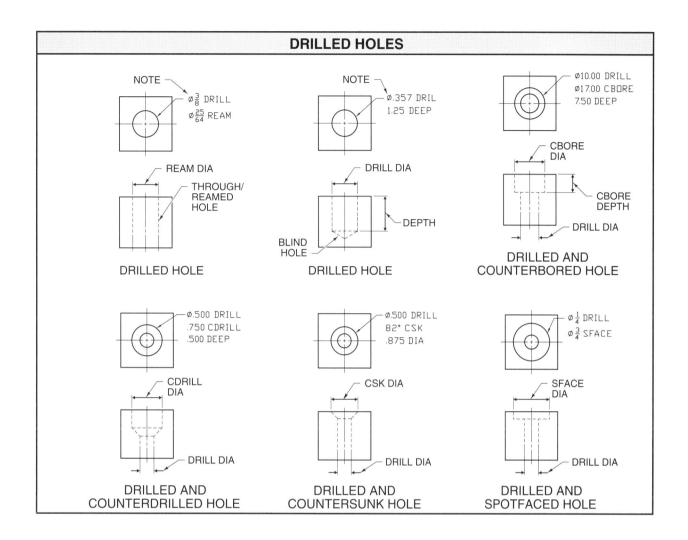

DRILL SIZES*

SIZE	DRILL DIAMETER	SIZE	DRILL DIAMETER	SIZE	DRILL DIAMETER	SIZE	DRILL DIAMETER	SIZE	DRILL DIAMETER	SIZE	DRILL DIAMETER
1	.2280	17	.1730	33	.1130	49	.0730	65	.0350	81	.0130
2	.2210	18	.1695	34	.1110	50	.0700	66	.0330	82	.0125
3	.2130	19	.1660	35	.1100	51	.0670	67	.0320	83	.0120
4	.2090	20	.1610	36	.1065	52	.0635	68	.0310	84	.0115
5	.2055	21	.1590	37	.1040	53	.0595	69	.0292	85	.0110
6	.2040	22	.1570	38	.1015	54	.0550	70	.0280	86	.0105
7	.2010	23	.1540	39	.0995	55	.0520	71	.0260	87	.0100
8	.1990	24	.1520	40	.0980	56	.0465	72	.0250	88	.0095
9	.1960	25	.1495	41	.0960	57	.0430	73	.0240	89	.0091
10	.1935	26	.1470	42	.0935	58	.0420	74	.0225	90	.0087
11	.1910	27	.1440	43	.0890	59	.0410	75	.0210	91	.0083
12	.1890	28	.1405	44	.0860	60	.0400	76	.0200	92	.0079
13	.1850	29	.1360	45	.0820	61	.0390	77	.0180	93	.0075
14	.1820	30	.1285	46	.0810	62	.0380	78	.0160	94	.0071
15	.1800	31	.1200	47	.0785	63	.0370	79	.0145	95	.0067
16	.1770	32	.1160	48	.0760	64	.0360	80	.0135	96	.0063

*in in.

LETTER SIZES*

SIZE	DRILL DIAMETER	SIZE	DRILL DIAMETER	SIZE	DRILL DIAMETER	SIZE	DRILL DIAMETER	SIZE	DRILL DIAMETER
A	.234	G	.261	L	.290	Q	.332	V	.377
B	.238	H	.266	M	.295	R	.339	W	.386
C	.242	I	.272	N	.302	S	.348	X	.397
D	.246	J	.277	O	.316	T	.358	Y	.404
E	.250	K	.281	P	.323	U	.368	Z	.413
F	.257								

*in in.

METRIC DRILL SIZES

DRILL DIAMETER						DRILL DIAMETER					
mm	in.	mm	in.	mm	in.	mm	in.	mm	in.	mm	in.
.40	.0157	1.95	.0768	4.70	.1850	8.00	.3150	13.20	.5197	25.50	1.0039
.42	.0165	2.00	.0787	4.80	.1890	8.10	.3189	13.50	.5315	26.00	1.0236
.45	.0177	2.05	.0807	4.90	.1929	8.20	.3228	13.80	.5433	26.50	1.0433
.48	.0189	2.10	.0827	5.00	.1969	8.30	.3268	14.00	.5512	27.00	1.0630
.50	.0197	2.15	.0846	5.10	.2008	8.40	.3307	14.25	.5610	27.50	1.0827
.55	.0217	2.20	.0866	5.20	.2047	8.50	.3346	14.50	.5709	28.00	1.1024
.60	.0236	2.25	.0886	5.30	.2087	8.60	.3386	14.75	.5807	28.50	1.1220
.65	.0256	2.30	.0906	5.40	.2126	8.70	.3425	15.00	.5906	29.00	1.1417
.70	.0276	2.35	.0925	5.50	.2165	8.80	.3465	15.25	.6004	29.50	1.1614
.75	.0295	2.40	.0945	5.60	.2205	8.90	.3504	15.50	.6102	30.00	1.1811
.80	.0315	2.45	.0965	5.70	.2244	9.00	.3543	15.75	.6201	30.50	1.2008
.85	.0335	2.50	.0984	5.80	.2283	9.10	.3583	16.00	.6299	31.00	1.2205
.90	.0354	2.60	.1024	5.90	.2323	9.20	.3622	16.25	.6398	31.50	1.2402
.95	.0374	2.70	.1063	6.00	.2362	9.30	.3661	16.50	.6496	32.00	1.2598
1.00	.0394	2.80	.1102	6.10	.2402	9.40	.3701	16.75	.6594	32.50	1.2795
1.05	.0413	2.90	.1142	6.20	.2441	9.50	.3740	17.00	.6693	33.00	1.2992
1.10	.0433	3.00	.1181	6.30	.2480	9.60	.3780	17.25	.6791	33.50	1.3189
1.15	.0453	3.10	.1220	6.40	.2520	9.70	.3819	17.50	.6890	34.00	1.3386
1.20	.0472	3.20	.1260	6.50	.2559	9.80	.3858	18.00	.7087	34.50	1.3583
1.25	.0492	3.30	.1299	6.60	.2598	9.90	.3898	18.50	.7283	35.00	1.3780
1.30	.0512	3.40	.1339	6.70	.2638	10.00	.3937	19.00	.7480	35.50	1.3976
1.35	.0531	3.50	.1378	6.80	.2677	10.20	.4016	19.50	.7677	36.00	1.4173
1.40	.0551	3.60	.1417	6.90	.2717	10.50	.4134	20.00	.7874	36.50	1.4370
1.45	.0571	3.70	.1457	7.00	.2756	10.80	.4252	20.50	.8071	37.00	1.4567
1.50	.0591	3.80	.1496	7.10	.2795	11.00	.4331	21.00	.8268	37.50	1.4764
1.55	.0610	3.90	.1535	7.20	.2835	11.20	.4409	21.50	.8465	38.00	1.4961
1.60	.0630	4.00	.1575	7.30	.2874	11.50	.4528	22.00	.8661	40.00	1.5748
1.65	.0650	4.10	.1614	7.40	.2913	11.80	.4646	22.50	.8858	42.00	1.6535
1.70	.0669	4.20	.1654	7.50	.2953	12.00	.4724	23.00	.9055	44.00	1.7323
1.75	.0689	4.30	.1693	7.60	.2992	12.20	.4803	23.50	.9252	46.00	1.8110
1.80	.0709	4.40	.1732	7.70	.3031	12.50	.4921	24.00	.9449	48.00	1.8898
1.85	.0728	4.50	.1772	7.80	.3071	12.80	.5039	24.50	.9646	50.00	1.9685
1.90	.0748	4.60	.1811	7.90	.3110	13.00	.5118	25.00	.9843		

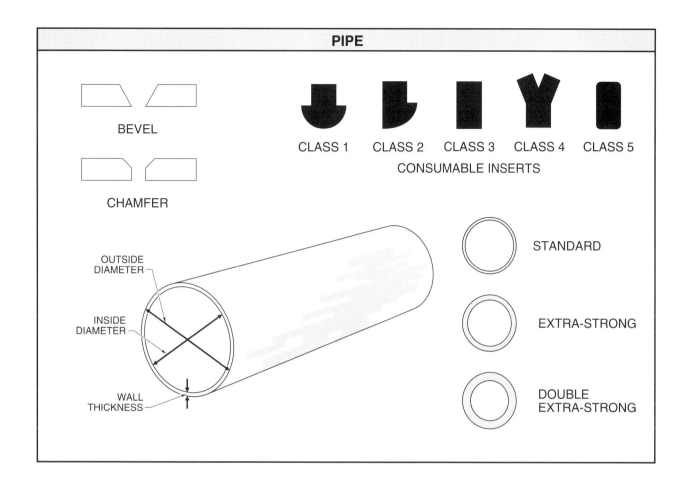

NOMINAL ID (IN.)	OD (BW GAUGE)	INSIDE DIAMETER (BW GAUGE)			NOMINAL WALL THICKNESS		
		STD	XS	XXS	SCHEDULE 40	SCHEDULE 60	SCHEDULE 80
⅛	0.405	0.269	0.215		0.068	0.095	
¼	0.540	0.364	0.302		0.088	0.119	
⅜	0.675	0.493	0.423		0.091	0.126	
½	0.840	0.622	0.546	0.252	0.109	0.147	0.294
¾	1.050	0.824	0.742	0.434	0.113	0.154	0.308
1	1.315	1.049	0.957	0.599	0.133	0.179	0.358
1¼	1.660	1.380	1.278	0.896	0.140	0.191	0.382
1½	1.900	1.610	1.500	1.100	0.145	0.200	0.400
2	2.375	2.067	1.939	1.503	0.154	0.218	0.436
2½	2.875	2.469	2.323	1.771	0.203	0.276	0.552
3	3.500	3.068	2.900	2.300	0.216	0.300	0.600
3½	4.000	3.548	3.364	2.728	0.226	0.318	
4	4.500	4.026	3.826	3.152	0.237	0.337	0.674
5	5.563	5.047	4.813	4.063	0.258	0.375	0.750
6	6.625	6.065	5.761	4.897	0.280	0.432	0.864
8	8.625	7.981	7.625	6.875	0.322	0.500	0.875
10	10.750	10.020	9.750	8.750	0.365	0.500	
12	12.750	12.000	11.750	10.750	0.406	0.500	

PIPE FITTINGS AND VALVES

	WELDED*	FLANGED	SCREWED		WELDED*	FLANGED	SCREWED		WELDED*	FLANGED	SCREWED
BUSHING				REDUCING FLANGE				AUTOMATIC BY-PASS VALVE			
CAP				BULL PLUG				AUTOMATIC REDUCING VALVE			
REDUCING CROSS				PIPE PLUG				STRAIGHT CHECK VALVE			
STRAIGHT-SIZE CROSS				CONCENTRIC REDUCER				COCK			
CROSSOVER				ECCENTRIC REDUCER				DIAPHRAGM VALVE			
45° ELBOW				SLEEVE				FLOAT VALVE			
90° ELBOW				STRAIGHT-SIZE TEE				GATE VALVE			
ELBOW—TURNED DOWN				TEE—OUTLET UP				MOTOR-OPERATED GATE VALVE			
ELBOW—TURNED UP				TEE—OUTLET DOWN				GLOBE VALVE			
BASE ELBOW				DOUBLE-SWEEP TEE				MOTOR-OPERATED GLOBE VALVE			
DOUBLE-BRANCH ELBOW				REDUCING TEE				ANGLE HOSE VALVE			
LONG-RADIUS ELBOW				SINGLE-SWEEP TEE				GATE VALVE			
REDUCING ELBOW				SIDE OUTLET TEE—OUTLET DOWN				GLOBE VALVE			
SIDE OUTLET ELBOW—OUTLET DOWN				SIDE OUTLET TEE—OUTLET UP				LOCKSHIELD VALVE			
SIDE OUTLET ELBOW—OUTLET UP				UNION				QUICK-OPENING VALVE			
STREET ELBOW				ANGLE CHECK VALVE				SAFETY VALVE			
CONNECTING PIPE JOINT				ANGLE GATE VALVE—ELEVATION				GOVERNOR-OPERATED AUTOMATIC VALVE			
EXPANSION JOINT				ANGLE GATE VALVE—PLAN							
LATERAL				ANGLE GLOBE VALVE—ELEVATION							
ORIFICE FLANGE				ANGLE GLOBE VALVE—PLAN							

* A • may be used instead of the "X" to represent a welded joint.

The American Society of Mechanical Engineers

ELECTRODE CHARACTERISTICS

TYPE	AWS CLASS	CURRENT TYPE	WELDING POSITION	WELD RESULTS	ELECTRODE GROUP
Mild Steel	E-6010	DCR (DC-EP)	All	Deep penetration, flat beads	Fast-freeze
	E-6011	AC			
	E-6012	DCS (DC-EN)	All	Shallow penetration, good bead contour, minimum spatter, for poor fit-up	Fill-freeze
	E-6013	AC (DC-EN) (DC-EP)			
	E-6020	DCR (DC-EP) DCS (DC-EN) AC	F, H	High deposition, deep groove single pass welds	Fast-fill
Iron Powder	E-6027	DCR (DC-EP) DCS (DC-EN) AC	F, H	High deposition, deep penetration	Fast-fill
	E-7014	DCR (DC-EP) DCS (DC-EN) AC	All	Low penetration, high speed	Fill-freeze
	E-7024	DCR (DC-EP) DCS (DC-EN) AC	F, H	High deposition, single and multiple passes	Fast-fill
Low Hydrogen	E-7016	DCR (DC-EP)	All	Welding of high-sulfur and high-carbon steels that tend to develop porosity and crack under weld bead	Fill-freeze
	E-7018	DCR (DC-EP) AC	All		
	E-7028	DCR (DC-EP) AC	F, H		Fast-fill

DCR — direct current reverse polarity
DCS — direct current straight polarity
AC — alternating current

F — flat, H — horizontal
DC-EP — direct current electrode positive
DC-EN — direct current electrode negative

MILD STEEL ELECTRODE SELECTION CHART*

	ELECTRODE CLASS										
	E6010	E6011	E6012	E6013	E7014	E7016	E7018	E6020	E7024	E6027	E7028
Groove butt welds, flat (< ¼″)	5	5	3	8	9	7	9	10	9	10	10
Groove butt welds, all positions (< ¼″)	10	9	5	8	6	7	6	**	**	**	**
Fillet welds, flat or horizontal	2	3	8	7	9	5	9	10	10	9	9
Fillet welds, all positions	10	9	6	7	7	8	6	**	**	**	**
Current***	DCR	AC DCR	DCS AC	AC DC	DC AC	DCR AC	DCR AC	DC AC	DC AC	AC DC	DCR AC
Thin material (¼″)	5	7	8	9	8	2	2	**	7	**	**
Heavy plate or highly restrained joint	8	8	8	8	8	10	9	8	7	8	9
High-sulfur or off-analysis steel	**	**	5	3	3	9	9	**	5	**	9
Deposition rate	4	4	5	5	6	4	6	6	10	10	8
Depth of penetration	10	9	6	5	6	7	7	8	4	8	7
Appearance, undercutting	6	6	8	9	9	7	10	9	10	10	10
Soundness	6	6	3	5	7	10	9	9	8	9	9
Ductility	6	7	4	5	6	10	10	10	5	10	10
Low-temperature impact strength	8	8	4	5	8	10	10	8	9	9	10
Low spatter loss	1	2	6	7	9	6	8	9	10	10	9
Poor fit-up	6	7	10	8	9	4	4	**	8	**	4
Welder appeal	7	6	8	9	10	6	8	9	10	10	9
Slag removal	9	8	6	8	8	4	7	9	9	9	8

*AWS — Rating is on a comparative basis of same size electrodes with 10 as the highest value. Ratings may change with size.

**Not recommended.

***DCR — direct current reverse, electrode positive; DCS — direct current straight, electrode negative; AC — alternating current; DC — direct current, either polarity.

AISI-SAE DESIGNATION SYSTEM

Numbers and Digits	Type of steel and/or nominal alloy content

Carbon steels

10xx — Plain carbon (1% Mn max)

11xx — Resulfurized

12xx — Resulfurized and rephosphorized

15xx — Plain carbon (1.00% Mn to 1.65% Mn max)

Manganese steels

13xx — 1.75% Mn

Nickel steels

23xx — 3.5% Ni

25xx — 5% Ni

Nickel-chromium steels

31xx — 1.25% Ni; .65% Cr and .80% Cr

32xx — 1.75% Ni; 1.07% Cr

33xx — 3.50% Ni; 1.50% Cr and 1.57% Cr

34xx — 3.00% Ni; .77% Cr

Molybdenum steels

40xx — .20% Mo and .25% Mo

44xx — .40% Mo and .52% Mo

Chromium-molybdenum steels

41xx — .50% Cr, .80% Cr, and .95% Cr; .12% Mo, .20% Mo, .25% Mo, and .30% Mo

Nickel-chromium-molybdenum steels

43xx — 1.82% Ni; .50% Cr and .80% Cr; .25% Mo

43BVxx — 1.82% Ni; .50% Cr; .12% Mo and .25% Mo; .03% V min

47xx — 1.05% Ni; .45% Cr; .20% Mo and .35% Mo

81xx — .30% Ni; .40% Cr; .12% Mo

86xx — .55% Ni; .50% Cr; .20% Mo

87xx — .55% Ni; .50% Cr; .25% Mo

88xx — .55% Ni; .50% Cr; .35% Mo

93xx — 3.25% Ni; 1.20% Cr; .12% Mo

94xx — .45% Ni; .40% Cr; .12% Mo

97xx — .55% Ni; .20% Cr; .20% Mo

98xx — 1.00% Ni; .80% Cr; .25% Mo

Nickel-molybdenum steels

46xx — .85% Ni and 1.82% Ni; .20% Mo and .25% Mo

48xx — 3.50% Ni; .25% Mo

Chromium steels

50xx — .27% Cr, .40% Cr, .50% Cr, and .65% Cr

51xx — .80% Cr, .87% Cr, .92% Cr, .95% Cr, 1.00% Cr, and 1.05% Cr

Chromium steels

50xxx — .50% Cr ⎫

51xxx — 1.02% Cr ⎬ C 1.00% min

52xxx — 1.45% Cr ⎭

Chromium-vanadium steels

61xx — .60% Cr, .80% Cr, and .95% Cr; .10% V and .15% V min

Tungsten-chromium steel

72xx — 1.75% W; 0.75% Cr

Silicon-manganese steels

92xx — 1.40% Si and 2.00% Si; .65% Mn, .82% Mn, and .85% Mn; 0% Cr and .65% Cr

High-strength low-alloy steels

9xx — Various SAE grades

Boron steels

xxBxx — B denotes boron steel

Leaded steels

xxLxx — L denotes leaded steel

NDE EXAMINATION METHODS

METHOD	LETTER DESIGNATION
Acoustic emmision	AET
Electromagnetic	ET
Leak	LT
Magnetic particle	MT
Neutron radiographic	NRT
Penetrant	PT
Proof	PRT
Radiographic	RT
Ultrasonic	UT
Visual	VT

NDE EXAMINATION SYMBOL

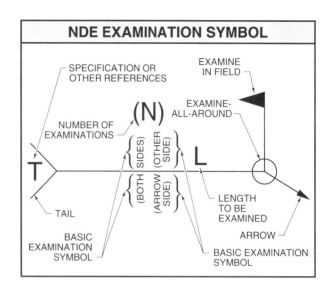

SPARK TEST

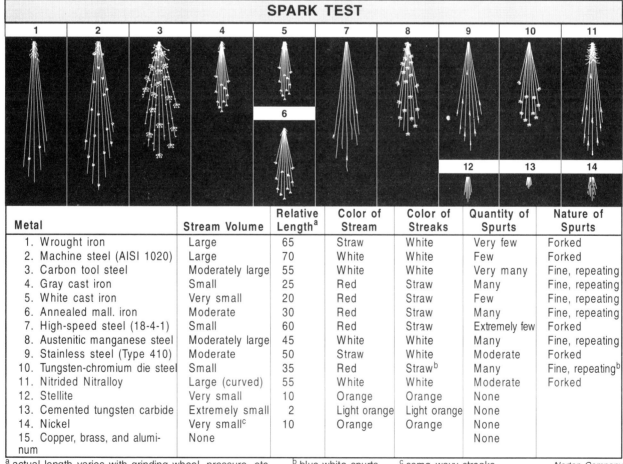

Metal	Stream Volume	Relative Length[a]	Color of Stream	Color of Streaks	Quantity of Spurts	Nature of Spurts
1. Wrought iron	Large	65	Straw	White	Very few	Forked
2. Machine steel (AISI 1020)	Large	70	White	White	Few	Forked
3. Carbon tool steel	Moderately large	55	White	White	Very many	Fine, repeating
4. Gray cast iron	Small	25	Red	Straw	Many	Fine, repeating
5. White cast iron	Very small	20	Red	Straw	Few	Fine, repeating
6. Annealed mall. iron	Moderate	30	Red	Straw	Many	Fine, repeating
7. High-speed steel (18-4-1)	Small	60	Red	Straw	Extremely few	Forked
8. Austenitic manganese steel	Moderately large	45	White	White	Many	Fine, repeating
9. Stainless steel (Type 410)	Moderate	50	Straw	White	Moderate	Forked
10. Tungsten-chromium die steel	Small	35	Red	Straw[b]	Many	Fine, repeating[b]
11. Nitrided Nitralloy	Large (curved)	55	White	White	Moderate	Forked
12. Stellite	Very small	10	Orange	Orange	None	
13. Cemented tungsten carbide	Extremely small	2	Light orange	Light orange	None	
14. Nickel	Very small[c]	10	Orange	Orange	None	
15. Copper, brass, and aluminum	None				None	

[a] actual length varies with grinding wheel, pressure, etc.   [b] blue-white spurts   [c] some wavy streaks   *Norton Company*

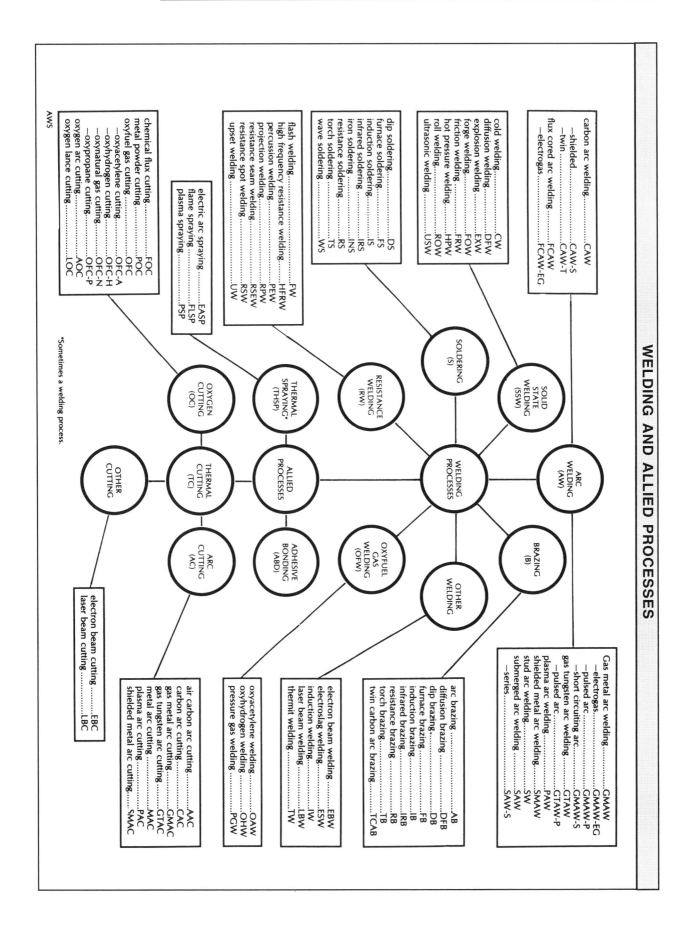

WELDING AND ALLIED PROCESSES

*Sometimes a welding process.

AWS

carbon arc welding............CAW
—shielded.............CAW-S
—twin..............CAW-T
flux cored arc welding.........FCAW
—electrogas............FCAW-EG

cold welding.............CW
diffusion welding...........DFW
explosion welding...........EXW
forge welding............FOW
friction welding...........FRW
hot pressure welding.........HPW
roll welding............ROW
ultrasonic welding..........USW

dip soldering.............DS
furnace soldering...........FS
induction soldering..........IS
infrared soldering..........IRS
iron soldering............INS
resistance soldering..........RS
torch soldering...........TS
wave soldering............WS

flash welding............FW
high frequency resistance welding......HFRW
percussion welding..........PEW
projection welding..........RPW
resistance seam welding........RSEW
resistance spot welding........RSW
upset welding............UW

electric arc spraying.........EASP
flame spraying............FLSP
plasma spraying............PSP

chemical flux cutting.........FOC
metal powder cutting.........POC
oxyfuel gas cutting..........OFC
—oxyacetylene cutting.........OFC-A
—oxyhydrogen cutting.........OFC-H
—oxynatural gas cutting........OFC-N
—oxypropane cutting.........OFC-P
oxygen arc cutting..........AOC
oxygen lance cutting..........LOC

SOLDERING (S)
SOLID STATE WELDING (SSW)
RESISTANCE WELDING (RW)
THERMAL SPRAYING* (THSP)
OXYGEN CUTTING (OC)
WELDING PROCESSES
ALLIED PROCESSES
ARC WELDING (AW)
OTHER CUTTING
THERMAL CUTTING (TC)
BRAZING (B)
ARC CUTTING (AC)
ADHESIVE BONDING (ABD)
OXYFUEL GAS WELDING (OFW)
OTHER WELDING

electron beam cutting.........EBC
laser beam cutting..........LBC

air carbon arc cutting........AAC
carbon arc cutting..........CAC
gas metal arc cutting.........GMAC
gas tungsten arc cutting........GTAC
metal arc cutting..........MAC
plasma arc cutting..........PAC
shielded metal arc cutting.......SMAC

oxyacetylene welding.........OAW
oxyhydrogen welding.........OHW
pressure gas welding.........PGW

electron beam welding.........EBW
electroslag welding..........ESW
induction welding..........IW
laser beam welding..........LBW
thermit welding...........TW

arc brazing.............AB
diffusion brazing...........DFB
dip brazing.............DB
furnace brazing...........FB
induction brazing..........IB
infrared brazing...........IRB
resistance brazing..........RB
torch brazing............TB
twin carbon arc brazing........TCAB

Gas metal arc welding..........GMAW
—electrogas.............GMAW-EG
—pulsed arc............GMAW-P
—short circuiting arc.........GMAW-S
gas tungsten arc welding........GTAW
—pulsed arc............GTAW-P
plasma arc welding..........PAW
shielded metal arc welding.......SMAW
stud arc welding...........SW
submerged arc welding.........SAW
—series..............SAW-S

SMAW

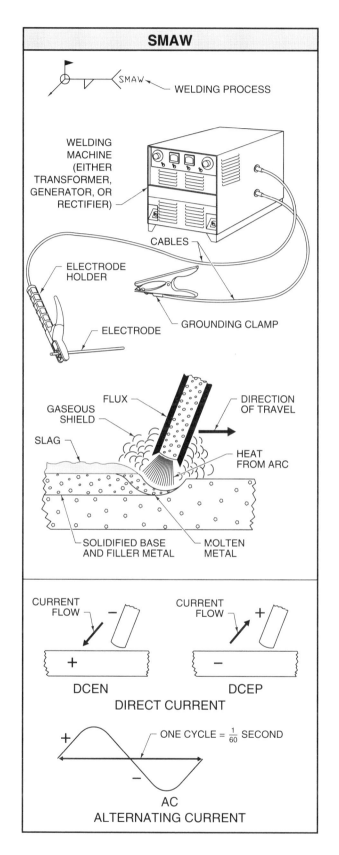

WELDING PROCESS

WELDING MACHINE (EITHER TRANSFORMER, GENERATOR, OR RECTIFIER)

CABLES

ELECTRODE HOLDER

GROUNDING CLAMP

ELECTRODE

FLUX

GASEOUS SHIELD

SLAG

DIRECTION OF TRAVEL

HEAT FROM ARC

SOLIDIFIED BASE AND FILLER METAL

MOLTEN METAL

CURRENT FLOW −

CURRENT FLOW +

DCEN

DCEP

DIRECT CURRENT

ONE CYCLE = $\frac{1}{60}$ SECOND

AC ALTERNATING CURRENT

FCAW

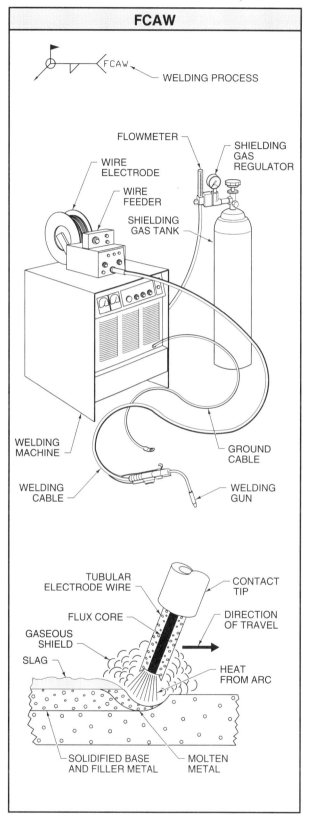

WELDING PROCESS

FLOWMETER

WIRE ELECTRODE

WIRE FEEDER

SHIELDING GAS REGULATOR

SHIELDING GAS TANK

WELDING MACHINE

GROUND CABLE

WELDING CABLE

WELDING GUN

TUBULAR ELECTRODE WIRE

CONTACT TIP

FLUX CORE

DIRECTION OF TRAVEL

GASEOUS SHIELD

SLAG

HEAT FROM ARC

SOLIDIFIED BASE AND FILLER METAL

MOLTEN METAL

GMAW

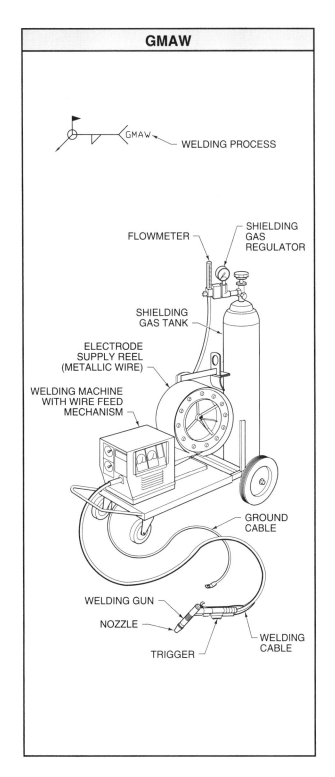

WELDING PROCESS

FLOWMETER

SHIELDING GAS REGULATOR

SHIELDING GAS TANK

ELECTRODE SUPPLY REEL (METALLIC WIRE)

WELDING MACHINE WITH WIRE FEED MECHANISM

GROUND CABLE

WELDING GUN

NOZZLE

TRIGGER

WELDING CABLE

OFW

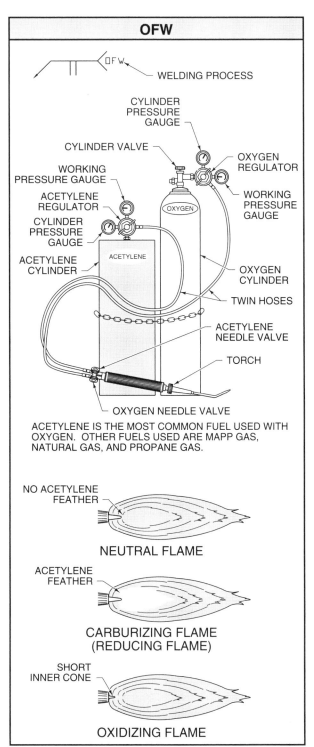

WELDING PROCESS

CYLINDER PRESSURE GAUGE

CYLINDER VALVE

OXYGEN REGULATOR

WORKING PRESSURE GAUGE

ACETYLENE REGULATOR

WORKING PRESSURE GAUGE

CYLINDER PRESSURE GAUGE

OXYGEN

ACETYLENE CYLINDER

ACETYLENE

OXYGEN CYLINDER

TWIN HOSES

ACETYLENE NEEDLE VALVE

TORCH

OXYGEN NEEDLE VALVE

ACETYLENE IS THE MOST COMMON FUEL USED WITH OXYGEN. OTHER FUELS USED ARE MAPP GAS, NATURAL GAS, AND PROPANE GAS.

NO ACETYLENE FEATHER

NEUTRAL FLAME

ACETYLENE FEATHER

CARBURIZING FLAME (REDUCING FLAME)

SHORT INNER CONE

OXIDIZING FLAME

GTAW

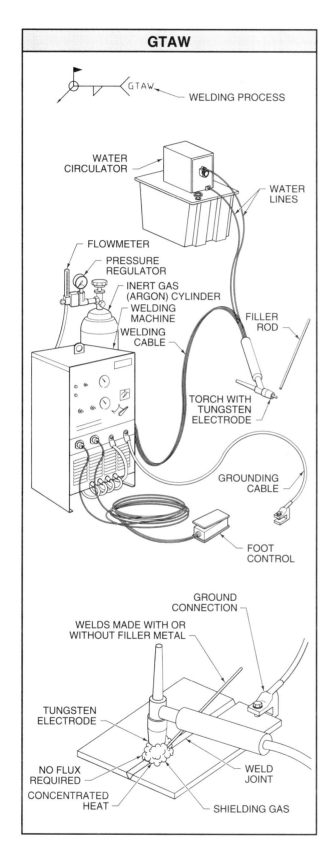

WELDING PROCESS

WATER CIRCULATOR

WATER LINES

FLOWMETER

PRESSURE REGULATOR

INERT GAS (ARGON) CYLINDER

WELDING MACHINE

WELDING CABLE

FILLER ROD

TORCH WITH TUNGSTEN ELECTRODE

GROUNDING CABLE

FOOT CONTROL

GROUND CONNECTION

WELDS MADE WITH OR WITHOUT FILLER METAL

TUNGSTEN ELECTRODE

NO FLUX REQUIRED

CONCENTRATED HEAT

WELD JOINT

SHIELDING GAS

RW

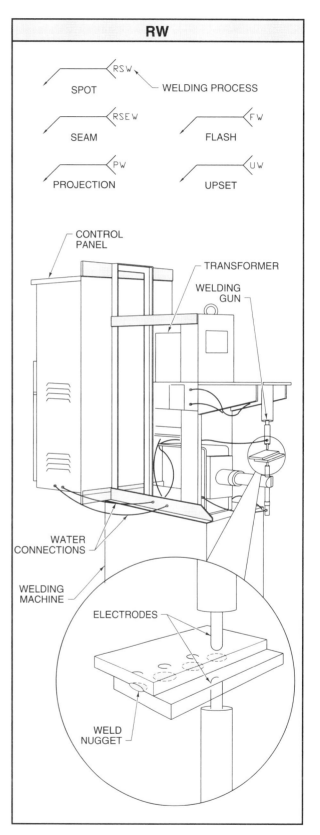

SPOT — WELDING PROCESS

SEAM

PROJECTION

FLASH

UPSET

CONTROL PANEL

TRANSFORMER

WELDING GUN

WATER CONNECTIONS

WELDING MACHINE

ELECTRODES

WELD NUGGET

B AND S

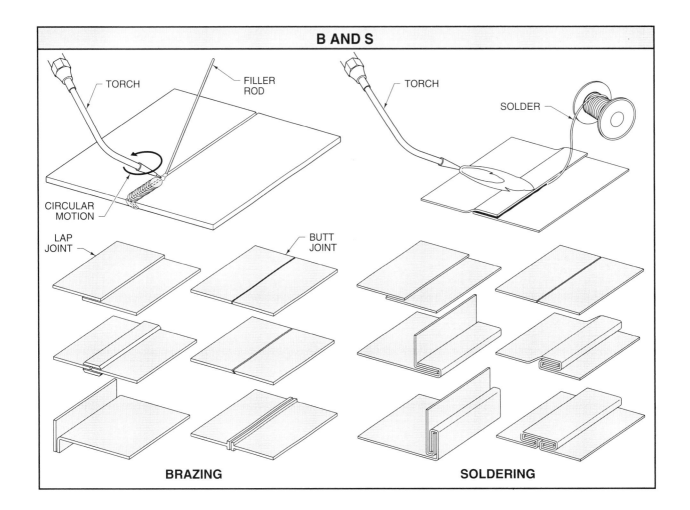

TORCH

FILLER ROD

CIRCULAR MOTION

TORCH

SOLDER

LAP JOINT

BUTT JOINT

BRAZING

SOLDERING

Form A.7.1

SUGGESTED
WELDING PROCEDURE SPECIFICATION (WPS)

Identification _____

Date _____ Revision _____

Company name _____

Supporting PQR no.(s) _____ Type - Manual ( )   Semi-Automatic ( )

Welding process(es) _____ Machine ( )   Automatic ( )

Backing: Yes ( )   No ( )

Backing material (type) _____

Material number _____ Group _____ To material number _____ Group _____

Material spec. type and grade _____ To material spec. type and grade _____

Base metal thickness range: Groove _____ Fillet _____

Deposited weld metal thickness range _____

Filler metal F no. _____ A no. _____

Spec. no. (AWS) _____ Flux tradename _____

Electrode-flux (Class) _____ Type _____

Consumable insert: Yes ( )   No ( )   Classifications _____

Shape _____

Position(s) of joint _____ Size _____

Welding progression: Up ( )   Down ( )   Ferrite number (when reqd.) _____

PREHEAT: **GAS:**

Preheat temp., min _____ Shielding gas(es) _____

Interpass temp., max _____ Percent composition _____
(continuous or special heating, where
applicable, should be recorded)   Flow rate _____
                                          Root shielding gas _____

POSTWELD HEAT TREATMENT: Trailing gas composition _____

Temperature range _____ Trailing gas flow rate _____

Time range _____

Tungsten electrode, type and size _____

Mode of metal transfer for GMAW: Short-circuiting ( )   Globular ( )   Spray ( )

Electrode wire feed speed range: _____

Stringer bead ( )   Weave bead ( )   Peening: Yes ( )   No ( )

Oscillation _____

Standoff distance

Multiple ( )   or single electrode ( )

Other _____

	Filler metal				Current			
Weld layer(s)	Process	Class	Dia.	Type & polarity	Amp range	Volt range	Travel speed range	
								e.g., Remarks, comments, hot wire addition, technique, torch angle, etc.

Approved for Production by _____
                                          Employer

Note: Those items that are not applicable should be marked N.A.

Form A.7.2   **SUGGESTED**   **Page 1 of 2**
                PROCEDURE QUALIFICATION RECORD (PQR)

WPS no. used for test _____   Welding process(es) _____

Company _____   Equipment type and model (sw) _____

JOINT DESIGN USED (2.6.1)   **WELD INCREMENT SEQUENCE**

Single ( )   Double weld ( )   **POSTWELD HEAT TREAMTENT (2.6.6):**

Backing material _____   Temp. _____

Root opening _____ Root face dimension _____   Time _____

Groove angle _____ Radius (J-U) _____   Other _____

Back gouging: Yes ( ) No ( ) Method _____   **GAS (2.6.7)**

BASE METALS (2.6.2)   Gas type(s) _____

Material spec._____ To _____   Gas mixture percentage

Type or grade _____ To _____   Flow rate _____

Material no. _____ To material no. _____   Backing gas _____ Flow rate _____

Group no. _____ To group no. _____   Root shielding gas

Thickness _____   EBW vacuum ( ) Absolute pressure ( )

Diameter (pipe) _____   **ELECTRICAL CHARACTERISTICS (2.6.8)**

Surfacing: Material _____ Thickness _____   Electrode extension _____

Chemical composition _____

Other _____   Standoff distance

FILLER METALS (2.6.3)   Transfer mode (GMAW) _____

Weld metal analysis A no. _____   Electrode diameter tungsten _____

Filler metal F no. _____   Type tungsten electrode _____

AWS specification _____   Current: AC ( ) DCEP ( ) DCEN ( ) Pulsed ( )

AWS classification _____   Heat input _____

Flux class _____ Flux brand _____   EBW: beam focus current _____ Pulse freq. _____

Consumable insert: Spec. _____ Class. _____   Filament type _____ Shape ___ Size _____

Supplemental filler metal spec. _____ Class. _____   Other _____

Non-classified filler metals _____   **TECHNIQUE (2.6.9)**

Consumable guide (ESW) Yes ( ) No ( )   Oscillation frequency _____Weave width _____

Supplemental deoxidant (EBW) _____   Dwell time _____

POSITION (2.6.4)   String or weave bead _____ Weave width _____

Position of groove _____ Fillet _____   Multi-pass or single pass (per side) _____

Vertical progression: Up ( )   Down ( )   Number of electrodes _____

                Peening _____

PREHEAT (2.6.5)   Electrode spacing _____

Preheat temp., actual min _____   Arc timing (SW) _____ Lift ( )

Interpass temp., actual max _____   PAW: Conventional ( ) Key hole ( )

                Interpass cleaning:

Pass no.	Filler metal size	Amps	Volts	Travel speed (ipm)	Filler metal wire (ipm)	Slope induction	Special notes (process, etc.)

Note: Those items that are not applicable should be marked N.A.

Form A.7.2 **Page 2 of 2**

TENSILE TEST SPECIMENS: SUGGESTED PROCEDURE QUALIFICATION RECORD PQR No.

Type: _____ Tensile specimen size: _____ Area: _____

Groove ( )   Reinforcing bar ( )   Stud welds ( )

Tensile test results: (Minimum required UTS _____ psi)

Specimen no.	Width, in.	Thickness, in.	Area, in.^{2}	Max load lbs	UTS, psi	Type failure and location

GUIDED BEND TEST SPECIMENS - SPECIMEN SIZE: _____

Type	Result	Type	Result

MACRO-EXAMINATION RESULTS: Reinforcing bar ( )   Stud ( )
1. _____ 4. _____
2. _____ 5. _____
3. _____

SHEAR TEST RESULTS - FILLETS:
1. _____ 3. _____
2. _____ 4. _____

IMPACT TEST SPECIMENS

Type: _____ Size: _____

Test temperature: _____

Specimen location: WM = weld metal;   BM = base metal;   HAZ = heat-affected zone

Test results:

Welding position	Specimen location	Energy absorbed (ft.-lbs.)	Ductile fracture area (percent)	Lateral expansion (mils)

IF APPLICABLE   **RESULTS**

Hardness tests: ( ) Values _____   Acceptable ( )   Unacceptable ( )

Visual (special weldments 2.4.2) ( )   Acceptable ( )   Unacceptable ( )

Torque ( ) psi   Acceptable ( )   Unacceptable ( )

Proof test ( ) Method _____   Acceptable ( )   Unacceptable ( )

Chemical analysis ( )   Acceptable ( )   Unacceptable ( )

Non-destructive exam ( ) Process _____   Acceptable ( )   Unacceptable ( )

Other _____   Acceptable ( )   Unacceptable ( )

Mechanical Testing by (Company) _____ Lab No. _____

We certify that the statements in this Record are correct and that the test welds were prepared, welded, and tested in accordance with the requirements of the American Welding Society Standard for Welding Procedure and Performance Qualification (AWS B2.1-83).

Qualifier: _____   Reviewed by: _____

Date: _____   Approved by: _____
                                          Employer

Form A.7.3

SUGGESTED
PERFORMANCE QUALIFICATION TEST RECORD

Name _____ Identification _____ Welder ( ) Operator ( )

Social security number: _____ Qualified to WPS no. _____

Process(es) _____ Manual ( ) Semi-Automatic ( ) Automatic ( ) Machine ( )

Test base metal specification _____ To _____

Material number _____ To _____

Fuel gas (OFW) _____

AWS filler metal classification _____ F no. _____

Backing: Yes ( ) No   ( ) Double ( ) or Single side ( )
Current: AC   ( ) DC   ( ) Short-circuiting arc (GMAW) Yes ( ) No ( )
Consumable insert: Yes ( ) No   ( )
Root shielding: Yes ( ) No   ( )

TEST WELDMENT   **POSITION TESTED**   **WELDMENT THICKNESS (T)**

GROOVE:
  Pipe   1G ( ) 2G ( ) 5G ( ) 6G ( ) 6GR ( ) Diameter(s) _____ (T) _____
  Plate   1G ( ) 2G ( ) 3G ( ) 4G ( )   (T) _____
  Rebar   1G ( ) 2G ( ) 3G ( ) 4G ( )   Bar size _____ Butt   ( )
                                                                       Spliced butt ( )

FILLET:
  Pipe ( ) 1F ( ) 2F ( ) 3F ( ) 4F ( ) 5F ( ) Diameter _____ (T) _____
  Plate ( ) 1F ( ) 2F ( ) 3F ( ) 4F ( )   (T) _____

  Other (describe) _____

Test results: Remarks
_____

  Visual test results   N/A ( )   Pass ( )   Fail ( )
  Bend test results   N/A ( )   Pass ( )   Fail ( )
  Macro test results   N/A ( )   Pass ( )   Fail ( )
  Tension test   N/A ( )   Pass ( )   Fail ( )
  Radiographic test results N/A ( )   Pass ( )   Fail ( )
  Penetrant test   N/A ( )   Pass ( )   Fail ( )

QUALIFIED FOR:
PROCESSES
GROOVE: **THICKNESS**
  Pipe   1G ( ) 2G ( ) 5G ( ) 6G ( ) 6GR ( ) (T) Min _____ Max _____ Dia _____
  Plate 1G ( ) 2G ( ) 3G ( ) 4G ( )   (T) Min _____ Max _____
  Rebar 1G ( ) 2G ( ) 3G ( ) 4G ( )   Bar size Min _____ Max _____

FILLET:
  Pipe   1F ( ) 2F ( ) 4F ( ) 5F ( )   (T) Min _____ Max _____
  Plate 1F ( ) 2F ( ) 3F ( ) 4F ( )   (T) Min _____ Max _____
  Rebar 1F ( ) 2F ( ) 3F ( ) 4F ( )   Bar size Min _____ Max _____

Weld cladding ( )   Position(s) _____ T Min _____ Max _____ Clad Min _____

Consumable insert ( )   Backing type ( )
Vertical Up ( )   Down ( )
Single side ( )   Double side ( )   No backing ( )
Short-circuiting arc ( )   Spray arc ( )   Pulsed arc ( )
Reinforcing bar - butt ( ) or Spliced butt ( )

The above named person is qualified for the welding process(es) used in this test within the limits of essential variables including materials and filler metal variables of the AWS Standard for Welding Procedure and Performance Qualification (AWS B2.1).

Date tested _____   Signed by _____
                                                         Qualifier

American Welding Society

GLOSSARY

A

acetylene: Colorless gas that is highly combustible when mixed with oxygen. Unstable at pressures above 15 psi. Used in oxyacetylene welding. See *oxyacetylene welding.*

actual throat: Shortest distance from the face of a fillet weld to the weld root after welding. See *weld face, fillet weld,* and *weld root.*

adhesion: Joining together of dissimilar metals by capillary action. See *capillary action.*

adhesive bonding: Joining of parts with an adhesive placed between faying (mating) surfaces.

alloy: Metal that consists of more than one chemical element, with at least one of the elements being a pure metal.

alternating current (AC): Electric current having alternating positive and negative values. See *current.*

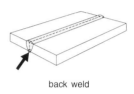

alternating current

ampere (amp or A): Unit of measure for electricity that expresses the quantity or number of electrons flowing through a conductor per unit of time. See *conductor.*

angle: 1. In plane figures, the intersection of two lines. **2.** In building steel, L-shaped structural steel of two equal or unequal widths. See *structural steel.*

annealing: Heat treatment process that softens a metal by heating it to a suitable temperature, holding it at that temperature, and cooling it at a suitable rate. See *metal.*

arc voltage: Voltage present after an arc is struck.

axis: Straight line around which a geometric figure is generated.

B

back (transverse) pitch: Distance from the center of one row of rivets to the center of the adjacent row of rivets. See *rivet.*

backing symbol: Supplementary symbol indicated by a rectangle on the opposite side of the groove weld symbol on the reference line. See *supplementary symbol.*

backing weld: Weld deposited in the weld root opposite the face of the weld on the other side of the joint member. Deposited before the weld on the opposite side of the part. See *back weld.*

backstep welding: Welding in which individual passes are made in the opposite direction of the weld.

back weld: Weld deposited in the weld root opposite the face of the weld on the other side of the joint member. Deposited after the weld on the opposite side of the part. See *backing weld.*

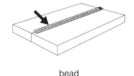

back weld

bar: Round-, square-, or rectangular-shaped structural steel. See *structural steel.*

base metal: Material to be welded.

bead: Narrow layer or layers of metal deposited on the base metal as an electrode melts. See *base metal* and *electrode.*

bead

beam: I-shaped structural steel. See *structural steel.*

bending strength: Quality which resists forces from causing a metal to bend or deflect in the direction in which the load is applied. See *metal.*

bending stress: Stress caused by equal forces acting perpendicular to the horizontal axis of an object. See *stress* and *axis.*

bevel: Sloped edge of an object running from surface to surface.

blind hole: Drilled hole that does not pass through.

blind rivet: Rivet with a hollow shank that joins two parts with access from one side. See *rivet* and *shank.*

brazing: Group of welding processes in which metal is joined by heating the filler metal at temperatures greater than 840°F, but less than the melting point of the base metal. See *filler metal* and *base metal.*

brazing symbol: Graphic symbol that shows braze locations and specifications on prints.

break line: Line that shows internal features or avoids showing continuous features.

brittleness: Lack of ductility in a metal. See *ductility.*

broken-out section: Partial section view which appears to have been broken out of the object. See *section view.*

butt joint: Weld joint formed when two joint members, located approximately in the same plane, are positioned edge to edge. See *weld joint.*

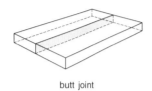
butt joint

C

capillary action: Force by which a liquid in contact with a solid is distributed between faying surfaces. See *faying surface.*

carburizing: Case-hardening process for low-carbon steels that uses an environment with sufficient carbon potential and a temperature above the upper critical temperature. See *case hardening.*

carburizing (reducing) flame: Oxyfuel flame with an excess of fuel.

case hardening: Process of hardening low-carbon or mild steels by adding carbon, nitrogen, or a combination of carbon and nitrogen to the outer surface, forming a hard, thin outer shell.

cast: Metal heated to its liquid state and poured into a mold, where it cools and resolidifies.

casting alloy: Alloy poured into a sand or permanent metal mold. See *alloy* and *metal.*

chamfer: Sloped edge of an object running from surface to side. See *edge.*

channel: C-shaped structural steel used in conjunction with other structural shapes as support members or combined to serve as an I beam. See *structural steel.*

Charpy: Impact test specimen supported horizontally between two anvils with the pendulum allowed to strike opposite the notch.

chemical properties: Properties of metals that are directly related to molecular composition and pertaining to the chemical reactivity of metals and the surrounding environment.

chemical test: Metal identification test using chemicals which react when placed on certain types of metals. See *metal.*

chip test: Metal identification test that identifies metal by the shape of its chips. See *metal.*

coefficient of thermal expansion: Unit change in the length of a material caused by changing the temperature 1°F.

cold worked: Metal that is hammered, rolled, or drawn through a die. See *metal.*

color test: Metal identification test that identifies metals by their color. See *metal.*

combined weld symbols: Weld symbols used when the weld joint, weld type, and welding operation require more information than can be specified with one weld symbol. See *weld symbol, weld joint,* and *weld type.*

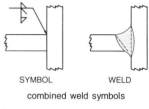
SYMBOL   WELD
combined weld symbols

compressive strength: Ability of a metal to resist being crushed. See *metal.*

compressive stress: Stress caused by two equal forces acting on the same axial line to crush an object. See *stress* and *axis.*

concave: Curved inward.

conductor: Any material through which electricity flows easily.

constant pitch: Standard screw thread series with a set number of threads per inch regardless of diameter. See *standard series.*

constant potential: Generation of a stable voltage regardless of the amperage output produced by the welding power supply.

consumable insert: Spacer that provides proper opening of weld joint and becomes part of the filler metal during welding. See *weld joint.*

consumable insert symbol: Supplementary symbol indicated by a square on the opposite side of a groove weld on reference line. See *supplementary symbol* and *groove weld.*

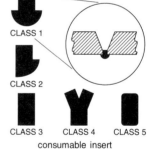
CLASS 1   CLASS 2   CLASS 3   CLASS 4   CLASS 5
consumable insert

contour symbol: Supplementary symbol indicated by a horizontal line or arc parallel to the weld symbol, which specifies the shape of the completed weld. See *supplementary symbol* and *weld symbol.*

convex: Curved outward.

corner-flange weld: Flange weld with one joint member bent. See *flange weld.*

corner joint: Weld joint formed when two joint members are positioned at an approximate 90° angle with the weld joint at the outside of the joint members. See *weld joint.*

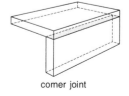
corner joint

corrosion: Combining metals with elements in the environment that leads to deterioration of the metal. See *metal.*

counterbored hole: Enlarged and recessed hole with square shoulders.

counterdrilled hole: Hole with a cone-shaped opening below the outer surface.

countersink: Tool that produces a countersunk hole. See *countersunk hole.*

countersunk hole: Hole with a cone-shaped opening or recess at the outer surface. See *countersink.*

cover pass: Final weld pass deposited. See *weld pass.*

crater: Depression in the base metal made by the welding heat source. See *base metal.*

creep: Slow progressive strain that causes metal to fail. See *strain.*

critical temperature: Temperature above which steel must be heated so it will harden when quenched.

cryogenic properties: Ability of a metal to resist failure when subjected to very low temperatures. See *metal.*

crystal: Solid composed of atoms arranged in a pattern that is repetitive in three dimensions.

cubic foot: 1'-0" × 1'-0" × 1'-0" or 1728 cu in.

cubic inch: 1" × 1" × 1" or its equivalent.

current: Movement of electrons through a conductor. See *conductor.*

cutting plane line: Line that shows where an object is imagined to be cut in order to view internal features.

cyaniding: Process of hardening low-carbon steel by heating it in sodium cyanide or potassium cyanide.

D

depth of fusion: Distance from the fusion face to the weld interface. See *fusion face* and *weld interface.*

destructive testing: Any type of testing that damages the test part (specimen).

direct current (DC): Electric current which flows in one direction. See *current.*

direct current electrode negative (DCEN): Flow of current from electrode (–) to work (+). See *electrode.*

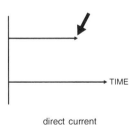

direct current

direct current electrode positive (DCEP): Flow of current from work (–) to electrode (+). See *electrode.*

distortion: Change in the original shape of the metal as metal expands when heated and contracts when cooled.

double-bevel-groove weld: Groove weld having joint members beveled on both sides with the weld made from both sides. See *groove weld.*

double-fillet weld: Fillet weld that has filler metal deposited on both sides. See *fillet weld.*

double-flare-bevel-groove weld: Groove weld having two radiused joint members with the weld made from both sides. See *groove weld.*

double-flare-V-groove weld: Groove weld having radiused joint members with the weld made from both sides. See *groove weld.*

double-J-groove weld: Groove weld having joint members grooved in a J shape on both sides with the weld made from both sides. See *groove weld.*

double-square-groove weld: Groove weld having square-edged joint members with the weld made from both sides. See *groove weld.*

double-U-groove weld: Groove weld having joint members grooved in a U shape on both sides with the weld made from both sides. See *groove weld.*

double-V-groove weld: Groove weld having joint members angled on both sides with the weld made from both sides. See *groove weld.*

drill: Round hole in a material produced by a twist drill.

ductility: Ability of a metal to stretch, bend, or twist without breaking or cracking.

duty cycle: Length of time (expressed as a percentage) that a welding machine can operate at its rated output within a 10 minute period. See *output.*

dynamic electricity: Electricity in motion in an electric current. See *current.*

E

edge: Intersection of two surfaces.

edge-flange weld: Flange weld with both joint members bent. See *flange weld.*

edge joint: Weld joint formed when the edges of two joint members are joined. See *weld joint.*

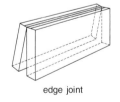

edge joint

effective throat: Shortest distance from the face of a fillet weld to the weld root, minus any convexity after welding. See *weld face, fillet weld,* and *weld root.*

elastic deformation: Ability of a metal to return to its original size and shape after loading and unloading.

elastic limit (yield): Last point at which a material can be deformed and still return to its original shape.

electrical circuit: Path taken by an electric current when flowing through a conductor from one terminal of the power source to the other. See *current* and *conductor.*

electrical properties: Ability of a metal to conduct or resist electricity or the flow of electrons.

electrode: Coated metal wire which forms and cleans the weld bead. See *weld bead*.

electrode holder: Hand-held device that holds the electrode securely at the required angle for maximum access to the weld area. See *electrode*.

F

face reinforcement: Filler metal which extends above the surface of the joint member on the side of the joint on which welding was done. See *filler metal*.

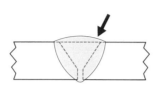

face reinforcement

fast-fill electrode: Iron powder electrode that has a soft arc and fast deposit rate. See *electrode*.

fast-freeze electrode: Electrode that produces a snappy, deep penetrating arc and fast-freezing deposits. See *electrode*.

fatigue strength: Property of a metal to resist various kinds of rapidly alternating stresses. See *metal* and *stress*.

faying surface: Part of the joint member which is in full contact prior to welding. See *capillary action*.

ferrous metal: Any metal with iron as a major alloying element.

field rivet: Rivet placed in the field. See *rivet*.

field weld symbol: Supplementary symbol indicated by a triangular flag rising from the intersection of the arrow and reference line, which specifies the welding operation is to be completed in the field at the location of final installation. See *supplementary symbol*.

field weld symbol

file test: Metal identification test in which a file is used to indicate the hardness of steel compared with that of the file. See *metal*.

filler metal: Metal deposited during the welding process. See *metal*.

filler pass: Weld pass that fills remaining portion of the weld after the root pass and hot pass. See *weld pass, root pass,* and *hot pass*.

fillet: Rounded interior corner.

fillet weld: Weld type made in the cross-sectional shape of a triangle. See *weld type*.

fillet weld leg: Distance from the joint root to the weld toe. See *joint root*.

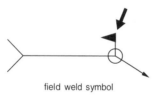

fillet weld

fillet weld leg size: Dimension from the root of a weld to the toes of a weld after welding. See *fillet weld leg*.

fill-freeze electrode: Electrode that has a moderately forceful arc and deposit rate. The rate is between those of the fast-freeze and fast-fill electrodes. See *electrode*.

fitting: Standard connection used to join two or more pieces of pipe.

fit-up: Positioning of pipe with other pipe or fittings before welding.

fixture: Device used to maintain the correct positional relationship between joint members required by print specifications.

flange height: Distance from the point of tangency on the flange of a flange weld to the edge of the flange before welding. See *flange weld*.

flange radius: Radius of the joint member(s) requiring edge preparation in a flange weld. See *flange weld*.

flange weld: Weld type made of light-gauge metal with one or both joint members bent at approximately 90°. See *weld type*.

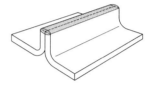

flange weld

flange weld thickness: Cross-sectional distance of a flange weld from the weld face to the weld root. See *flange weld, weld face,* and *weld root*.

flux: Coating on the electrode. See *electrode*.

flux cored arc welding (FCAW): Welding process that uses an arc shielded by gas from within the electrode. See *electrode*.

forged: Metal formed by a mechanical or hydraulic press with or without heat. See *metal*.

fracture test: Metal identification test that breaks the metal sample to check for ductility and grain size. See *metal* and *ductility*.

fusion: Melting together of filler metal and base metal. See *filler metal* and *base metal*.

fusion face: Surface of the base metal that is melted during welding. See *fusion*.

G

gas metal arc welding (GMAW): Welding process with a shielded gas arc between a continuous wire electrode and the weld metal. See *electrode*.

gas-shielded flux cored arc welding (FCAW-G): Flux cored arc welding process variation in which shielding gas is provided through a gas nozzle, in addition to that obtained from electrode flux.

gas tungsten arc welding (GTAW): Welding process in which shielding gas protects the arc between a tungsten electrode and the weld area. See *electrode*.

globular transfer: Metal transfer in which molten metal from a consumable electrode is spread across the arc in large drops. See *metal* and *electrode*.

grain: Individual crystal in a metal that has multiple crystals. See *metal* and *crystal.*

grain structure: Pattern of the grains in a metal. See *grain* and *metal.*

grip: Effective holding length of a rivet. See *rivet.*

groove face: Surface of the joint member included in the groove of the weld.

groove weld: Weld type made in the groove of the pieces to be welded. See *weld type.*

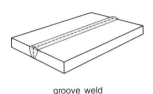
groove weld

grounding device (ground): Connection between welding cable and weld parts in the welding circuit.

H

hardfacing: Application of filler metals which provide a coating to protect the base metal from wear caused by impact, abrasion, erosion, or from other wear. See *filler metal* and *base metal.*

hardness: Ability of a metal to resist indentation. See *metal.*

heat-affected zone: Area of base metal in which the mechanical properties and structure are affected by the welding process. See *base metal* and *mechanical properties.*

high-carbon steel: Steel with a carbon range of 0.45% to 0.75%.

hot pass: Weld pass that penetrates deeply into the root pass and the root face of the joint. See *weld pass, root pass,* and *root face.*

I

impact load: Load that is applied suddenly or intermittently. See *load.*

impact strength: Ability of a metal to resist loads that are applied suddenly and often at high velocity. See *metal.*

inclusion: Impurity or foreign substance forced in a molten puddle during the welding process.

incomplete fusion: Discontinuance of a weld where complete fusion does not occur between the weld metal and the fusion faces of the joint of the joint members. See *fusion.*

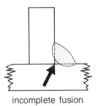
incomplete fusion

inert gas: Gas that does not readily combine with other elements.

input: Electrical requirements for operating a welding machine.

intermittent fillet welds: Short sections of fillet welds applied at specified intervals on the weld part. See *fillet weld.*

interpass temperature: Weld area temperature between passes of a multiple-pass weld. See *weld pass.*

Izod: Impact test specimen that is supported as a vertical cantilever beam and struck on the free end projecting above the holding vise.

J

joint root: Part of a joint to be welded where the members are the closest to each other.

L

lap joint: Weld joint formed when two joint members are lapped over one another. See *weld joint.*

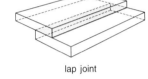
lap joint

large rivets: Rivets with a shank of ½″ or greater in diameter. See *rivet* and *shank.*

load: External force applied to an elastic body that causes stress in a material. See *stress.*

low-carbon steel: Steel with a carbon range of 0.05% to 0.30%.

M

machinable electrode: Electrode whose deposits are soft and ductile enough so they can easily be machined after welding. See *electrode.*

magnetic test: Metal identification test that checks for the presence of iron in a metal. See *metal.*

malleability: Ability of a metal to be deformed by compressive forces without developing defects. See *metal.*

margin: Distance from the edge of a plate to the centerline of the nearest row of rivets. See *plate* and *rivet.*

material safety data sheet (MSDS): Printed material used to relay chemical hazard information.

mechanical properties: Properties that describe the behavior of metals under applied loads. See *metal.*

medium-carbon steel: Steel with a carbon range of 0.30% to 0.45%.

melting point: Amount of heat required to melt a given amount of metal. See *metal.*

melt-through symbol: Supplementary symbol indicated by a darkened radius on the reference line opposite the weld symbol specified. Filler metal deposited on one side must completely penetrate through to the other side of the weld. See *supplementary symbol*.

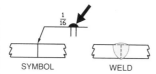

SYMBOL   WELD

melt-through symbol

metal: Material consisting of one or more chemical elements having crystalline structure, high thermal and electrical conductivity, the ability to be deformed when heated, and high reflectivity.

metallizing: Spray coating procedure where finely divided particles of metal are deposited on worn surfaces. See *metal*.

microstructure: Microscopic arrangement of the components within a metal. See *metal*.

modulus of elasticity: Ratio of stress to strain within the elastic limit. See *stress, strain,* and *elastic limit*.

N

neutral flame: Oxyfuel flame with a balanced mixture of oxygen and fuel.

nitriding: Subcritical case-hardening process that introduces nitrogen into the surface of a steel. See *case hardening*.

nondestructive examination (NDE) symbol: Symbol that specifies examination methods and requirements to verify weld quality.

nondestructive testing: Any type of testing that leaves the test part undamaged.

nonferrous metal: Pure metal, other than iron or metals with iron as a major alloying element. See *metal*.

nonmachinable electrode: Electrode with a mild steel core containing a heavy coating which melts at low temperatures, allowing the use of low welding current. See *electrode*.

non-threaded fasteners: Devices that join or fasten parts together without threads.

O

open-circuit voltage: Voltage produced when a welding machine is running and no welding is being done.

optical properties: Color of a metal and how it reflects light. See *metal*.

output: Maximum amperage and voltage of a welding machine. See *ampere* and *volt*.

overlapping: Extending weld metal beyond the weld toes or weld root. See *weld toe* and *weld root*.

oxidation: Combination of metal and oxygen into metal oxides. See *metal*.

oxidizing flame: An oxyfuel flame with an excess of oxygen.

oxyacetylene welding (OAW): Oxyfuel welding with acetylene. See *oxyfuel welding* and *acetylene*.

oxyfuel welding (OFW): Welding process that uses oxygen combined with a fuel to sustain a flame that generates the heat necessary for welding.

P

padding: Process of building worn surfaces by depositing several layers of beads. See *bead*.

parent metal: Metal to be welded. See *metal*.

pass: Each layer of beads deposited on the base metal. See *bead* and *base metal*.

penetration: Depth of fusion with the base metal. See *fusion* and *base metal*.

physical properties: Thermal, electrical, optical, magnetic, and general properties of metal. See *metal*.

pipe: Round-shaped structural steel. See *structural steel*.

pipe-jig: Device which holds sections of pipe or fittings before tack welding. See *pipe* and *tack weld*.

pitch: Distance between corresponding points on adjacent thread forms.

plastic deformation: Failure of a metal to return to its original size and shape after being loaded and unloaded. See *metal* and *load*.

plate: ³⁄₁₆″ or more thick structural steel used to cover large expanses of a structure. See *structural steel*.

plug weld: Weld type made in the cross-sectional shape of a hole in one of the joint members. See *weld type*.

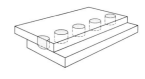

plug weld

plug weld size: Diameter of the hole through the joint member at the faying surface of the weld joint. See *faying surface* and *weld joint*.

polarity: Direction of current in a circuit. See *current*.

porosity: Cavity or cavities in the weld metal or weld interface caused by trapped gas. See *metal*.

positioner: Mechanical device that supports and moves joint members for maximum loading, welding, and unloading efficiency.

postheating: Application of heat to the weld part after welding to facilitate a controlled cooling rate.

preheating: Application of heat to the base metal before welding to reduce the temperature difference between the weld metal and the surrounding base metal. See *base metal*.

primary weld: Weld that is an integral part of a structure and that directly transfers the load. See *load.*

projection weld: Weld type produced by confining fusion of molten base metal using heat and pressure with a preformed dimple or projection in one joint member prior to welding. See *weld type, fusion,* and *base metal.*

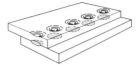

projection weld

proportional limit: Maximum stress a material can withstand without permanent deformation. See *stress.*

pure metal: Metal that consists of one chemical element. See *metal.*

R

radiographic examination: Testing of welds for weld defects and strength requirements using X rays. See *X ray.*

reaming: Enlarging and improving the surface quality of a hole.

rectifier: Welding machine that produces AC or DC.

reducing (carburizing) flame: Oxyfuel flame with an excess of fuel. See *carburizing flame.*

reduction: Loss or removal of oxygen during the welding process.

reinforcement: Amount of weld metal that is piled up above the surface of the pieces being joined.

resistance: Opposition to the flow of electrons.

resistance welding (RW): Welding processes in which welding occurs from the heat obtained by resistance to the flow of current through the workpieces. See *current.*

reverse polarity: An arc welding circuit in which the electrode is connected to the positive terminal. See *electrode.*

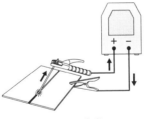

reverse polarity

right angle: Angle that contains 90°. See *angle.*

ripple: Shape within the deposited bead that is caused by movement of the welding heat source. See *bead.*

rivet: Cylindrical metal pin with a preformed head. See *metal.*

rivet pitch: Distance from the center of one rivet to the center of the next rivet in the same row. See *rivet.*

root edge: Weld face that comes to a point and has no width. See *weld face.*

root face: Surface of the groove next to the root.

root opening: Distance between joint members at the root of the weld before welding.

root pass: Initial weld pass that provides complete penetration through the thickness of the joint member. See *weld pass* and *penetration.*

root reinforcement: Filler metal which extends above the surface of the joint on the opposite side of the joint on which welding was done. See *filler metal.*

root surface: Surface of the weld on the opposite side of the joint on which welding was done.

S

screw thread series: Groups of diameter-pitch combinations. See *pitch.*

seam weld: Weld type produced by confining fusion of molten base metal using heat and pressure for a series of continuous or overlapping successive spot welds on joint members. See *weld type, fusion,* and *spot weld.*

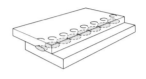

seam weld

secondary weld: Weld used to hold joint members and subassemblies together.

section view: Interior view of an object through which a cutting plane has been passed. See *cutting plane line.*

segregation: Separation of elements comprising the base metal. See *base metal.*

self-shielded flux cored arc welding (FCAW-S): Flux cored arc welding process variation in which shielding gas is provided exclusively by the flux within the electrode.

shank: Cylindrical body of a rivet. See *rivet.*

sheared plate: Plate that is rolled between horizontal and vertical rollers and trimmed on all edges.

shearing stress: Stress caused by two equal and parallel forces acting upon an object from opposite directions. See *stress.*

shear strength: Ability of a metal to withstand two equal forces acting in opposite directions. See *metal.*

sheet: 3/16″ or less structural steel used to cover large expanses of a structure. See *structural steel.*

shielded metal arc welding (SMAW): Arc welding process in which the arc is shielded by the decomposition of the electrode covering. See *electrode.*

shop rivet: Rivet placed in the shop. See *rivet.*

short-circuit transfer: Metal transfer in which molten metal from a consumable electrode is deposited during repeated short circuits. See *metal* and *electrode.*

single-bevel-groove weld: Groove weld having one joint member beveled with the weld made from that side. See *groove weld.*

single-fillet weld: Fillet weld having filler metal deposited on one side. See *fillet weld* and *filler metal.*

single-flare-bevel-groove weld: Groove weld having one straight and one radiused joint member with the weld made from one side. See *groove weld.*

single-flare-V-groove weld: Groove weld having radiused joint members with the weld made from one side. See *groove weld.*

single-J-groove weld: Groove weld having joint members grooved in a J shape on one side with the weld made from that side. See *groove weld.*

single-square-groove weld: Groove weld having square-edged joint members with the weld made from one side. See *groove weld.*

single-U-groove weld: Groove weld having joint members grooved in a U shape on one side with the weld made from that side. See *groove weld.*

single-V-groove weld: Groove weld having both joint members angled on the same side with the weld made from that side. See *groove weld.*

slag inclusions: Small particles of slag (cooled flux) trapped in the weld metal which prevent complete penetration. See *metal* and penetration.

slot weld: Weld type made in the cross-sectional shape of a slot (elongated hole) in one of the joint members. See *weld type.*

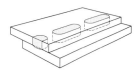

slot weld

small rivet: Rivet with slot weld a shank of $7/16''$ or less in diameter. See *rivet* and *shank.*

soldering: Group of welding processes in which metal is joined by heating the filler metal at temperatures less than 840°F and less than the melting point of the base metal. See *filler metal* and *base metal.*

solidification temperature: Temperature at which the atoms of a metal assume their characteristic crystal structure. See *metal* and *crystal.*

space lattice: Uniform pattern produced by lines connected through the atoms.

spacer symbol: Supplementary symbol indicated by a rectangle centered on reference line. See *supplementary symbol.*

spark test: Metal identification test that identifies metals by the shape, length, and color of a spark emitted from contact with a grinding wheel. See *metal.*

special series: Screw thread series with combinations of diameter and pitch not in the standard screw thread series. See *screw thread series* and *pitch.*

specifications: Documents that supplement working drawings with written instructions giving additional information.

spotface: Flat surface machined at a right angle to a drilled hole. See *right angle.*

spot weld: Weld type produced by confining the fusion of molten base metal using heat and pressure without

preparation to the joint members. See *weld type* and *fusion.*

spray transfer: Metal transfer in which molten metal from a consumable electrode is sprayed across the arc in small drops. See *metal* and *electrode.*

staggered intermittent fillet welds: Intermittent fillet welds that have a staggered pitch and are applied to both sides of a weld joint. See *intermittent fillet welds* and *pitch.*

standard series: Screw thread series of coarse (UNC/UNRC), fine (UNF/UNRF), and extra-fine (UNEF/UNREF) graded pitches and eight series with constant pitches. See *screw thread series* and *pitch.*

static electricity: Electricity at rest.

static load: Load that remains constant. See *load.*

straight polarity: An arc welding circuit in which the electrode is connected to the negative terminal. See *electrode.*

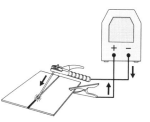

strain: Deformation per unit length of a solid under stress. See *stress.*

stress: Effect of an external force applied upon a solid material.

straight polarity

stress relieving: Process of heating a metal to a suitable temperature, holding it at that temperature to reduce residual stresses, and cooling it slowly to minimize the development of new residual stresses. See *metal* and *stress.*

structural steel: Steel used in the erection of a structure.

stud weld: Weld type made by joining threaded studs with other parts using heat and pressure. See *weld type.*

supplementary symbol: Symbol used on welding symbols to further define the operation to be completed. See *welding symbol.*

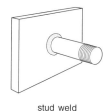

stud weld

surface feature: A part of a surface where change occurs.

surfacing: Applying filler metals which have similar characteristics to the base metal. See *filler metal* and *base metal.*

surfacing weld: Weld type in which weld beads are deposited on a surface to increase the dimensions of the part or to add special properties to the weld part. See *weld type* and *bead.*

T

tack weld: Weld that joins the joint members at random points to keep the joint members from moving out of their required positions.

tail: Part of a welding symbol included when a specific welding process, specification, or procedure must be indicated. See *welding symbol*.

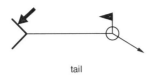

tail

tee: T-shaped structural steel made of I beams cut to specifications by mill or suppliers. See *structural steel*.

tensile strength: Ratio of the maximum load to the original cross-sectional area. See *load*.

tensile stress: Stress that is caused by two equal forces acting on the same axial line to pull an object apart. See *stress*.

theoretical throat: Distance from the face of a fillet weld to the root before welding. See *fillet weld, weld face*, and *weld root*.

thermal conductivity: Rate which metal transmits heat. See *metal*.

thermal expansion: Expansion of a metal when subjected to heat. See *metal*.

thermal properties: One of the physical properties of metal. Includes melting point, thermal conductivity, and thermal expansion and contraction. See *physical properties* and *metal*.

threaded fasteners: Devices such as nuts and bolts that join or fasten parts together with threads.

through hole: Drilled hole passing completely through the material.

T-joint: Weld joint formed when two joint members are positioned approximately 90° to one another in the form of a T.

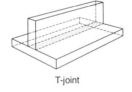

T-joint

torch test: Metal identification test that can be used to identify a metal by its color change with the application of heat, its melting point, and its behavior in the molten state.

torque: Product of the applied force (P) times the distance (L) from the center of application.

torsional strength: Ability of a metal to withstand forces that cause it to twist. See *metal*.

torsional stress: Stress caused by two forces acting in opposite twisting motions. See *stress*.

toughness: Combination of strength and ductility of metals. See *ductility*.

transformer: Welding machine that produces AC only.

travel angle: Angle less than 90° of the electrode in relation to a perpendicular line from the weld and the direction of the weld. See *electrode*.

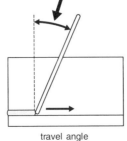

travel angle

travel speed: Speed at which the electrode is moved across the weld area. See *electrode*.

tubing: Round-, square-, or rectangular-shaped structural steel. See *structural steel*.

U

undercutting: Creating a groove that is not completely filled by weld metal in the base metal during the welding process. See *base metal*.

union: Fitting consisting of three parts having threads and flanges which draw together when tightened.

universal plate: Plate that is rolled between horizontal and vertical rollers and trimmed only on the ends. See *plate*.

V

variable load: Load that varies with time and rate, but without the sudden change that occurs with an impact load. See *impact load*.

very-high carbon steel: Steel with a carbon range of 0.75% to 1.7%.

volt (V): Unit of measure for electricity that expresses the electrical pressure differential between two points in a conductor. See *conductor*.

voltage drop: Voltage developed across a component by the flow of current through the resistance or impedance of the component. See *current* and *resistance*.

W

weaving: Welding technique used to increase the width and volume of the bead. See *bead*.

weld-all-around symbol: Supplementary symbol indicated by a circle at the intersection of the arrow and reference line, which specifies that the weld extends completely around the joint. See *supplementary symbol*.

weld bead: Weld that results from a weld pass. See *weld pass*.

weld contour: Cross-sectional shape of the completed weld face. See *weld face*.

weld cracks: Linear discontinuities that occur in the base metal, weld interface, or the weld metal. See *base metal* and *weld interface*.

weld defects: Undesirable characteristics of a weld which may cause the weld to be rejected.

weld face: Exposed surface of weld, bounded by the weld toes of the side on which welding was done. See *weld toe*.

weld finish: Method used to achieve the surface finish. See *base metal*.

welding symbol: Graphic symbol that shows weld locations and specifications on prints.

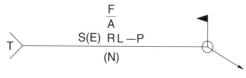

welding symbol

weld interface: Area where filler metal and base metal mix together. See *filler metal* and *base metal.*

weld joint: Physical configuration of the joint members to be joined.

weld leg: Size of fillet welds made in lap or T-joints. See *fillet weld, lap joint,* and *T-joint.*

GROOVE WELDS			
SQUARE	SCARF	V	BEVEL
‖	⁄⁄	∨	⌵
U	J	FLARE-V	FLARE-BEVEL

OTHER WELDS				
FILLET	PLUG OR SLOT	STUD	SPOT OR PROJECTION	SEAM
BACK OR BACKING	SURFACING	FLANGE		
		EDGE	CORNER	

weld symbol

weld pass: Single progression of welding along a joint.

weld root: Area where filler metal intersects base metal opposite weld face. See *filler metal, base metal,* and *weld face.*

weld symbol: Graphic symbol which defines the cross-sectional shape of a weld.

weld throat: Distance through the center of the weld from the face to the root. See *weld face* and *weld root.*

weld toe: Intersection of the base metal and the weld face. See *base metal* and *weld face.*

weld type: Cross-sectional shape of the filler metal after welding. See *filler metal.*

weld width: Distance from toe to toe across the face of the weld. See *weld toe* and *weld face.*

work angle: Angle less than 90° of the electrode in relation to the workpiece. See *electrode.*

wrought alloy: Alloy that contains one or more alloying elements and possesses a high tensile strength. See *alloy* and *tensile strength.*

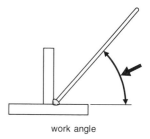

work angle

X

X ray: Electromagnetic radiation with a very short wavelength.

Y

yield (elastic limit): Last point at which a material can be deformed and still return to its original shape. See *elastic limit.*

INDEX

X

Z